2019

INTERNATIONAL DESIGN EXCHANGE PROJECT

The 10th Hidden Space Project

REMEDITAION 再媒介

01
The Cass London Metropolitan University
The Sir John Cass Faculty of Art, Architecture and Design
Interior Design BA (Hons)
UK

02
Dongyang Mirae University
Dept. of Interior Environmental Design, in School of Living and Environmental Engineering
Rep. of KOREA

GRASSROOTS
INDEPENDENCE
MUSIC
EVENTS
CONVENIENCE
INDIVIDUALITY
TEAM WORK

_ 18 Projects

contents

01 The Cass London Metropolitan University _UK

02 Dongyang Mirae University _ Rep. of KOREA

INTRODUCTION

PROJECTS

1

The Cass London Metropolitan University
The Sir John Cass Faculty of Art, Architecture and Design
Interior Design BA (Hons)

The Sir John Cass Faculty of Art, Architecture and Design
41-47 Commercial Road
Lonodon E1 1LA
Switchboard 020 7423 0000
www.thecass.com

Studio **2**

STATE of INDEPENDENCE

LIVE PROJECT

8h shift Romanian Pavilion Venice Biennale 2014

Preamble:

The last few years has seen a significant growth of Co-working spaces that operate in the adapted shells of buildings whose original purpose has become redundant. The premise of operating together with shared facilities seems to be a good economic model where different businesses can share the costs and in turn network, expanding business opportunities. Many of these Co-worker spaces are owned and run by other corporate businesses which have used the opportunity to brand the interiors with their own corporate identity, which at a glance would seem business like, friendly and fun. However, what of the individual company, what of their identity how does this manifest itself as separate and distinct entity? One such business is the WIRE, a music and performance magazine that supports and reflects the work of grass roots musicians and performance artists across the world. They describe themselves as 'pathologically independent'. It is essential that the WIRE occupy their 'own space' in order to keep their original values and philosophies true as much through the space that they occupy as in the copy that they write.

The WIRE have in the past considered sharing workspace with like-minded performance- interested companies, so does this ask the question whether sharing space with compatible, or other businesses attuned to your genre is a way forward? Or does the grass roots approach suggest that independence is independent and that this provides the edge, self- determination that allows unconventional behaviours and interactions, and opportunities to occur, where companies are open to outside approaches by having an absolutely zero allegiance to others?

The building for our "pathologically independent inhabitants' is Stoke Newington Town Hall. Originally built in 1936, it was named the Civic Centre' the first time that this term had been used. 1936 was a period of renewed ideals of the future; the building followed the art and design precepts of the time by being in parts a solid example of Art Deco. It started with three main ideas to govern, educate and gather its community within the Council Chambers, the Library and the Assembly Hall.
Today the Town Hall is all but redundant, only 20% of its space is used, mainly for Weddings and similar gatherings. London Borough of Hackney council has recently spent some money to refurbish the Assembly rooms but is now asking for ideas as to how to use, adapt and design the rest of the space.

지난 몇 년 동안 원래 목적이 중복 된 건물의 개조 된 껍질에서 작동하는 공동 작업 공간이 크게 성장했습니다. 공유 시설과 함께 운영한다는 전제는 다른 비즈니스가 비용을 공유하고 네트워크를 통해 비즈니스 기회를 확대 할 수 있는 좋은 경제 모델로 보입니다. 이러한 공동 작업 공간의 대부분은 다른 회사 비즈니스가 소유하고 운영하며, 회사의 정체성으로 인테리어를 브랜딩 할 수 있는 기회를 한눈에 볼 수 있습니다. 그러나 개별 회사는 무엇이고, 그들의 정체성은 무엇입니까? 이것이 어떻게 분리되고 뚜렷한 독립체로서 나타납니까? 그러한 사업 중 하나는 WIRE, 전세계의 인디 음악가 및 공연 예술가의 작품을 지원하고 반영하는 음악 및 공연 잡지. 그들은 스스로를 "병리학적으로 독립" 된 것으로 묘사합니다. WIRE는 자신이 쓴 사본에서 차지하는 공간을 통해 원래의 가치와 철학을 그대로 유지하기 위해 자체 공간을 차지하는 것이 중요합니다.

WIRE는 과거에 같은 생각을 가진 연주에 관심이 있는 회사와 공유 작업 공간을 고려해 왔으므로 호환 가능한 또는 다른 장르에 맞게 조정된 다른 비즈니스와 공간을 공유하는 것이 앞으로 나아갈 수 있을까? 또는 풀뿌리 접근 방식은 독립성이 독립적이며, 이것이 틀에 박힌 행동, 상호 작용을 허용하는 자기 결정, 기회가 발생할 수 있음을 암시하는가?

병리학적으로 독립적인 주민을 위한 건물은 스토크 뉴 잉턴 시청입니다. 원래 1936 년에 지어졌으며 처음으로 이 용어가 사용된 시민 센터로 명명되었습니다. 1936은 미래에 대한 새로운 이상의 시기였습니다. 그 건물은 Art Deco의 견고한 사례로 부분적으로 예술과 당시의 교훈을 따랐습니다. 이 회의는 시의회 회의소, 도서관 및 총회 홀을 통해 지역 사회를 지배, 교육 및 수집하기 위한 세 가지 주요 아이디어로 시작되었습니다.

This is a **Live** project, Hackney Borough Council planning service, which consists of Conservation, Urban Design and Sustainability and Growth Teams will deliver the over-arching brief and be your main client.

- London Borough of Hackney council has recently spent some money to refurbish the Assembly rooms but is at a loss as to how to use, adapt and design the rest of the space. Hackney Borough Council have asked us to investigate how this large series of spaces could be used for the good of the community, its business, progress and pleasure.

Brief:
This project asks you to:

- Re-design a number of connected or adjoining spaces within the grade II listed Town Hall for the good of the community, its enterprise, education and enjoyment.
- You are to rediscover or re-invent the meaning of Civic centre.

The main user is the WIRE magazine; they require you to design their workspace together with associated and supportive spaces

- Gathering space
- Archive
- Gallery
- Radio station
- Small performance space for bands, poets, vocal performers

Tony Herrington, the publisher of the WIRE will deliver the brief and will therefore form the premise of the Interior client. Because of its connection to grass roots music and performance the studio will visit Berlin for their Study Trip. It is also hoped that you visit music and performance venues in London.

You will be required to:

- Translate the identity of the WIRE magazine together with their sense of 'pathological independence' through the designed space.
- Explore how to mediate between the grade II listed art deco architecture and decoration and the identities and environment of the incoming WIRE magazine.

As the main brief is concerned with workspace design, you will have the opportunity to work with Hitch Mylius, a contemporary furniture company.

You will be required to:

- Design a one-off piece of furniture/ installation that describes the WIRE's 'pathological Independence'.
- Work with Hitch Mylius to understand how to specify furniture in the context of workspace and the WIRE magazine.

Later in the year you will be asked to present your designs to Hitch Mylius who will then exhibit the chosen projects at Clerkenwell Design Week in May 2019.

Studio 2 is always committed to sustainable and well-being outcomes were possible and as such the British Wool Organisation will present the many attributes of using wool in many different forms within an interior. They are giving us unspun wool to work with.

Workshops

In the course of the year as you will be exploring the Stoke Newington environment, historic notions and materiality of Art Deco and the music magazine publishing and performance we have added several workshops to support your explorations.

- Cyanotypes
- Screen printing
- Veneering and Surface finishes
- Working with wool
- Furniture specification and FF& E

Pod Casting

The studio will be regularly working with pod casting as a way to document the whole process and to actively discuss ideas and approaches. It will also help to understand acoustic environments and radio presentation.

개요

이것은 Hackney Borough Council 계획 서비스로, 보존, 도시 설계, 지속가능성 및 성장 팀으로 구성된 라이브 프로젝트로서, 중요한 개요를 전달하고 여러분의 주요 고객이 될 것이다. Hackney Council의 London Borough는 최근 국회 회의실을 재정비하는 데 약간의 돈을 썼지만 나머지 공간을 어떻게 사용하고, 적응하고, 설계해야 할지 막막하다. Hackney Borough Council은 우리에게 어떻게 이 거대한 일련의 공간들이 지역사회의 이익, 사업, 발전과 쾌락을 위해 사용될 수 있는지 조사해 줄 것을 요청했다.

이 프로젝트에서 요구하는 사항:

· Grade Ⅱ 의 연결되거나 인접하는 여러 공간을 지역사회의 이익, 기업, 교육 및 즐거움을 위해 재설계한다.

· 시민 센터의 의미를 재발견하거나 재발명해야 한다.

주요 사용자는 WIRE 매거진이며, WIRE 매거진은 관련되고 지원되는 공간과 함께 작업 공간을 설계할 것을 요구한다.

· 공간 수집
· 보관
· 갤러리
· 라디오 방송국
· 밴드, 시인, 보컬 연주자를 위한 작은 공연 공간

WIRE의 출판인 토니 헤링턴은 그 개요를 전달할 것이며 따라서 인테리어 클라이언트의 전제를 형성할 것이다. 인디 음악과 공연과의 연관성 때문에 스튜디오는 그들의 학습 여행을 위해 베를린을 방문할 것이다. 또한 당신이 music의 음악과 공연장을 방문하기를 희망한다.

다음 요구 사항:

· WIRE 매거진의 정체성을 '병리학적 독립성'에 대한 감각과 함께 설계된 공간을 통해 번역한다.

· Grade Ⅱ I에 열거된 Art Deco건축과 장식, 그리고 들어오는 WIRE 잡지의 정체성과 환경 사이에서 중재하는 방법을 탐구한다.

작업 공간 디자인에 관한 것이므로, 당신은 현대 가구 회사인 Hitch Mylius와 함께 일할 기회를 갖게 될 것이다.

다음 요구 사항:

· WIRE의 '병리적 독립성'을 설명하는 일회성 가구/설비를 설계한다.

· Hitch Mylius와 협력하여 작업 공간 및 WIRE 매거진 맥락에서 가구를 지정하는 방법을 이해한다.

연말에 당신은 당신의 디자인을 Hitch Mylius에게 제시하도록 요청받을 것이다. 그는 2019년 5월에 Clerkenwell Design Week에서 선택된 프로젝트를 전시할 것이다.

스튜디오 2는 항상 지속 가능하고 웰빙 결과가 가능했으며, 이와 같이 영국 양모 기구는 실내에서 많은 다양한 형태로 양모를 사용하는 많은 속성을 보여줄 것이다. 그들은 우리에게 작업할 수 있는 스펀지 털실을 주고 있다.

워크샵

당신이 Stoke Newington 환경, Art Deco의 역사적 개념과 중요성, 그리고 음악 잡지 출판과 공연을 탐험할 때 탐험을 지원하기 위해 몇 개의 워크샵을 추가했다.

· 시아노 타입
· 화면 인쇄
· 베니어링 및 표면 마감
· 모직 작업
· 가구 사양 및 FF&E

팟 캐스팅

스튜디오는 전체 과정을 문서화하고 아이디어와 접근방식을 적극적으로 논의하기 위한 방법으로 포드 캐스팅과 함께 정기적으로 작업할 예정이다. 그것은 또한 음향 환경과 라디오 프리젠테이션을 이해하는데 도움이 될 것이다.

Kaye Newman MA FRSA

CASS: Head of Interiors
Course Leader: Interior Design BA(Hons)
Acting Course Leader: Interior Design & Decoration BA(Hons)
CASS Faculty of Art, Architecture & Design
London Metropolitan University

State of Independence

This year we were approached by the chief conservation officer from Hackney Borough Council, Tim Walder to take on Stoke Newington Town Hall as the location for our next project. The Town Hall was built in 1937, it was the first of its type to use the term Civic Centre. Sitting next to Stoke Newington library it also incorporated a large assembly room and presided over the Governance, Knowledge and Community gathering in its day.
The quadrant shaped building is detailed with beautiful art deco features and shows a high presence of occasion and hierarchy. Its imposing façade and roomy interior mainly stands empty now. Its services now digitised or centralised elsewhere it asks the question is what is a modern day Town Hall? who should it serve, what does its community need? So many of our London Town Hall's are being sold off to developers to re make into hotels as the interior accommodation is often quite grand but isn't this ignoring the potential of designing new public spaces which could also generate a income stream?

Possibilities of an income stream:
The last few years has seen a significant growth of Co-working spaces that operate in the adapted shells of buildings whose original purpose has become redundant. The premise of operating together with shared facilities seems to be a good economic model where different businesses can share the costs and in turn network, expanding business opportunities. Many of these Co-worker spaces are owned and run by other corporate businesses who have used the opportunity to brand the interiors with a corporate identity, which in one glance would seem business like, friendly and fun. However, what of the individual company, what of their identity, how does this manifest itself as separate and distinct entity. We invited the WIRE, a music and performance magazine that supports and reflects the work of grass roots musicians and performance artists across the world to be a part of the project. They describe themselves as 'pathologically independent'. It was essential that the WIRE occupy their 'own space' in order to keep their original values and philosophies true as much through the space that they occupy as in the copy that they write.

The building has a grade 2 listing, initially the students were more than respectful of its heritage and found designing interventions, breaking through walls and therefore breaking through proportioned geometry quite an undertaking. However, once the idea of the community, their spatial requirements and idea of neighbourly connection started to resonate, the new spaces began to take on these new ideals as well respectfully showing the art deco features.

The Conservation team have initiated two public consultations within the Stoke Newington neighbourhood, to ascertain a way forward from which nothing has decided. I am very pleased to say that following Tim Walder's visit to the Summer Show, he requested the portfolio of the students' featured in the book to present to the local councillors as he had recognised that the work offered so many ideas, that were both researched and visualised.

It is very pleasing to know that our students offer such excellent, well presented and professional work.

Kaye Newman
Course Organizer - 코스 기획자

State of Independence

올해 우리는 Hackney Borough Council의 최고 보존 책임자인 Tim Walder에 의해 다음 프로젝트의 장소로 Stoke Newington Town Hall을 맡게 되었다. 타운홀은 1937년에 지어졌으며, 시민센터라는 용어를 사용한 최초의 타입이었다. 스토크 뉴잉턴 도서관 옆에 앉아 그것은 또한 큰 회의실을 설립하고 그 시대에 거버넌스, 지식 그리고 지역사회 모임을 주재했다.

사분면 모양의 건물에는 아름다운 예술 디코 기능이 상세하게 표현되어 있으며, 때와 서열이 매우 높다. 그것의 당당한 파사드와 넓은 내부는 지금은 주로 텅 비어 있다. 그것의 서비스는 현재 디지털화되거나 다른 곳에 집중화되었다. 그것은 현대적인 타운홀이 무엇인가? 그것은 누구를 위해 봉사해야 하는가, 그것의 지역사회가 필요로 하는 것은 무엇인가? 그래서 우리 런던 시청의 상당수는 종종 실내 시설이 꽤 웅장하기 때문에 호텔로 개조하기 위해 개발자들에게 매각되고 있지만, 이것은 또한 소득을 창출할 수 있는 새로운 공공 공간을 설계 할 가능성을 무시하는 것이 아닌가?

지난 몇 년 동안 원래의 목적이 중복된 건물의 개조된 조개 껍질에서 운영되는 공동 작업 공간이 크게 성장했다. 공유시설과 함께 운영한다는 전제는 서로 다른 사업체가 비용을 분담하고, 차례로 네트워크화하여 사업기회를 확대할 수 있는 좋은 경제모델로 보인다. 많은 직장 동료들은 기업 정체성을 가진 인테리어 브랜드를 사용했던 다른 기업체들에 의해 소유되고 운영되는데, 한 눈에 보면 친근하고 재미있는 사업처럼 보일 것이다. 그러나 개별 기업의 어떤 것, 그들의 정체성, 이것이 어떻게 분리되고 뚜렷한 실체로 그 자체를 드러낼 수 있는가. 우리는 전세계의 인디 뮤지션과 공연예술인들의 작품을 지원하고 반영하는 음악 및 공연잡지 WIRE를 이 프로젝트에 참여하도록 초청했다. 그들은 스스로를 '병적으로 독립적인' 존재라고 묘사한다. WIRE는 그들이 쓰는 사본만큼 그들이 점유하는 공간을 통해 그들의 원래 가치와 철학을 진실되게 유지하기 위해 그들의 '소유 공간' 을 차지하는 것이 필수적이었다.

그 건물은 grade 2으로 되어 있는데, 처음에는 학생들이 그 유산을 존중하는 것 이상이었고, 간섭을 디자인하고, 벽을 깨고, 그에 따라 비례적인 기하학을 꽤 큰 사업으로 돌파하는 것을 발견했다. 그러나, 일단 공동체의 생각, 그들의 공간적 요건, 그리고 이웃의 연결에 대한 생각이 반향을 일으키기 시작하자, 새로운 공간들은 예술적 디코 특징을 정중하게 보여줄 뿐만 아니라 이러한 새로운 이상을 떠맡기 시작했다.

보존 팀은 아무것도 결정된 것이 없는 앞으로 나아갈 길을 찾기 위해 스토크 뉴잉턴 지역 내에서 두 가지 공공 협의를 시작했다. 나는 팀 월더가 서머 쇼에 방문한 후, 그가 이 책에 수록된 학생들의 포트폴리오를 지역 의원들에게 제시해 줄 것을 요청했다는 것이 매우 기쁘다. 그는 이 작품이 연구되고 시각화된 많은 아이디어를 제공한다는 것을 알고 있었기 때문이다.

우리 학생들이 이렇게 훌륭하고, 잘 표현되고, 전문적인 일을 한다는 것을 알게 되어 매우 기쁘다.

Annabelle Roennberg

Interior Architecture & Design
BA(Hons)

Studio Studio 2:
The state of independence
Stoke Newington Town Hall

DESIGN PERFORMANCE COMMUNITY **WIRE**

As part of the project, a community in Stoke Newington is to be found together in a new way. The 80% of the unused Town Hall will be brought back to live with lots of music and dance. The Wire, a 'pathological' independent Music Magazine will find a new creative Venue in the Town Hall. It will use the power of the Town Hall walls and the former decision-making to transform their magic into a new form of creative communication. With this the Wire will awaken the old spirit to a new life. Tradition and history meet contemporary forms of creation, yet the functional significance of the space remains intact.

In addition an artist organisation is opened, which is also situated on the ground floor. The performance organisation will hold regular dancing classes, Body Conditioning, Independent dance and artist development. With one goal in mind – to bring everyone in the Stoke Newington community together, from young to old.

이 프로젝트의 일환으로 스토크 뉴잉턴의 한 지역사회는 새로운 방식으로 함께 발견될 예정이다. 사용되지 않은 시청의 80%는 많은 음악과 춤으로 다시 살아날 것이다. '병리학적' 독립 음악 잡지인 'The WIRE' 는 시청에서 새로운 창작 명소를 발견하게 될 것이다. 그것은 시청 벽과 이전의 의사 결정의 힘을 이용하여 그들의 마법을 새로운 형태의 창조적 의사소통 방식으로 바꿀 것이다. 이것으로 WIRE는 옛 정신을 새로운 삶으로 각성시킬 것이다. 전통과 역사는 현대적 창조의 형태를 충족하지만, 그 공간의 기능적 의의는 그대로 남을 것이다.

또한 1층에 위치한 예술인 조직이 개설된다. 공연 조직은 정기적으로 댄스 수업, 바디 컨디셔닝, 인디펜던트 댄스 및 아티스트 개발을 개최할 것이다. 한 가지 목표를 염두에 두고, 스토크 뉴잉턴 지역 사회의 모든 사람들을 젊은 층에서 노인층까지 하나로 모으는 것이다.

LOCATION

The complex is located in Stoke Newington, borough of Hackney and the Clissold ward. It is encircled by Church Street to the south, Clissold Park and St. Mary's Church to the west of Lordship Terrace to the north and Edwards Lane to the east.
Various bus connections and the Overground which is 1km away by foot can be reached by public transport to the Town Hall.
In Stoke Newington are living a diverse community. Jewish people, families, artistic people. The Church Street where the Town hall is located is a successful independent shopping and food economy, the wider area is home of adults aged 30-64.

이 단지는 해크니와 클리솔드 병동인 스토크 뉴잉턴에 위치해 있다. 남쪽으로는 처치 스트리트, 남쪽으로는 클리솔드 파크, 그리고 세인트로 둘러싸여 있다. 메리의 교회는 북쪽으로는 경의 선 테라스 서쪽, 동쪽으로는 에드워즈 레인의 서쪽이다. 도보로 1km 거리에 있는 다양한 버스 연결과 육상에 대중교통을 통해 타운홀로 갈 수 있다. 스토크 뉴잉턴은 다양한 공동체를 살고 있다. 유대인, 가족, 예술인. 타운홀이 위치한 교회거리는 성공적인 독립적인 쇼핑과 음식 경제로, 넓어진 지역은 30-64세의 성인들의 집이다.

1

2

3

1 **TOWN HALL**

The Complex is located in Stoke Newington ward, the north-west corner of Hackney. It falls within the Stoke Newington Church Street Local Shopping Centre.

이 콤플렉스는 해크니의 북서쪽 코너인 스토크 뉴잉턴 병동에 위치해 있다. 그것은 스토크 뉴잉턴 교회 거리 지역 쇼핑 센터 안에 있다.

2 **CLISSOLD PARK**

Clissold Park is an open space in Stoke Newington, in the London Borough of Hackney. It is bounded by Greenway Close (to the north), Stoke Newington Church Street (to the south) and Green Lanes (west) and Queen Elizabeth's Walk (east).

클리솔드 공원은 해크니의 런던 보루에 있는 스토크 뉴잉턴에 있는 열린 공간이다. 그린웨이 클로즈(북쪽), 스토크 뉴잉턴 교회 거리(남쪽)와 그린 레이네스(서쪽), 엘리자베스 여왕의 산책로(동쪽)와 경계를 이룬다.

3 **St. MARY'S CHURCH**

The buildings as they stand today are used in many different ways both for worship and in support of the local community and the most vulnerable in Stroke Newington.

오늘날 그들이 서 있는 건물들은 예배와 지역사회 그리고 스트로크 뉴잉턴에서 가장 취약한 지역사회 모두를 위해 많은 다양한 방법으로 사용된다.

BUS STATIONS

HISTORY

The Town and Assembly Hall municipal buildings on Stoke Newington Church Street were built between 1935-1937 by J Reginald Truelove. It used to be a government building and stopped being useful for that, what means a lot less people working in there so the whole complex turned in to being 80% unused. The mission is to secure a more sustainable future for this important building, to optimise current uses and bringing new uses into the Complex.

스토크 뉴잉턴 교회 거리의 타운과 의회 홀 시립 건물들은 J Reginald Truelove에 의해 1935-1937년 사이에 지어졌다. 그것은 정부 건물이었고 이곳의 사용은 그리 썩 유용하지 않았다. 그것은 그 곳에서 일하는 사람들이 훨씬 적다는 것을 의미하며, 그래서 전체 단지는 80%는 사용되지 않는 것으로 변했다. 우리의 임무는 이 중요한 건물에 대한 보다 지속 가능한 미래를 확보하고, 현재의 사용을 최적화하고, 새로운 용도를 단지 안으로 가져오는 것이다.

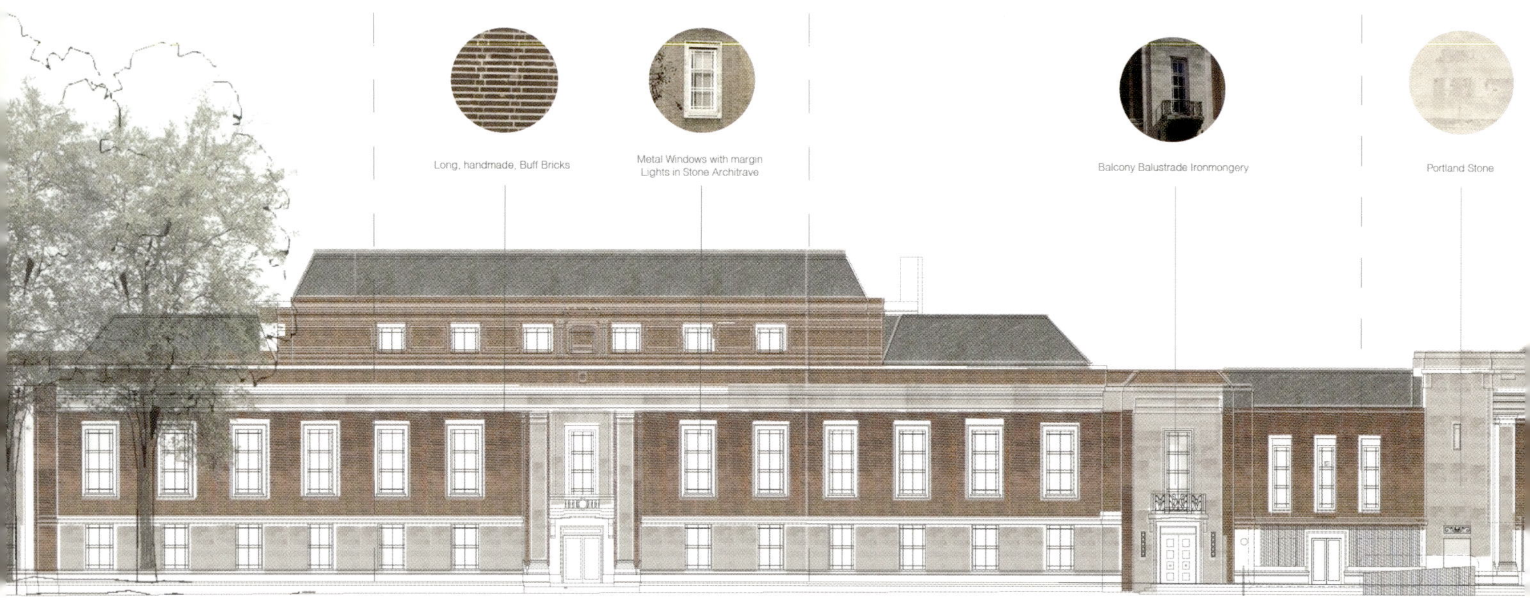

Existing Part West Elevation - Facing Clissold Park [1]
Existing Part West Elevation - Facing Old St Mary's Church [2]
Existing Part West Elevation - Facing Old St Mary's Church [3]

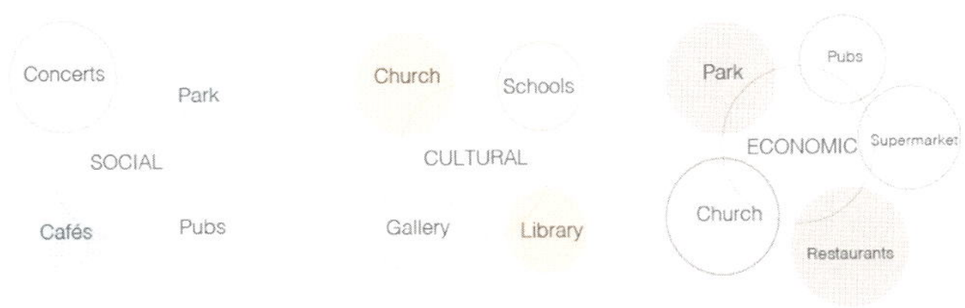

CLOSING MUSIC PLACES

Years ago the Lucky Seven which found its place in the Church Street for 8 years and the Vortex Jazz Bar which were opened in 1984 had been popular musical points in Stoke Newington. Due to the rent increase and expiring rental contracts, these had to close.

음악 장소의 폐쇄

몇 년 전, 8년 동안 교회 거리에서 자리를 잡은 럭키 세븐과 1984년에 문을 연 보텍스 재즈 바는 스토크 뉴잉턴에서 인기 있는 음악점이었다. 임대료 인상과 임대 계약 만료로 인해, 이것들은 문을 닫아야 했다.

EXISTING FACADE AND MATERIALS

The main level of the council offices is a segmental curve trough ninety degrees. The Town Hall is a symmetrical composition. The material of the basement facade, as well as the end pavilions and the wide central section set behind huge pillars, are out of Portland Stone. The centrepiece of this symmetrical composition is the entrance, which consists of a double door in bronze and a marbleised architrave. All windows are made of metal, have margin lights and are embedded in stone architrave edging.

기존 파사드 및 재료

구의회 사무실의 주요 수준은 구십도 정도의 구획 곡선이다. 시청은 대칭적인 구성이다. 지하 면의 재료는 물론 끝의 정관과 거대한 기둥 뒤에 세워진 넓은 중앙 부분이 포틀랜드 스톤에서 벗어났다. 이 대칭 구성의 중심부는 청동으로 된 이중 문과 대리석으로 된 아치트레이브로 구성된 입구다. 모든 창문은 금속으로 만들어졌고, 여백 조명이 있고, 돌 아치트레이브 에지 안에 내장되어 있다.

THE CLIENT

Our client the WIRE is a music magazine which had its first edition in 1982. During this time, it specialized mainly in jazz music. Originally the wire comes from the Soho music scene. But as the Soho scene changed, the Wire in the East moved in which the creative and musical life takes place. Today it is a independent music magazine which does not want to specialize in any kind of music but offers a wide range of alternative, underground, Jazz, Hip Hop, RnB. They do not care how popular the musicians are, they care about how they play their Music attitude: many cultures, individuals, international

우리의 고객인 WIRE는 1982년에 초판을 낸 음악잡지다. 이 시기에는 주로 재즈음악에 특화되었다. 원래 WIRE는 소호 음악계에서 온 것이다. 그러나 소호 장면이 바뀌면서 동양에서의 와이어는 창조적이고 음악적인 삶이 이루어지는 곳으로 이동했다. 오늘날 그것은 어떤 종류의 음악도 전문적으로 다루고 싶지 않지만 다양한 대안, 즉 지하, 재즈, 힙합, R&B를 제공하는 독립적인 음악 잡지다. 그들은 음악가들이 얼마나 인기 있는지 신경쓰지 않고, 그들이 어떻게 그들의 음악적 태도를 연주하는지, 즉 많은 문화, 개인, 국제적인 것에 관심을 갖는다.

Current Office: In the office the WIRE Team work together at a long central table. (Promotes communication, exchanges views become easier) People communicate better, everyone's opinion is important and everyone should come together and talk. They also have a separate space for the graphic designer. Speaking of graphic designers: The layout of WIRE is very simple. With this they want to have the layout not to distract from the topic or the photography.
Overall, the WIRE is a natural usual identity magazine which is completely independent and which should be the most important point in our project. DO NOT BE MAINSTREAM

현재 사무실: 사무실에서 WIRE 팀은 긴 중앙 테이블에서 함께 일한다. (소통을 제안하고, 의견을 교환하는 것이 쉬워진다.) 사람들은 의사소통을 더 잘 하고, 모두의 의견이 중요하며, 모두가 함께 모여 이야기를 나누어야 한다. 그들은 또한 그래픽 디자이너를 위한 별도의 공간을 가지고 있다. 그래픽 디자이너에 대해 말하자면, WIRE의 레이아웃은 매우 간단하다. 이것으로 그들은 주제나 사진에서 주의를 딴 데로 돌리지 않는 레이아웃을 갖기를 원한다.
전반적으로, WIRE는 완전히 독립적이고 우리 프로젝트에서 가장 중요한 포인트가 되어야 하는 자연적인 통상적인 아이덴티티 잡지다. 주류가 되길 원하지 않는다.

As part of the project, a community in Stoke NEWington is to be found together in a NEW way. The 80% of the unused Town Hall will be brought back to live with lots of music and dance. The Wire, a 'pathological' independent Music Magazine will find a new creative Venue in the Town Hall. It will use the power and the decision making walls of the Town Hall to transform their magic into a new form of creative communication. With this the Wire will awaken the old spirit to a new life. Tradition and history meet modern and contemporary forms of creation, yet the functional significance of the space remains intact.

The Wire will sell their magazines, t-shirts in their new shop. In addition an artist organisation is opened this will hold regular dancing classes, Body Conditioning, Independent dance and artist development. With one goal in mind – to bring everyone in the Stoke Newington community.

이 프로젝트의 일환으로 스토크 뉴잉턴의 한 지역사회는 새로운 방식으로 함께 발견될 예정이다. 사용되지 않은 시청의 80%는 많은 음악과 춤으로 다시 살아날 것이다. '병적' 독립 음악잡지인 'THE WIRE'는 시청에서 새로운 창작 명소를 발견하게 될 것이다. 그것은 그들의 마법을 새로운 형태의 창조적 소통으로 바꾸기 위해 시청의 힘과 결정을 이용할 것이다. 이것으로 WIRE는 옛 정신을 새로운 삶으로 각성시킬 것이다. 전통과 역사는 현대와 현대 창조의 형태를 충족하지만, 공간의 기능적 의의는 그대로 남아 있다.
WIRE는 그들의 잡지와 티셔츠를 새 가게에서 판매할 것이다. 또한 예술가들로 주관하는 여러 정기적으로 댄스 수업, 바디 컨디셔닝, 인디펜던트 댄스 및 아티스트 개발을 진행한다. 한 가지 목표를 염두에 두고 – 스토크 뉴잉턴 지역 사회의 모든 사람들을 새로운 사람들과 섞이게 하는 것이다.

Blending the old with the new

The South- West facade of the Town Hall is entirely made up of coloured glass fins which reflected the colours of the historic building. Laminated collared glass fins were fixed on a wall system. They prvide a sense of privacy for the WIRE Workers. The Grätzel cells are part of an integral and multi-layered energy concept of alternative energy resources that heat or cool the Town Hall. The entire hot water for the use of the toilet facilities and showers of the dance school on the solar cells installed on the roof and by the Grätzel cells or produced via a heat recovery system of the heated indoor air.The Grätzel cells are used to convert light energy into electrical energy. It is an application of bionics, which is also called electrochemical dye-solar cell in its function. This solar cell is named after Michael Grätzel from Switzerland. The electrochemical dye solar cell does not use a semiconductor material to absorb light, but organic dyes, for example, the leaf dye chlorophyll. The solar cells that are attached to the roof, you will not see from the perspective of a passers, so the aesthetics of the building is not affected.

시청의 남서부 정면은 전적으로 역사적 건물의 색깔을 반영하는 색의 유리 지느러미로 이루어져 있다. 적층 콜러드 글라스 핀은 벽 시스템에 고정되었다. 그들은 WIRE 노동자들에게 사생활 의식을 자랑한다. Grazel 전지는 타운 홀을 가열하거나 냉각시키는 대체 에너지 자원의 일체적이고 다층적인 에너지 개념의 부분이다. 지붕에 설치된 태양전지 및 그렉첼 세포에 의해 또는 가열된 실내공기의 열회수 시스템을 통해 생산되는 댄스스쿨의 화장실 시설과 샤워기의 사용을 위한 전체 온수. 그래젤 세포는 빛 에너지를 전기 에너지로 변환하는데 사용된다. 생체역학을 응용한 것으로, 그 기능에서 전기화학적 염료-솔라세포라고도 한다. 이 태양 전지는 스위스에서 온 마이클 그래젤의 이름을 따서 명명되었다. 전기화학 염료 태양전지는 반도체 물질을 사용하여 빛을 흡수하는 것이 아니라 유기 염료, 예를 들어 잎 염료 엽록소를 사용한다. 지붕에 붙어 있는 태양전지는 행인의 입장에서 볼 수 없기 때문에 건물의 미학에는 영향이 없다..

FURNITURE DESIGN

At the same place where in the past the rooms were separated by folding doors, a rotatable shelf is created here, which not only serves as a room divider, but also as a storage space and a separate workplace.

How to identify the WIRE with the independent piece of furniture

The long clean lines which go horizontally and vertically remind of clear lines which are also used in the WIRE ethic and content. The shelf can rotate.
This is reflected that it does not commit itself. Furthermore it reminds that they do not write about a particular genre of music or value how well known the musicians are.
The solid glazed frame is meant to show that they are a strong and independent music magazine. The single place in the shelf is for Tony, one of the owners of the WIRE. I wanted to take over the concept that he had a separate place at his old office. Therefore he can decide if he wants to work independently or join the group at the big table.

가구디자인

과거에는 접이식 문으로 방을 분리했던 바로 그 곳에는 이곳에 회전식 선반이 만들어지는데, 이 선반은 객실 칸막이 역할을 할 뿐만 아니라 수납공간과 별도의 작업공간 역할을 한다.

독립적인 가구로 The WIRE를 식별하는 방법

수평과 수직으로 가는 길고 깨끗한 선은 WIRE 윤리 및 내용에서도 사용되는 명확한 선을 연상시킨다. 선반은 회전할 수 있다. 이것은 스스로 범하지 않는다는 것을 반영한다. 게다가 그것은 그들이 특정한 장르의 음악에 대해 쓰지 않거나 얼마나 잘 알려진 음악가들에 대해 소중하게 생각하지 않는다는 것을 상기시킨다. 견고한 유리 프레임은 그들이 강하고 독립적인 음악 잡지라는 것을 보여주기 위한 것이다. 선반에 있는 한 곳은 WIRE의 주인 중 한 명인 Tony를 위한 것이다. 나는 그가 예전 사무실에 따로 자리를 잡고 있다는 개념을 이어받고 싶었다. 따라서 그는 독립적으로 일할 것인지 아니면 큰 테이블에서 그 그룹에 참여할 것인지를 결정할 수 있다.

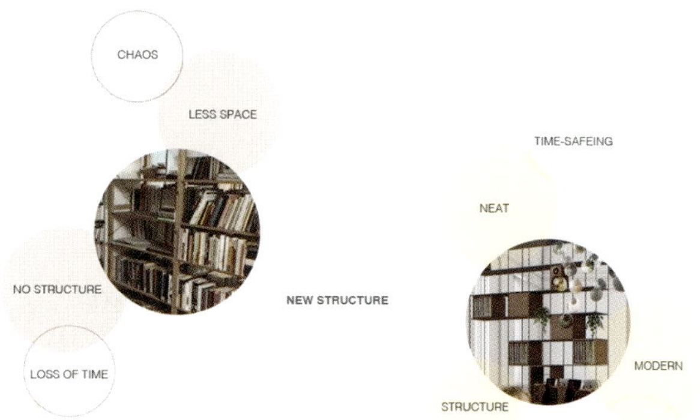

WIRE Office

The new WIRE office is designed according to the new requirements of the employees. The desire for more space, structure and order was implemented here in a more modern and sustainable way. The big wire table has a constants present that remain in the new spaces as well. The four meter high window openings enrich the office with natural light and create a pleasant working environment. The rotating shelf which is embedded in the new glazed frame structure creates a new organized storage spaced. The glass material, even though separating the prolonged room, keeps the long curved space in tact. The Bocci lamp creation is from a German designer. These are very suitable for high rooms and the colours of the glass spheres are reflected in the overall interior colour palette and also in the exterior.

새로운 WIRE 사무실은 직원들의 새로운 요구에 따라 설계되었다. 더 많은 공간, 구조, 질서에 대한 욕구가 보다 현대적이고 지속 가능한 방법으로 여기서 구현되었다. 큰 철사 테이블에는 새로운 공간에도 남아 있는 상수가 있다. 높이 4m의 창문 개구부는 자연광으로 사무실을 풍요롭게 하고 쾌적한 근무환경을 만들어 준다. 새로운 유리로 된 프레임 구조에 내장된 회전 선반은 새로운 정리된 저장 공간을 만든다. 유리로 된 물질은, 길게 뻗은 방을 분리해도, 길게 휘어진 공간을 재치있게 유지시켜 준다. 보키 램프는 독일의 디자이너가 만든 것이다. 이것들은 높은 방에 매우 적합하며 유리구체의 색상은 전체적인 내부 색상 팔레트와 외부에도 반영된다.

Reception, Private Room

The reception area which can also be used as a private room, is a very representative space. This is engulf by its floor-to-ceiling balcony door opening with a lot of natural light. The two cuved Hitch Mylius sofas facing each other are situated within the central part of the curvature of the Town Hall. The magazines covers on the wall are used for decoration.

개인방으로도 사용할 수 있는 접대공간은 매우 대표적인 공간이다. 이것은 많은 자연광으로 바닥에서 천장까지 이어지는 발코니 문틈에 휩싸여 있다. 서로 마주보고 있는 두 개의 큐브한 히치 밀리어스 소파는 시청 곡선의 중심부에 위치하고 있다. 벽에 붙어 있는 잡지 표지는 장식용으로 쓰인다.

Coloured Glass

Polyester Wrapl

Upgraded Existing
Concrete floor

WIRE Archive

The spiral-glass staircase connecting the WIRE Office with the Archive. The glass is not only used for the safety panel, but for the steps as well. It brings more daylight into the interior and through this it opens the slightly darker room on the ground floor. The staircase can be left halfway up and reach the gallery-like mezzanine floor. The glass panels are fixed to the newly used floor which serves as a gallery. Through this intervention the Archive becomes enlarged and the space is used to a maximum extend. It was a wish of the staff as little as possible to have around. Important is a large table, music and storage space for current magazines.

WIRE 보관소

WIRE 사무실과 Archive를 연결하는 나선유리 계단. 이 유리는 안전판뿐만 아니라 계단에도 사용된다. 그것은 실내로 더 많은 햇빛을 가져오고 이것을 통해 1층의 약간 어두운 방을 연다. 계단은 반쯤 올라가서 갤러리 같은 메자닌 바닥에 닿을 수 있다. 유리 패널은 갤러리 역할을 하는 새로 사용한 바닥에 고정되어 있다. 이 개입을 통해 보관소는 확대되고 공간이 최대한 확장되도록 사용된다. 곁에 있는 것이 가능한 한 적은 수의 스태프들의 소망이었다. 중요한 것은 현재 잡지를 위한 큰 테이블, 음악, 저장 공간이다.

Laminated Coloured Glass

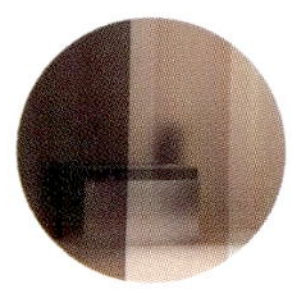

Black coated Metal

Upgraded Existing Herringbone Parquet

Performance/ Dance

The ballet room and the stretching room shine with their simplicity and their calm and sober colours. The intense ultramarine blue serves as a visual accent. These spaces were inspired by a Ballet School in Melbourne which also work with clear lines and colours.
The aim was to create modern spaces for a mix of traditional and contemporary performances that make one forget that they are in a historic Complex from the 30's. All rooms are equipped with a modern ceiling ventilation. Through the contrast of a modern design environment and classical ballet, a reference is once again made to the main concept proposal.

퍼포먼스/댄스

발레실과 스트레칭룸은 그들의 소박함과 차분하고 냉정한 색감으로 빛난다. 강렬한 울트라마린 블루는 시각적인 억양의 역할을 한다. 이 공간들은 멜버른에 있는 발레학교에서 영감을 받아 만들어진 것인데, 이 학교는 이러한 선명한 선과 색깔로 학교를 나타낸다. 목표는 30년대부터 역사적인 콤플렉스에 있다는 사실을 잊게 하는 전통과 현대 공연이 혼합된 현대적 공간을 만드는 것이었다. 모든 방에는 현대적인 천장 환기 장치가 설치되어 있다. 현대적인 디자인 환경과 고전 발레의 대조를 통해 다시 한 번 메인 컨셉트 제안에 대한 언급이 이루어진다.

Catherine Zaluski

Interior Architecture & Design
BA(Hons)

Studio Studio 2:
The state of independence
Stoke Newington Town Hall

SITE LOCATION
Stoke Newington Town Hall

Stoke Newington was designated a conservation area in 1983 and covers Clissold Park conservation area along side Church Street, including substantial parts of the High Street and all of Abney Park Cemetery. The conservation area was reviewed in 2004, the boundaries were revised. The area is mostly commercial with some residential and several listed buildings including the Town Hall.

스토크 뉴잉턴은 1983년에 보존 구역으로 지정되었고, 하이 스트리트와 애비 공원 묘지의 상당 부분을 포함한 클리솔드 공원 보존 지역을 교회 가와 나란히 덮고 있다. 보존 구역은 2004년에 검토되었다. 경계선은 개정되었다. 이 지역은 대부분 상업적으로 일부 주거지와 시청을 포함한 몇몇 상장된 건물들로 이루어져 있다.

The Town Hall and Assembly Hall are Grade II listed designed by John Reginald Truelove. The building was originally built in 1936 and is a great example of Art Deco. The Council currently use some of the building to house teams such as their Emergency Planning team, as do the public for weddings and other functions.
The Town Hall complex fronts Stoke Newington Church Street has Edwards Lane to the east, Lordship Terrace to the north with St Mary's Church (including the Grade II listed Old Churchand and adjoining Churchyard) and the Capital Ring footpath to the west.

타운 홀과 어셈블리 홀은 존 레지날드 트루엘로브가 디자인한 Grade II 이다. 이 건물은 원래 1936년에 지어졌으며 Art Deco의 훌륭한 예다. 의회는 현재 비상 계획 팀과 같은 팀들을 수용하기 위해 건물의 일부를 사용하고 있으며 결혼식이나 다른 행사에도 일반인들이 사용하고 있다.
타운홀 단지 앞면 스토크 뉴잉턴 교회가는 동쪽으로 에드워즈 레인, 북쪽으로 세인트 메리 교회와 함께 있는 영주 테라스, 서쪽으로 캐피털 링 산책로가 있다.

Front View

Back View

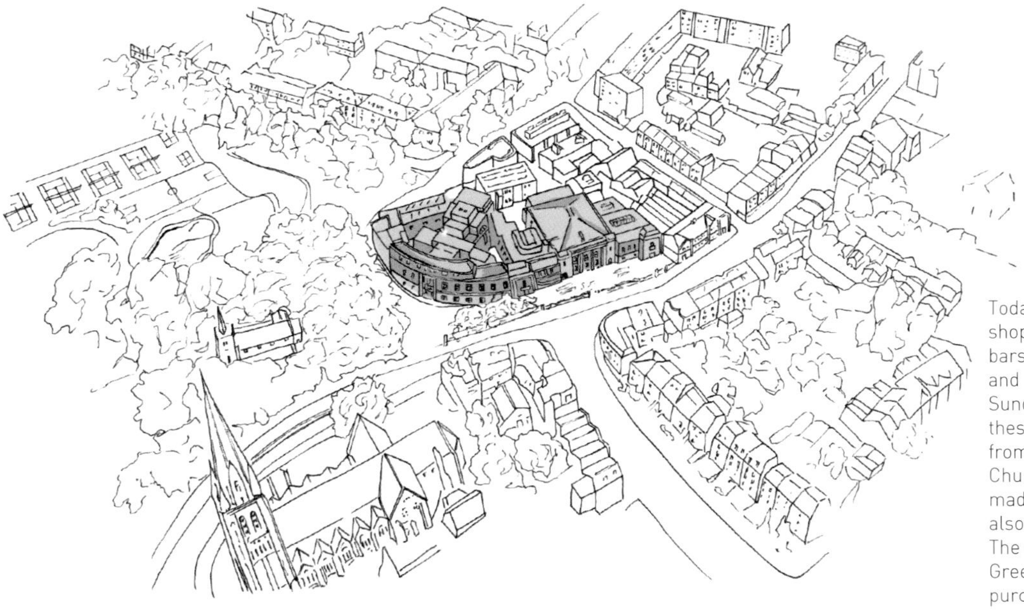

Today Church Street has a selection of independent shops including vintage and boutique clothes stores, bars, clubs and a wide choice of restaurants, café' and coffee shops, including a farmer's market every Sunday. For more major or well-known chain store these are found in the High Street just down the road from Church Street.
Church street is mainly a residential area and mainly made up of Victorian style terrace houses, there are also a few housing estates around the area.
The cost of renting is high but slightly cheaper on the Green Lanes side of Clissold Park. Property prices to purchase are expensive and out of the price range of most first-time buyers.

오늘날 교회가는 빈티지 및 부티크 의류 상점, 바, 클럽, 일요일마다 농부의 시장을 포함한 다양한 레스토랑, 카페, 커피숍을 포함한 독립적인 상점을 선정한다. 더 큰 체인점이나 잘 알려진 체인점들은 교회 거리에서 바로 아래 High street에서 발견된다.
교회가는 주로 주거지역이며 주로 빅토리아풍의 테라스 주택으로 이루어져 있으며, 그 주변에는 몇 채의 주택도 있다.
임대료는 비싸지만 클리솔드 공원의 그린 레이네스 쪽에서는 약간은 싼편이며 구입해야 할 부동산 가격은 비싸고 대부분의 최초 구매자들의 가격대를 벗어난다.

The area has a very mixed community, but a prominent population of Kurdish and Turkish, alongside wealthy young professionals, hipsters and people born and bred in Stoke Newington. Next to Stoke Newington Town Hall sits Clissold Park, popular with families and for lesuire activities.
Since the 1990's Stoke Newington has seen a transformation from a fairly rough area into one of the most desirable postcodes in north London.

이 지역은 매우 다양한 공동체를 가지고 있지만 부유한 젊은 전문가, 힙스터, 스토크 뉴잉턴에서 태어나고 자란 사람들과 함께 쿠르드족과 터키인들이 주된 인구층을 이루고 있다. 스토크 뉴잉턴 타운 홀 옆에는 가족과 레저 활동으로 인기 있는 클리솔드 공원이 있다.
1990년대부터 스토크 뉴잉턴은 우편번호 면에서 상당히 어려운 지역에서 북부 런던에서 가장 바람직한 우편번호 중 하나로 변모해 왔다.

PROJECT BRIEF

The focus for Studio 2 was to look at Stoke Newington's Town Hall which is a Grade II listed building, located in Church Street Stoke Newington. How to revitalise the unused connected and adjoining spaces within the building complex. Only 25% of the Town Hall is currently used and the brief was to re-design the connected or adjoining spaces within the Grade II building for its enterprise, education and enjoyment. To rediscover or reinvent the meaning of Civic Centre.
The brief also included the main client a company called The WIRE. The WIRE run an avant garde music magazine founded in 1982 by the jazz promoter Andy Wood and journalist Chrissie Murray. The investigation of the brief included a talk from one of the publishers from the Wire, outlining the background of the company, how they worked throughout their offices. The design of the new space would need to be flexible and adaptable, as they have around 12 to 14 staff members, some of these are part of the design team freelancers.

Studio2의 초점은 Church Street Street Stoke Newington에 위치한 Grade II 건물인 Stoke Newington의 타운 홀을 보는 것이었다. 건물 단지 내 미사용 연결 및 인접 공간 활성화 방안ol 주목적이다. 현재 시청의 25%만이 사용되고 있으며, 그 개요는 시민 센터의 의미를 재발견하기 위해서 Grade II 건물 내에 연결되거나 인접하는 공간을 기업, 교육 및 즐거움을 위해 재설계하는 것이었다.
개요에는 주요 고객인 The WIRE도 포함되어 있었다. WIRE는 1982년 재즈 프로모터 앤디 우드와 저널리스트 Chrissie Murray에 의해 창간된 아반트 정원 음악 잡지를 운영하고 있다.이 브리핑의 조사에는 WIRE의 출판사 중 한 명이 회사의 배경과 사무실 전체에 걸쳐 어떻게 일했는지를 설명하는 강연이 포함되었다.새로운 공간의 디자인은 유연하고 다용적이면서, 12명에서 14명 정도의 직원이 있기 때문이며, 이들 중 일부는 디자인 팀의 프리랜서들이다.

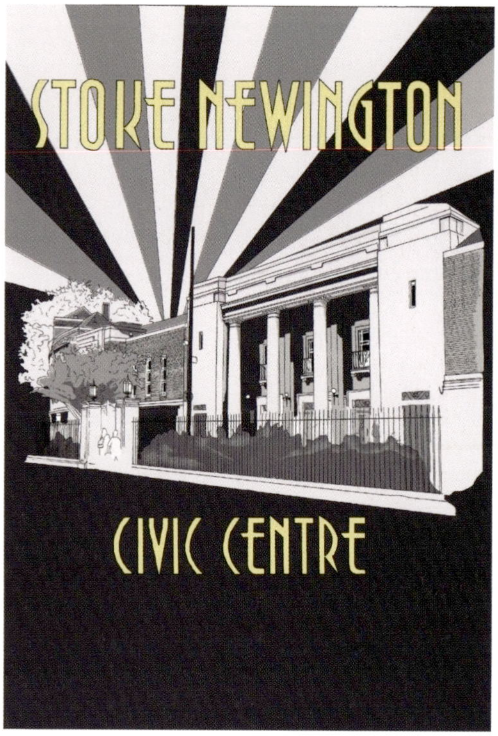

Advertisment Poster Design For Town Hall

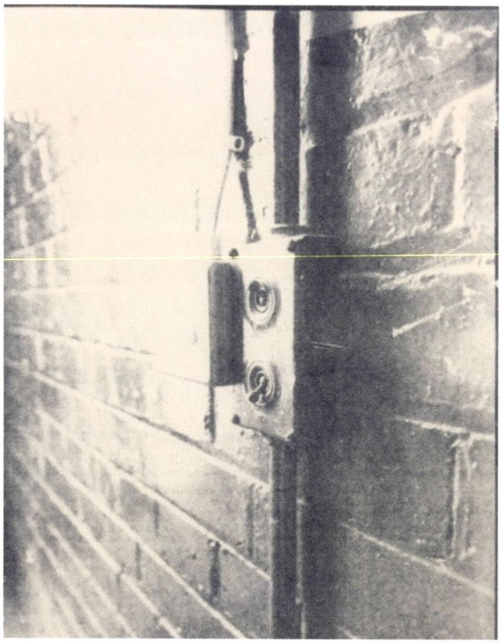
Light Switch Cyanotype

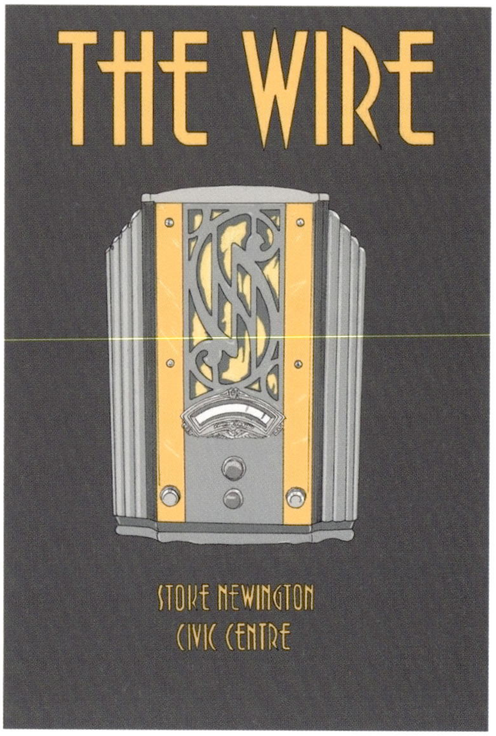

Advertisment Poster Design For The WIRE

Seating Booth Design

The shape of the booth came from research involving the original seating in the Council Chambers at the town hall. The original seating was curved and made from Australian Walnut. These were later replaced. Not only the shape of the original seating but the curve shape in the building has inspired circular and curved designs.

좌석 부스 디자인이 부스의 모양은 마을 회관에 있는 의회 회의실의 원래 좌석을 포함하는 연구에서 나온 것이다. 원래의 좌석은 구부러졌고 호주산 월넛으로 만들어졌다. 이것들은 나중에 교체되었다. 원래 좌석의 모양뿐만 아니라 건물의 곡선 모양도 원형과 곡선의 디자인에 영감을 주었다.

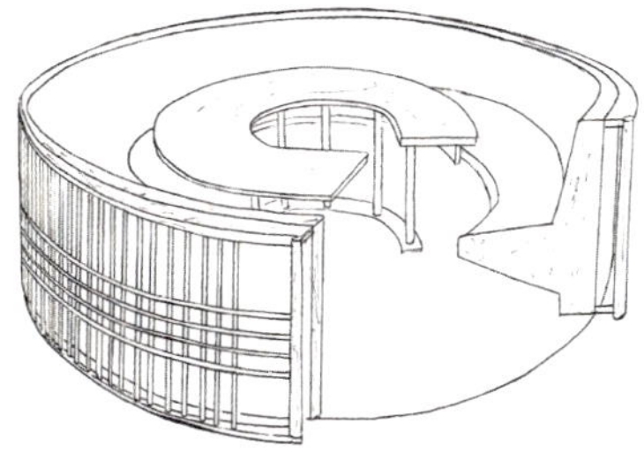

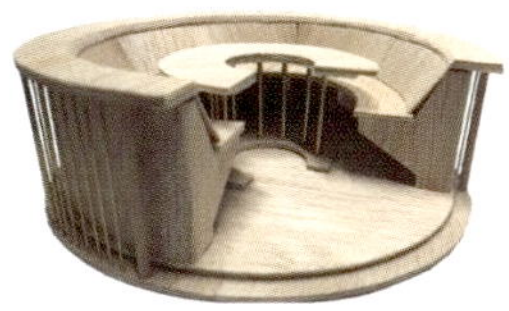

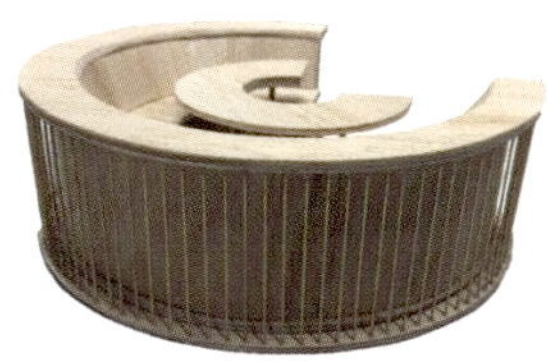

Lightwell Design

Within the lightwells, a number of these walkways could be installed.
Allowing quicker access within the building and the hubs in between could allow for extra outside seating areas. The walkways will be smaller in size and spaced to not obstruct the natural light.

라이트웰 안에는 이런 통로들을 여러 개 설치할 수 있었다. 건물 내부와 그 사이에 있는 허브에 더 빨리 접근할 수 있도록 하면 외부 좌석 구역이 추가로 허용될 수 있다. 산책로는 크기가 작아지고 자연광을 방해하지 않도록 간격을 둘 것이다.

DESIGN DEVELOPMENT

The intension was to look at the under used/vacant areas of the building and bring back into use, to include a commercial elements, educational and attract the public into the building by providing facilities that encourages them to engage. Making it an extension to the community.
Following all the research and investigation, the keys consideration in the individual brief was to use the first floor of the building as the public domain, add a cinema in the Council Chambers that have individual seating booths with additional seating in the original gallery area. To include an exhibition area, that could be used to display the work of local artist or hired out for other business users etc.

의도는 건물의 미사용/사용 지역을 살펴보고, 상업적 요소, 교육적 요소를 포함시키고, 그들이 참여하도록 장려하는 시설을 제공하여 대중을 건물 안으로 끌어들이는 것이었다. 지역사회로의 확장성.
모든 조사와 조사에 따라, 개별 브리핑의 주요 고려사항은 건물의 1층을 공공 영역으로 사용하고, 원래 갤러리 구역에 추가 좌석이 있는 개별 좌석 부스를 가진 Council Chamber에 영화를 추가하는 것이었다. 전시장을 포함하기 위해, 지역 예술가의 작품을 전시하거나 다른 사업 사용자들을 위해 고용할 수 있다.

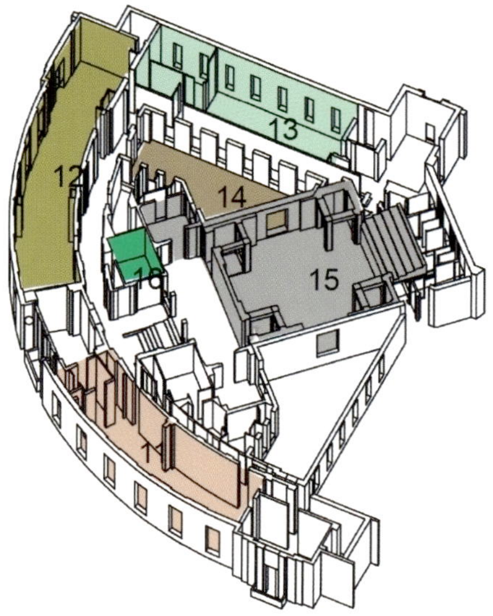

A Café and Lounge Bar would be well used by cinema goers in the evening and also by the general public visiting perhaps the exhibition area, and by the employee within the whole of the Town Hall.
The ground floor would mainly private and two sets of office space would be created, one designed for the requirement of the WIRE another to be rented/leased to currently unknown company.
The ground floor would also contain a radio room (local radio), Podcast rooms from hire and a recording studio. Apart from the radio room, the Podcast rooms and recording rooms could be used for educational / workshops.

카페와 라운지 바는 저녁에 영화를 보러 오는 사람들이나 전시장을 방문하는 일반 대중들, 그리고 시청의 전체 직원들에 의해 잘 이용될 것이다.
1층은 주로 사적인 공간이고 2개의 사무실 공간이 만들어질 것인데 하나는 현재 알려지지 않은 회사에 WIRE를 임대/사용하도록 설계되었다. 1층에는 라디오룸(현지 라디오), 팟캐스트 룸, 녹음 스튜디오도 들어설 것이다.
라디오 룸과는 별도로 팟캐스트 룸과 녹음실을 교육/작업장에 사용할 수 있다.

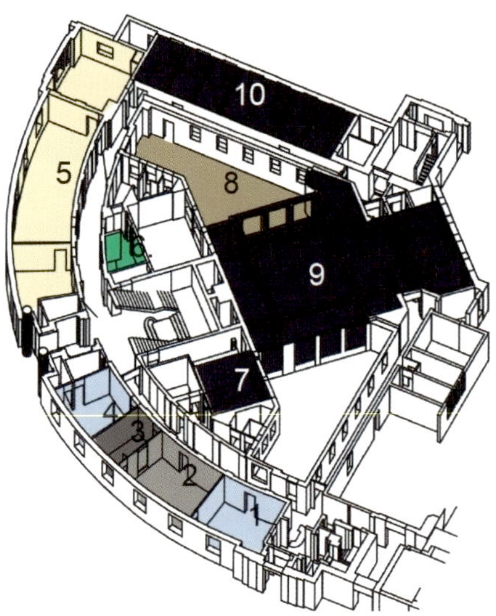

1. Podcast Room/Radio Station
2. Recording Studio
3. Recording Booth
4. Podcast Room
5. Office Space
6. New Lift
7. The WIRE Kitchen Area
8. Light Well Courtyard & New Walkway
9. The WIRE Main Office
10. The WIRE Storage Room
11. Cafe Area
12. Exhibition Space
13. Bar
14. Larger Walkway
15. Independent Cinema
16. New Lift

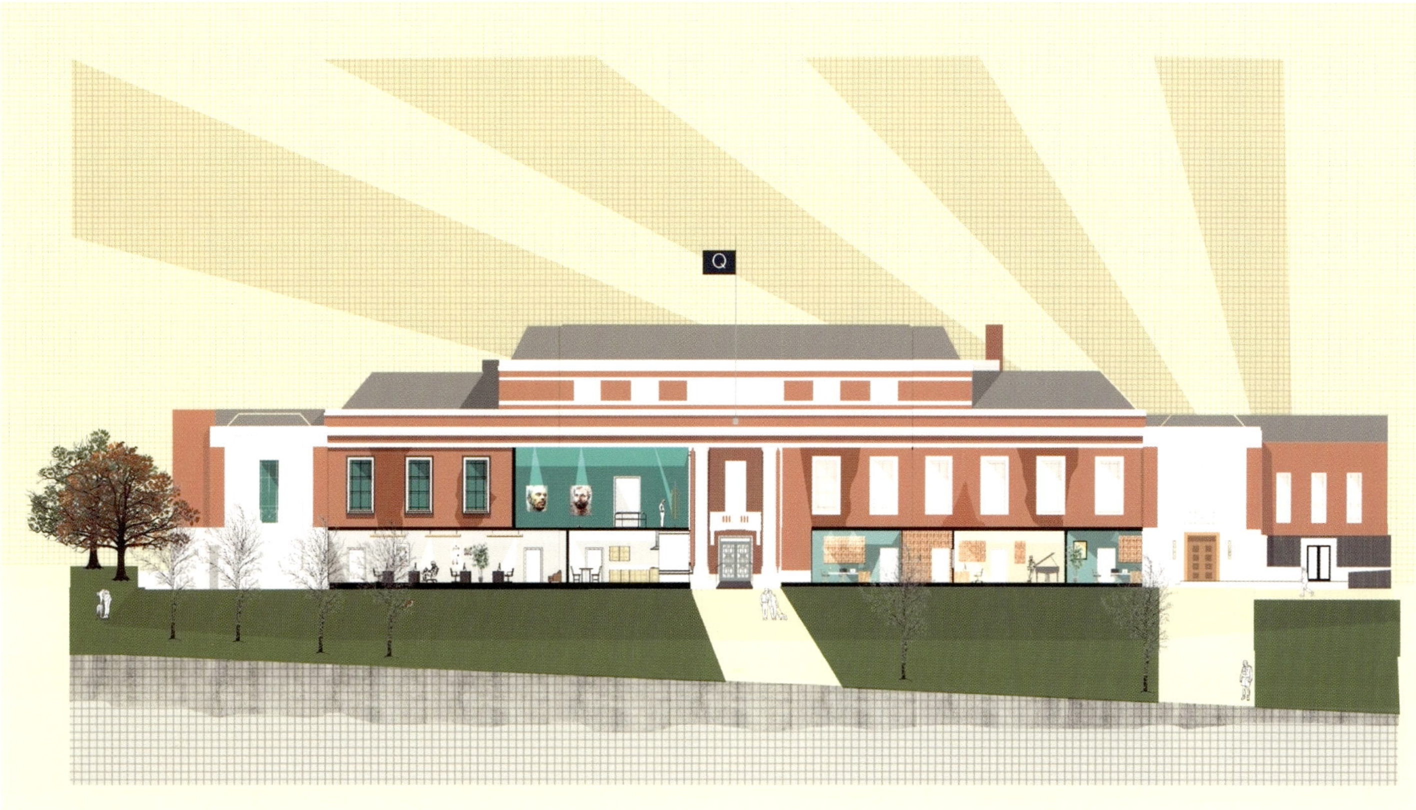

Front Elevation

Ground Floor

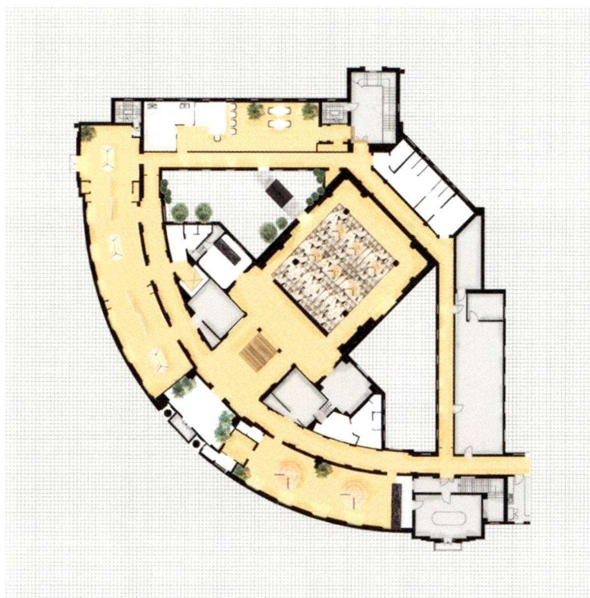

First Floor

The proposal provides for open spaces where the employees can grow their own vegetables, including other areas of planting that will need to be watered. Rain water harvesting will be installed in these areas allowing a sustainable way of maintaining the plants without the need to draw water from the main.
These areas are environmentally sustainable and a benefit to the employees within the building. Offers an opportunity to get out into natural daylight and pick fresh food to eat.

이 제안은 직원들이 물을 주어야 할 다른 모내기를 포함한 채소를 직접 재배할 수 있는 열린 공간을 제공한다. 이들 지역에는 빗물 채취가 설치돼 물을 끌어 올 필요 없이 식물을 유지할 수 있는 지속 가능한 방법이 마련된다.
이러한 영역은 환경적으로 지속가능하며 직원에게 혜택이 주어진다.건물 밖의 자연광을 받으면서 신선한 음식을 고를 수 있는 기회를 제공한다.

Exhibition Room

Front of Hall

Lightwell Section, Showing Cinema & WIRE office

Back Elevation

Charlotte Ponting

Interior Architecture & Design
BA(Hons)

Studio Studio 2:
The state of independence
Stoke Newington Town Hall

The focus for our studio was on reworking a once pivotal building from redundancy to new life whilst creating the opportunity for shared facilities for businesses and the community. The building we are designing within is Stoke Newington Town Hall. Originally built in 1936 functioning as a Civic Centre and Assembly Hall, it holds solid examples of Art Deco design and architecture and is Grade II listed. We aim to avoid the corporate business of co-working and instead considering individual brand identity, friendliness and fun.

우리 스튜디오의 초점은 한때의 중심이었던 건물을 이중화를 통한 새로운 생활로 재작업하는 것과 동시에 기업과 지역사회를 위한 공유시설을 만드는 데 있었다. 우리가 디자인하고 있는 건물은 스토크 뉴잉턴 타운 홀이다. 원래 1936년에 시민회관과 집회관으로 기능하면서 건축되었으며, 아트 데코 디자인과 건축의 확실한 예를 보여주는 Grade II의 건축물이다. 우리는 공동 작업이라는 기업 비즈니스와의 공동 작업의 이미지를를 피하고 대신 개인의 브랜드 아이덴티티, 친근감, 재미를 고려하는 것을 목표로 한다.

The business we are working with is The Wire, a music magazine supporting grass-roots musicians and performance artists across the world. The Wire describe themselves as 'pathologically independent' and want to keep their original values and philosophies throughout the space.

As well as working with The Wire, a new Civic Centre is to be proposed. Currently the building sits 80% vacant and is rarely used by the community so my goal is to bring in the community, creating a space for the local people to meet one and another and have somewhere to spend time in, hire out, eat and have fun.

우리가 함께 일하고 있는 사업은 전세계의 풀뿌리 음악가들과 공연 예술가들을 지원하는 음악 잡지인 The Wire이다. Wire는 스스로를 '병적으로 독립적' 이라고 묘사하며 공간 전체에 걸쳐 원래의 가치와 철학을 유지하기를 원한다.

WIRE와 함께 일하는 것뿐만 아니라 새로운 시민 센터도 제안될 것이다. 현재 그 건물은 80%의 공석이 있고 지역사회가 거의 사용하지 않기 때문에 우리의 목표는 지역사회를 끌어들여 지역 주민들이 서로 만날 수 있는 공간을 만들고, 고용하고, 먹고, 즐길 수 있는 장소를 만드는 것이다.

The building sits just south of Clissold Park which is currently separated by a fence and a disused road, extending the park through the Stoke Newington Town Hall will attract people to enter and use the building.

이 건물은 현재 울타리와 사용하지 않는 도로로 분리되어 있는 클리졸드 공원 바로 남쪽에 자리잡고 있으며, 스토크 뉴잉턴 타운 홀을 통해 공원을 확장하면 사람들이 이 건물에 들어가 이용하게 될 것이다.스토크 뉴잉턴 지역은 당신을 런던에서 캠브리지까지 인도하는 A10의 주요 고가도로를 가지고 있다. 마을회관이 자리잡고 있는 도로는 제과점, 카페, 부티크 가게와 펍들이 즐비한 거리인 Church Street다.

The area of Stoke Newington has a main high street which is the A10 leading you out of London through to Cambridge. The road which the town hall sits on is Church Street, a street lined with bakeries, cafés, boutique shops and pubs.

이 건물은 현재 울타리와 사용하지 않는 도로로 분리되어 있는 클리졸드 공원 바로 남쪽에 자리잡고 있으며, 스토크 뉴잉턴 타운 홀을 통해 공원을 확장하면 사람들이 이 건물에 들어가 이용하게 될 것이다.스토크 뉴잉턴 지역은 당신을 런던에서 캠브리지까지 인도하는 A10의 주요 고가도로를 가지고 있다. 마을회관이 자리잡고 있는 도로는 제과점, 카페, 부티크 가게와 펍들이 즐비한 거리인 Church Street다.

BUILDING MATERIALS

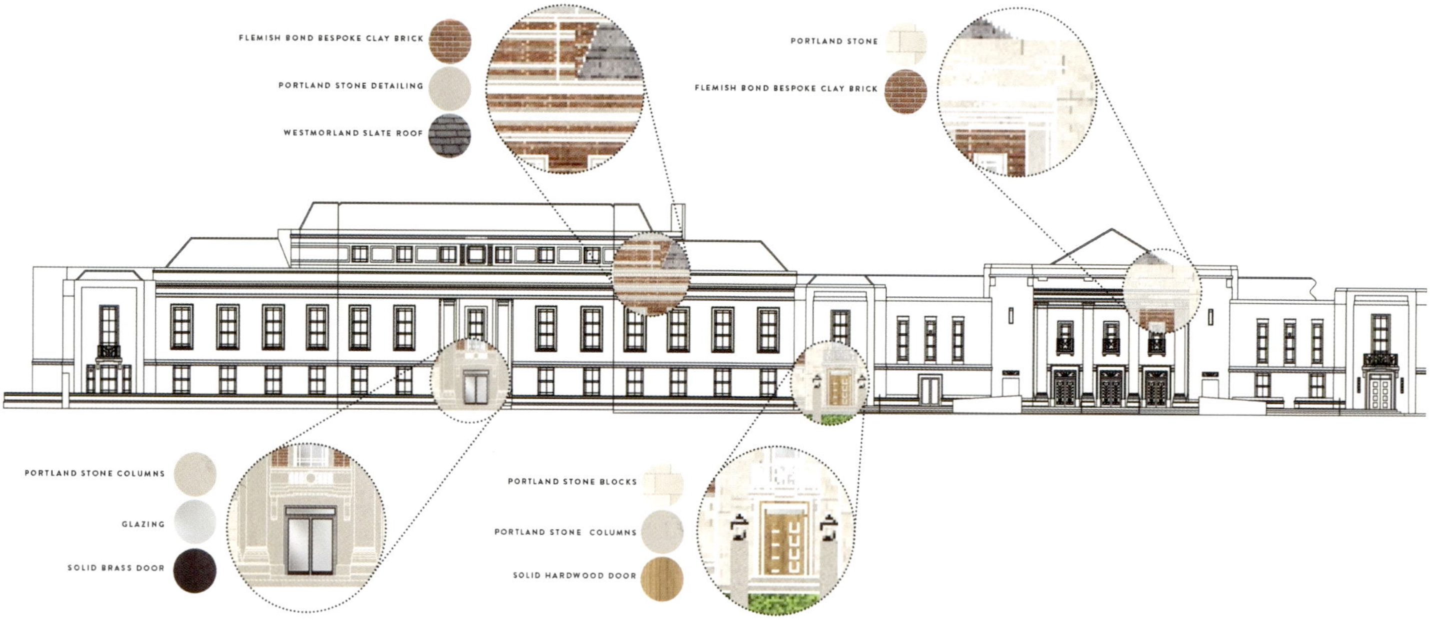

CONCEPT COLLAGE

THE WIRE OFFICE COLLAGE

Our client for this project was The Wire, a grass roots music magazine. We visited their current office in Hackney to see how they work and what was important to them in the work place. This is a collage of that main elements picked up from the visit.

이 프로젝트의 고객은 인디 음악잡지인 The Wire였다. 우리는 Hackney 에 있는 그들의 현 사무실을 방문하여 그들이 어떻게 일하고, 직장에서 그들에게 중요한 것이 무엇인지 알아보았다. 이것은 방문에서 알게 된 주요 요소들의 콜라주다.

CABINET ROOM DIVIDER

This works as both a room divider and a cabinet. The Wire hold a lot of memorabilia and books, old records, tapes and magazines so designing a piece of furniture to fit these items was ideal.

이것은 방 칸막이 겸 수납장 역할을 한다. WIRE는 많은 기념품, 책, 오래된 기록, 테이프, 잡지를 소장하고 있어서 이러한 물품에 맞는 가구들을 디자인하는 것이 이상적이었다.

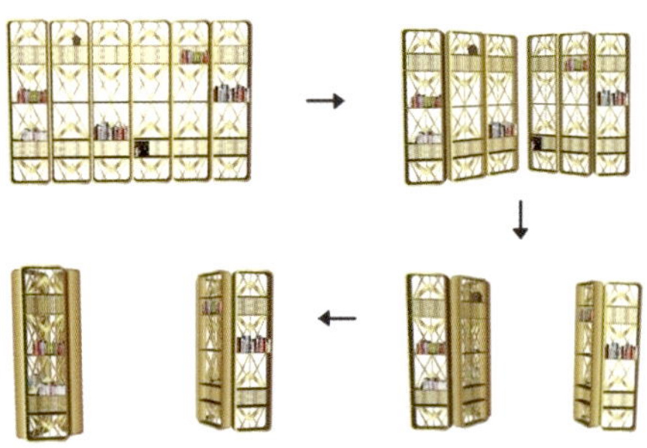

BRASS EFFECT MATERIAL

BRASS ROOM DIVIDER

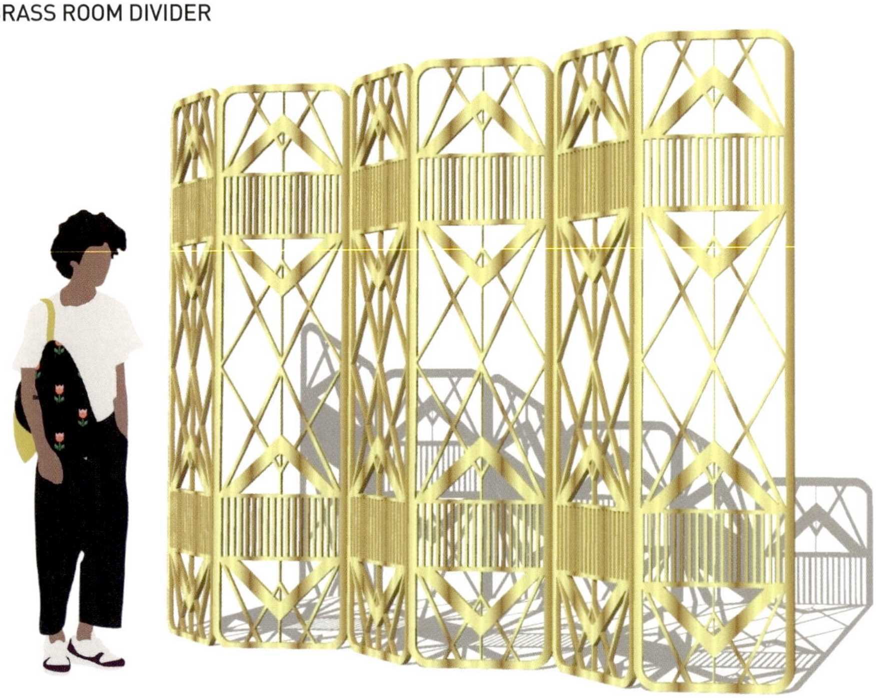

SOLID BRASS

This room divider is elegant, it will create patterns of shadows and light in the room it is used in. This can be used as a screen to divide spaces or as a pull around to make an area feel more private.

이 방 칸막이는 우아하고, 그것이 사용되는 방에 그림자와 빛의 패턴을 만들어 낼 것이다. 이것은 공간을 나누기 위한 스크린으로 쓰일 수도 있고, 한 지역을 좀 더 개인적으로 느끼도록 하기 위해 당김으로 사용될 수도 있다.

ACOUSTIC BAFFLE

BOILED WOOL

Using the same geometric pattern but the negative lines, I designed this acoustic baffle which can be hung from the ceiling to act as a 'cloud' or it can be used as a standing baffle/room divider. It is made out of British wool to absorb the sound and also act as an air purifier.

같은 기하학적 패턴이지만 음의 선을 사용하여, 천장에 매달아 '구름' 역할을 하거나, 혹은 입석 배플/방 칸막이 역할을 할 수 있는 음향 배플을 설계했다. 영국산 양모로 만들어 소리를 흡수하고 공기청정기 역할도 한다.

WIRE
WIRE
WIRE
WIRE
SMEG
DOORWAYS
WAYFINDING
PANELS ALONG WALLS
WOOL
MIRROR
STONE
LIGHT
BRASS
DOORS
MEMOR
LIGHTING ON WALLS.

Whilst designing and brainstorming I made sketches to visualise my ideas. These sketches show furniture ideas, sign-age for the building and rooms and testing room layouts.

디자인과 브레인스토밍을 하면서 아이디어를 시각화하기 위해 스케치를 그렸다. 이 스케치들은 가구 아이디어, 건물과 방 그리고 시험실 배치를 보여준다.

SECTIONAL VISUALS

These sections show most of the main rooms that have been redesigned in the building.

Section AA

This section shows the bar, exercise studio, archive/gallery and the roof garden.

Section BB

This section shows The Wire office, the 'community' room and another view of the exercise studio.

Section CC

This section shows the event room, a space which can be used for gigs, parties ,conferences etc.

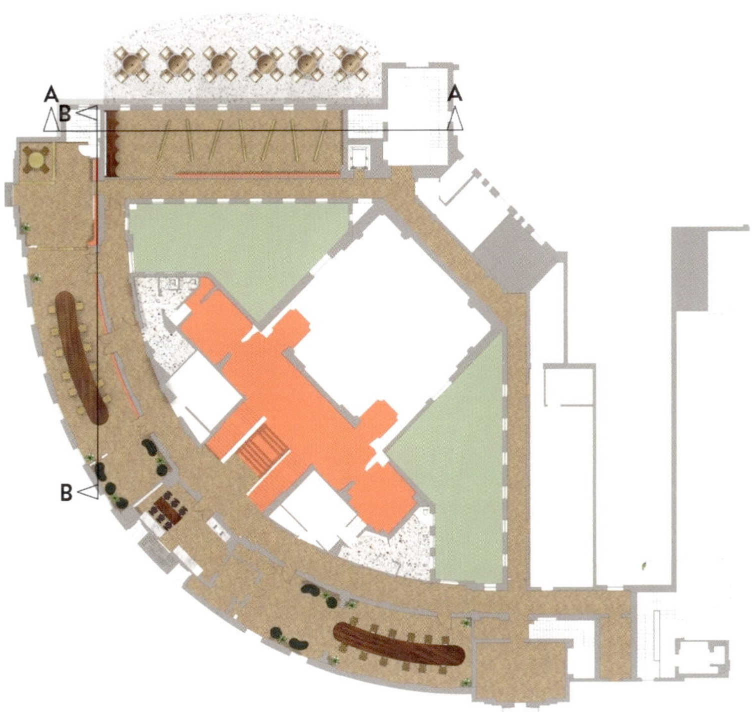

First floor plan showing materiality and layout.

FRONT ELEVATION

To make the building somewhere that people want to visit I have landscaped the front of the building to make it more appealing. The building currently has a big metal fence making it intimidating to enter. The intention is to extend Clissold Park which sits next to the building.

사람들이 방문하고 싶은 곳에 그 건물을 만들기 위해 나는 그 건물의 앞부분을 조경하여 그것을 더욱 매력적으로 만들었다. 그 건물에는 현재 커다란 금속 울타리가 있어 진입을 위협하고 있다. 그 의도는 건물 옆에 있는 클리솔드 공원을 확장하는 것이다.

STOKE
NEWINGTON
TOWN
HALL

ARCHIVE/GALLERY VISUAL

The Wire have many items and photographs stored away. To get these on display and show their great catalogue an Archive and Gallery space has been designed. The room has 'light stands' designed to fit in with the art deconess of the building. These lights can have the negatives placed on them to enable them to be seen. There will be shelves at one end to show all the items they have collected over the years and at the other end, an area to listen to their archive of music.

WIRE는 많은 물품과 사진들을 보관하고 있다. 이것들을 전시하고 그들의 훌륭한 카탈로그를 보여주기 위해 보관 및 갤러리 공간이 설계되었다. 그 방에는 그 건물의 예술적 단정에 맞게 디자인 된 '라이트 스탠드' 가 있다. 이 불빛들은 그것들을 볼 수 있게 하기 위해 부정적인 것들을 둘 수 있다. 한 쪽 끝에는 그들이 수년간 수집한 모든 아이템을 보여주는 선반이 있을 것이고 다른 쪽 끝에는 그들의 음악보관소를 들을 수 있는 영역이 있을 것이다.

BAR VISUAL

It was really important for my design to have somewhere for the community to gather and meet each other in a social environment. To meet this I designed a Bar/ Cafe space, somewhere which you can come for a coffee or a cocktail, a cake or a canapé. I have replaced windows with Crittall doors which will open up in the summer to extend the space outside to the nearby park.

지역사회가 모여 사회적 환경에서 서로 만날 수 있는 장소를 갖는 것이 정말 중요했다. 이를 위해 커피나 칵테일, 케이크나 카나페를 마시러 올 수 있는 바/카페 공간을 디자인했다. 외부공간을 근처 공원으로 확장하기 위해 여름에 열리는 크리톨 문으로 창문을 교체했다.

The materials used are tactile and intriguing. The colours will run throughout all the spaces.

FURNITURE

All the furniture used is a modern take on Art Deco, lots of colour, velvets and fringing. Having brass details is key throughout my design.

Eleanor Hopwood

Interior Architecture & Design
BA(Hons)

Studio Studio 2:
The state of independence
Stoke Newington Town Hall

TOWN HALL - AREA

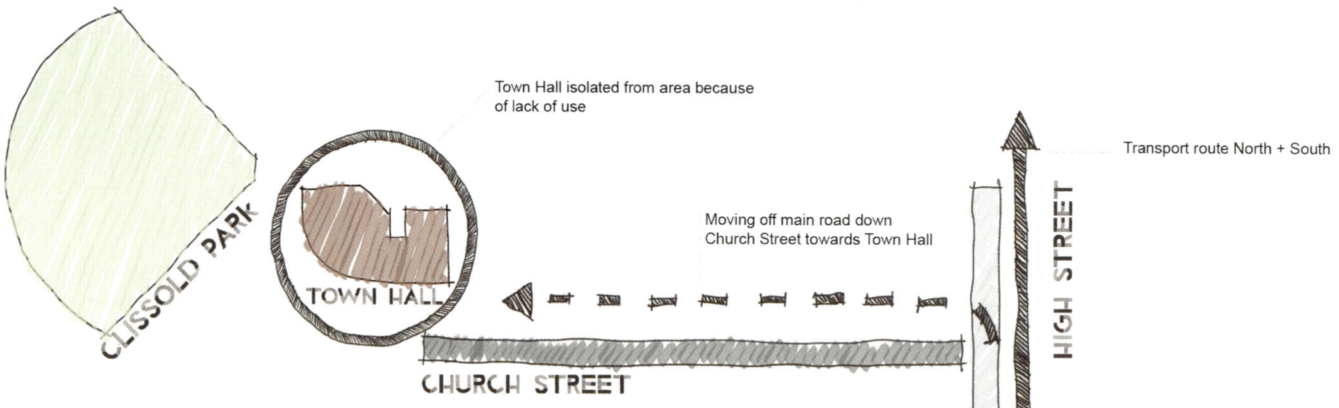

An abudant green area offering the community outdoor activities + historic venues.
Its proximity to the town hall could mean that visitors could be drawn inside after spending time outside - possibly continuing the green space through the building to make it a more cohesive relationship.

지역사회 야외활동 + 역사적 장소를 제공하는 풍부한 녹색 지역. 마을회관과의 근접성은 방문객들이 외부로부터 시간을 보낸 후에 안으로 끌어들일 수 있다는 것을 의미하고 좀 더 화합적인 관계를 만들기 위해서 건물을 통한 좀 더 화합적인 관계를 만들 수 있는 지속 가능한 녹색지역

The quieter of the two main streets in the area
Offering a number of independant cafes + shops it has a friendly + inviting feel with a lot of people meandering slowly exploring what is on offer.
It is very nice to see so many business owners clearly passionate about what they do by the carefully curated specialist shops.

지역내의 두 개의 주요 도로들 중에 훨씬 조용한 공간.
다수의 독립 카페 와 상점들을 제공하는 것은 많은 사람들이 천천히 그곳에서 제공되는 것을 친근하고 매력적인 느낌을 준다.
많은 사업주들이 세심하고 정성스럽게 마련한 전문 상점에서 그들이 무엇을 하고 있는 지를 열정적으로 보여주는 것은 매우 좋은 일이다.

Largest + most trafficky road in area, it serves as a typical high street with a lot of main brand shops.
Running north to south it is the main transport route, connecting the more southern parts of east London, Aldgate + Shoreditch with the Stoke Newington.

이 지역에서 가장 크고 밀도가 높은 도로로, 많은 주요 브랜드 상점들이 있는 전형적인 번화가 역할을 한다.
북에서 남으로 운행하는 것은 주요 교통로로서, 런던 동부의 더 남쪽 지역인 올드게이트 + 쇼어디치와 스토크 뉴잉턴을 연결한다.

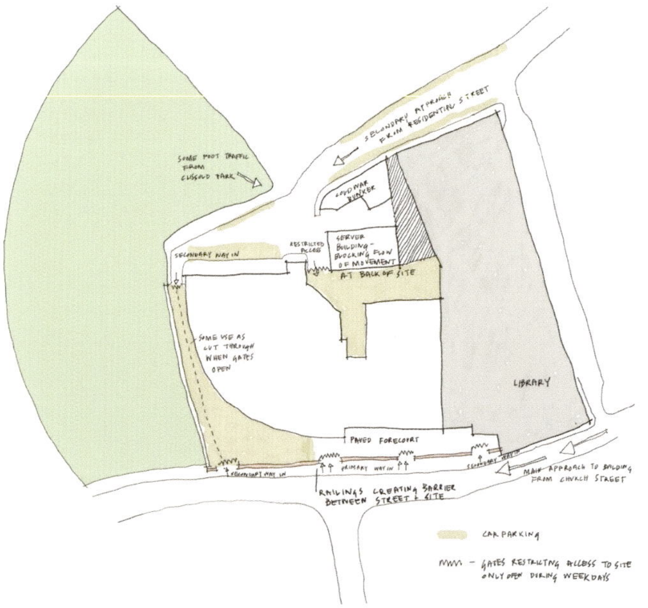

Originating as a medival settlement Stoke Newington still maintains a village feels even with it's close proximity to busier parts of east London.
There is a strong sense of community + caring for the needs of the local people, creating an independant, sensitive identity.
Stoke Newington Town Hall once at the heart of the community, housing the council offices now sits empty + isolated at the quiet end of Church Street.
Bringing community back into the building will reconnect it with its surroundings + the local people.

중세시대 삶의 터전을 기원으로 하는 스토크 뉴잉턴은 여전히 마을 다운 느낌을 유지하면서 동런던의 더 바쁜 지역들과 가깝다.
공동체 의식 과 지역민의 욕구를 배려하면서 독립적이고 남을 배려하는 세심한 정체성을 만들어 낸다.
스토크 뉴잉턴 타운 홀은 한때 지역 사회의 중심부였지만, 구의회 사무소를 수용하는 것은 현재 텅 빈 + 교회 거리의 조용한 끝에 고립되어 있다.
지역사회를 다시 건물 안으로 데려오는 것은 주변환경과 지역주민들을 다시 연결시킬 것이다.

STOKE NEWINGTON COMMUNITY

WHO ARE THEY + WHAT DO THEY NEED

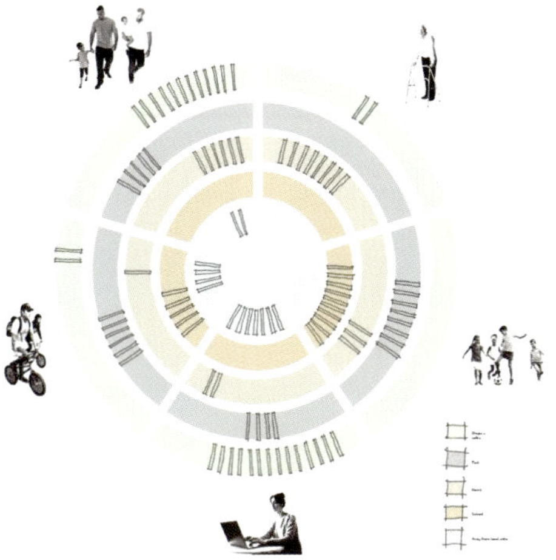

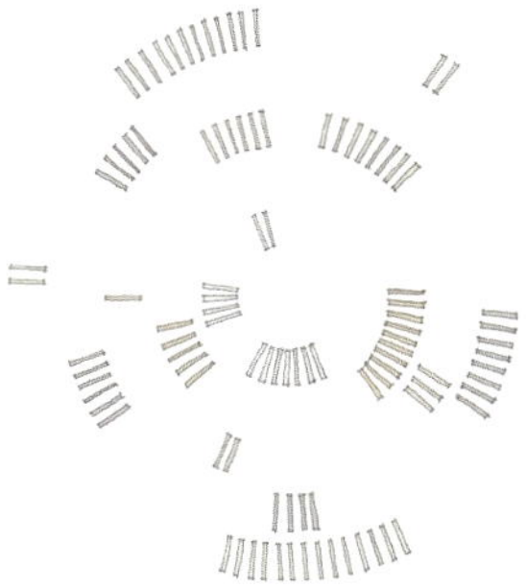

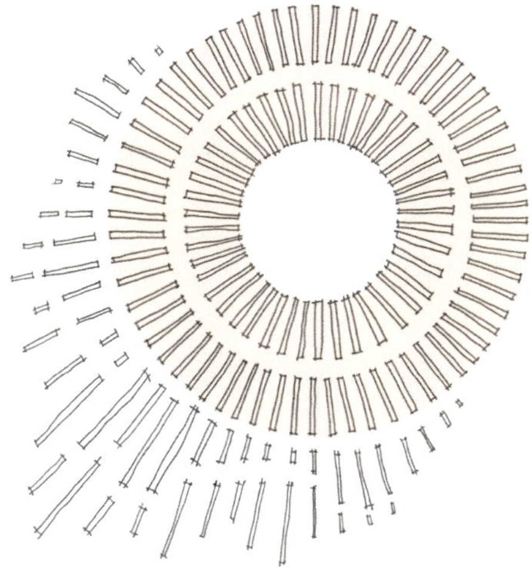

WHO ARE THE PEOPLE OF STOKE NEWINGTON

Where do they spend their time?

STOKE NEWINGTON PEOPLE

The community have places to spend their time but scattered over the area.

NEW COMMUNITY SPACE

Will bring them all together + give opportunities to explore + change perspectives.

COMMUNITY SPACE

Providing a space for the Stoke Newington Community that brings everyone together.
Allowing + encouraging people to explore, adventure + start to change their perspectives on how we see things through our everyday lives.
Bright + comforting environment that draws people in but also inspires.

모두를 하나로 묶는 스토크 뉴잉턴 커뮤니티를 위한 공간 제공.
사람들에게 호기심을 자극하고 장려함으로써 공동체의 사람들이 우리들이 일상생활을 통해 어떻게 사물을 보는 방법에 대한 그들의 관점을 바꾸기 시작한다.
사람을 끌어들이면서도 영감을 주는 밝고 편안한 환경.

A SPACE TO PLAY, THINK, REIMAGINE

THE WIRE MAGAZINE - IDENTITY

COMMUNICATION IS EVERYTHING | Spreading the music + sound experience to people everywhere

Fiercly independant in their grass roots approach, representing underground musicians from all over the world.
Quiet + sentive aesthetic - Communication through writing is most important with design secondary.
A small full time team supported by freelance writers.
Their passion for their work is very apparent - Also seen in the independant shops of Stoke Newington Church Street.

풀뿌리에서 극도로 독립적인 전세계의 지하 음악가를 대표하는 접근법.
차분하면서 섬세한 미학 – 글쓰기를 통한 의사소통은 디자인 부차적인 측면에서 가장 중요하다.
프리랜서 작가들에 의해서 운용되는 작은 단위의 하나의 팀.
그들의 일에 대한 그들의 열정은 누가 봐도 알 수 있을 만큼 명백하고 또한 스토크 뉴잉턴 교회 거리의 독립적인 상점에서도 볼 수 있다.

A SPACE TO HONOUR THE WIRE'S WORK CHAMPIONING MUSICS HIDDEN WONDERS

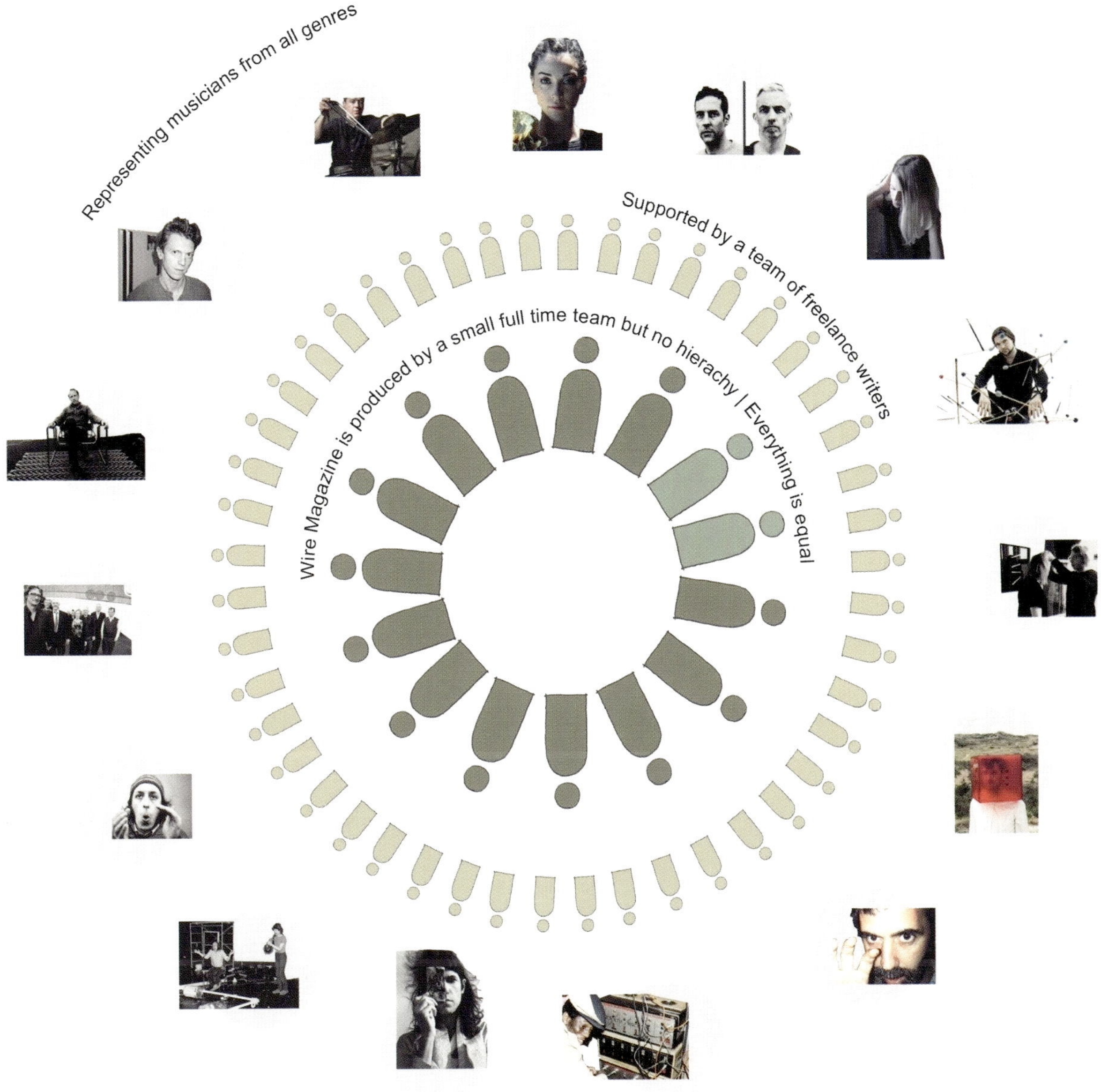

LOCATION

The Wire are very keen to stay in east London which has been there home since they moved from central London because of rising rent prices.

Currently located in Hackney Central only two miles from Stoke Newington + both contained within the London Borough of Hackney.

The wire는 집세 상승으로 인해 런던 중심부에서 이사를 온 이후로 줄곧 런던 동부에 머무르고 싶어한다. Hackney Central에 위치한 현재의 위치는 Stoke Newington과는 오로지 2마일 떨어진 곳이고 London Borough of Hackney내에 포함되어 있다.

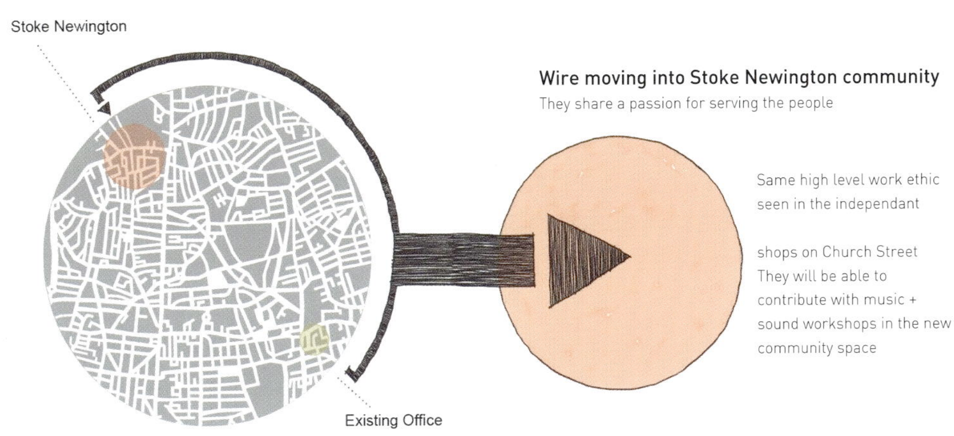

CIRCULATION + WAYFINDING

EXTERIOR

Who will be visiting the building?

How will the building be approached?

How will visitors locate the building?

GREEN SPACE

Improve the relationship between the town hall building + neighbouring Clissold Park by drawing green through the building to Church Street.

교회로에 이르는 건물 초록색으로 물들임으로써 마을회관 건물과 이웃한 클리솔드 공원의 관계 개선

SURROUNDING AREA WAYFINDING

Wayfinding totems introduced peirodically down Church Street to encourage people to continue exploring as far as the town hall.

길찾기 토템은 사람들이 마을회관까지 계속 호기심을 장려하기 위해 교회 가를 따라 주기적으로 도입된다

PROPOSED EXTERNAL WAYFINDING

Outdoor meeting area

Colourful floor treatment to direct visitors to back entrance

Indoor/outdoor meeting area

Mixed planted/paved area

Colourful floor treatment to direct visitors to front entrance

Railings removed to make site more approachable

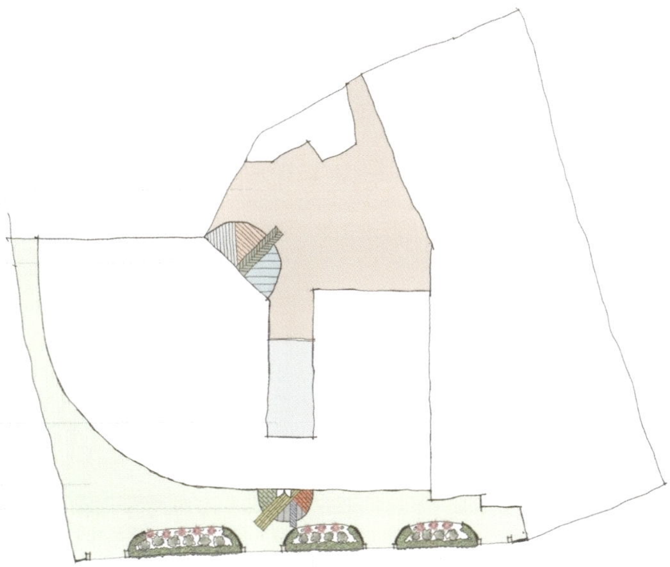

BUILDING PROGRAMME

- Assembly Hall - Wedding venue
- Library
- Entry corridor - Linking front entrance to back area
- Open community areas i.e. workshop, interactive etc
- Closed room community areas i.e. study rooms, meeting rooms etc
- Reception
- Outdoor meeting area
- Private office space including The Wire
- Co-working office space

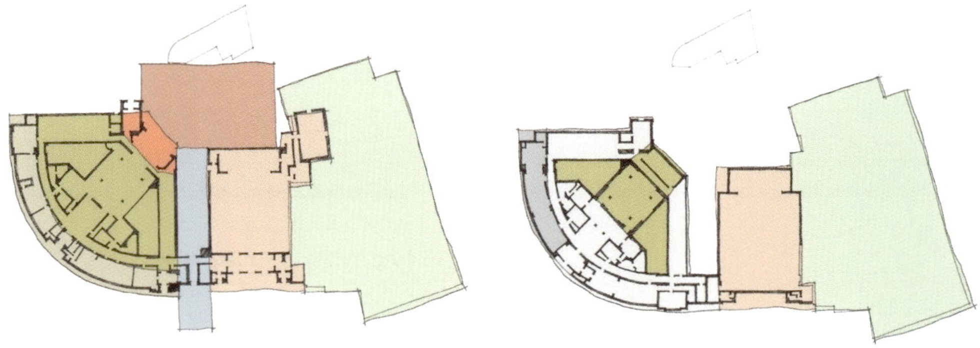

WIRE OFFICE + COMMUNITY SPACE LOCATION

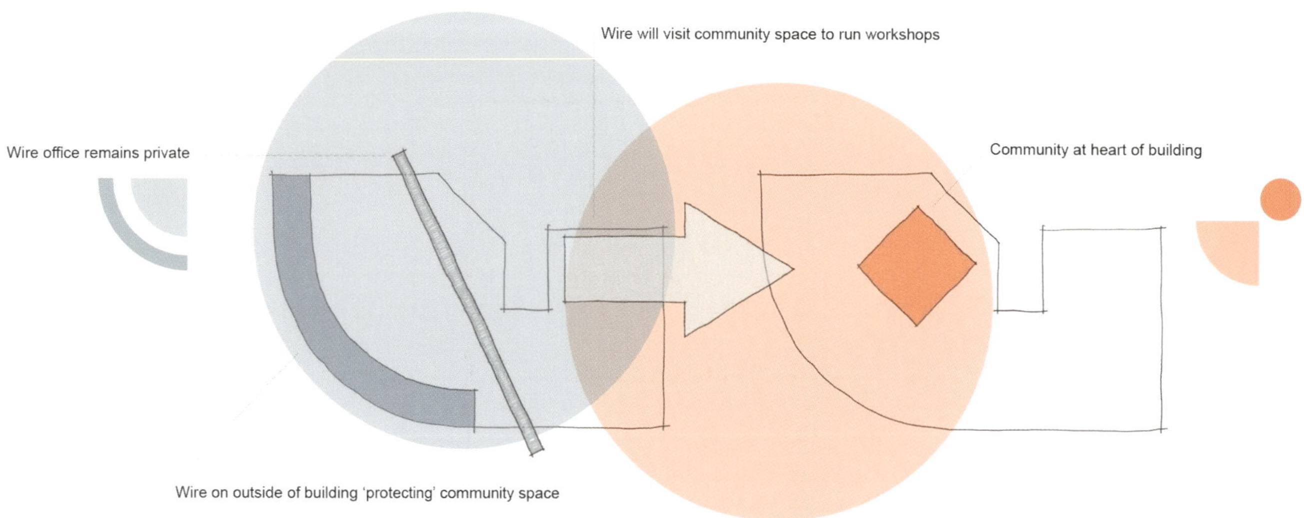

CIRCULATION + WAYFINDING

EXTERIOR

- Who will be visiting the building?
- How will the building be approached?
- How will visitors locate the building?

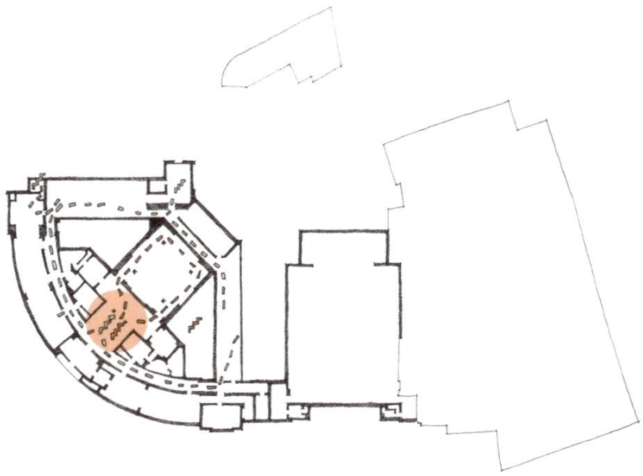

COMMUNITY VISITOR | Journey through building will be about discovering + exploring

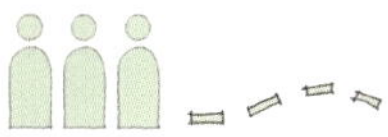

CO-WORKING OFFICE USER | Journey through building will be more direct to destination with some exploring

WIRE OFFICE USER | Journey through building will be very direct to office space

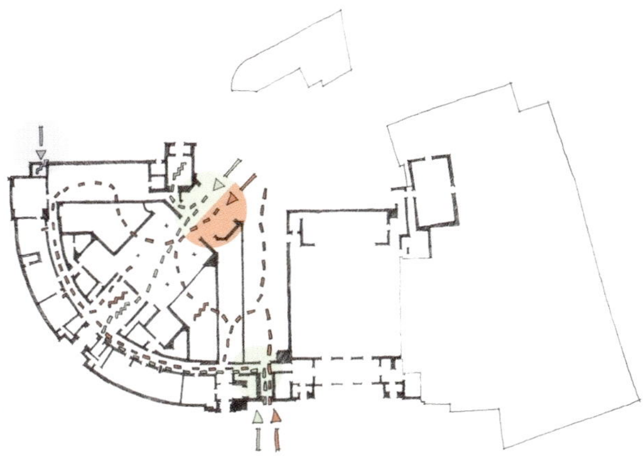

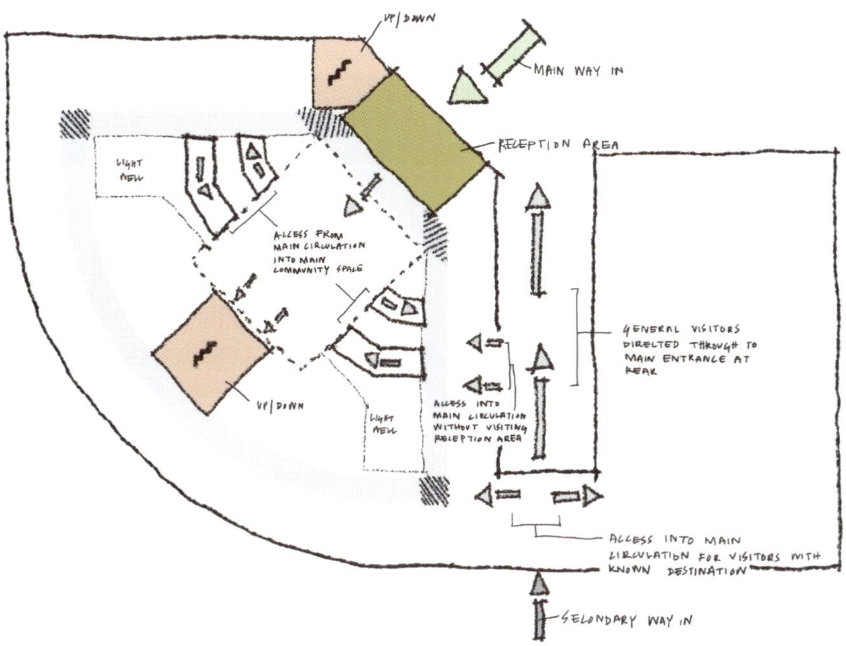

PROPOSED CIRCULATION

COMMUNITY - SPACE PLANNING + DESIGN

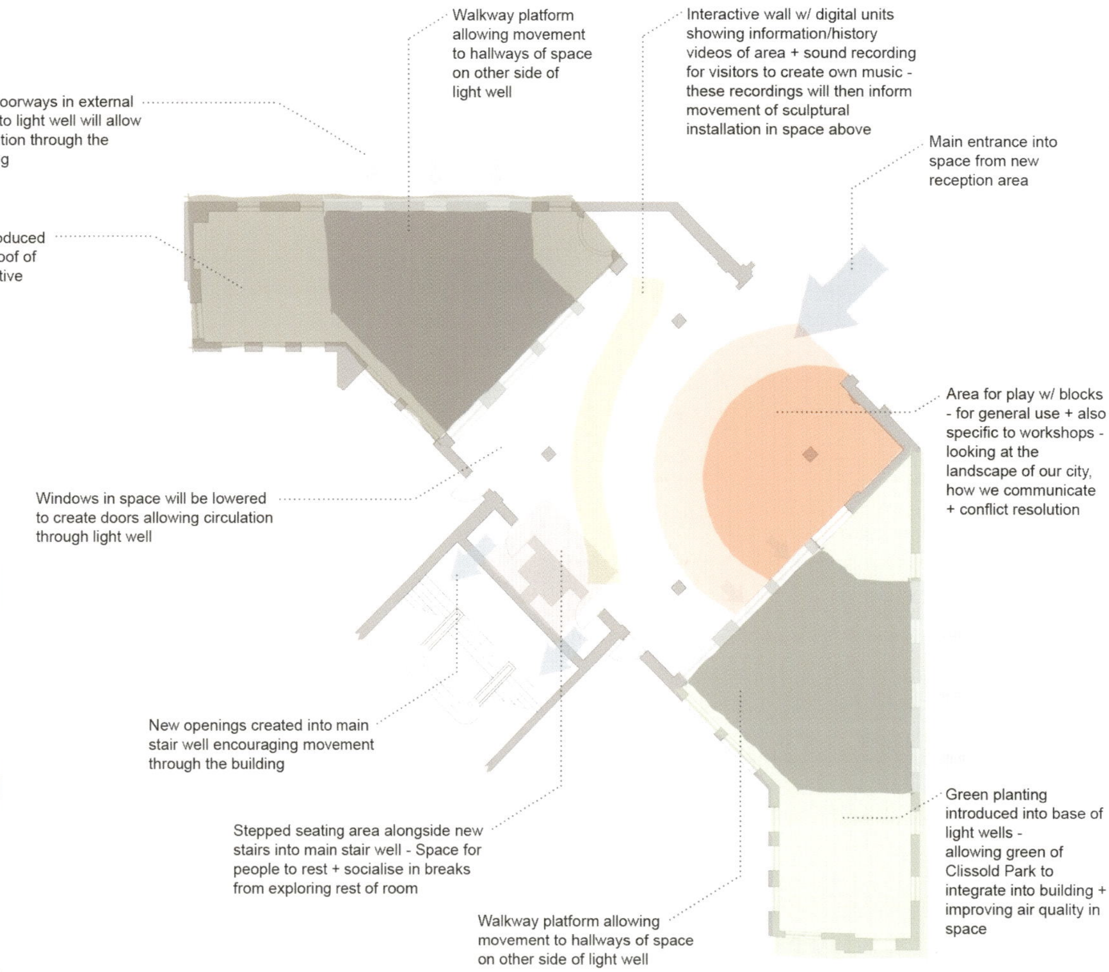

Interactive cultural space that inspires visitors

Flexible to allow for multiple functions - Workshops, music events, exhibition space

Allow room in concept for community to take ownership of space

WIRE OFFICE – SPACE PLANNING + DESIGN

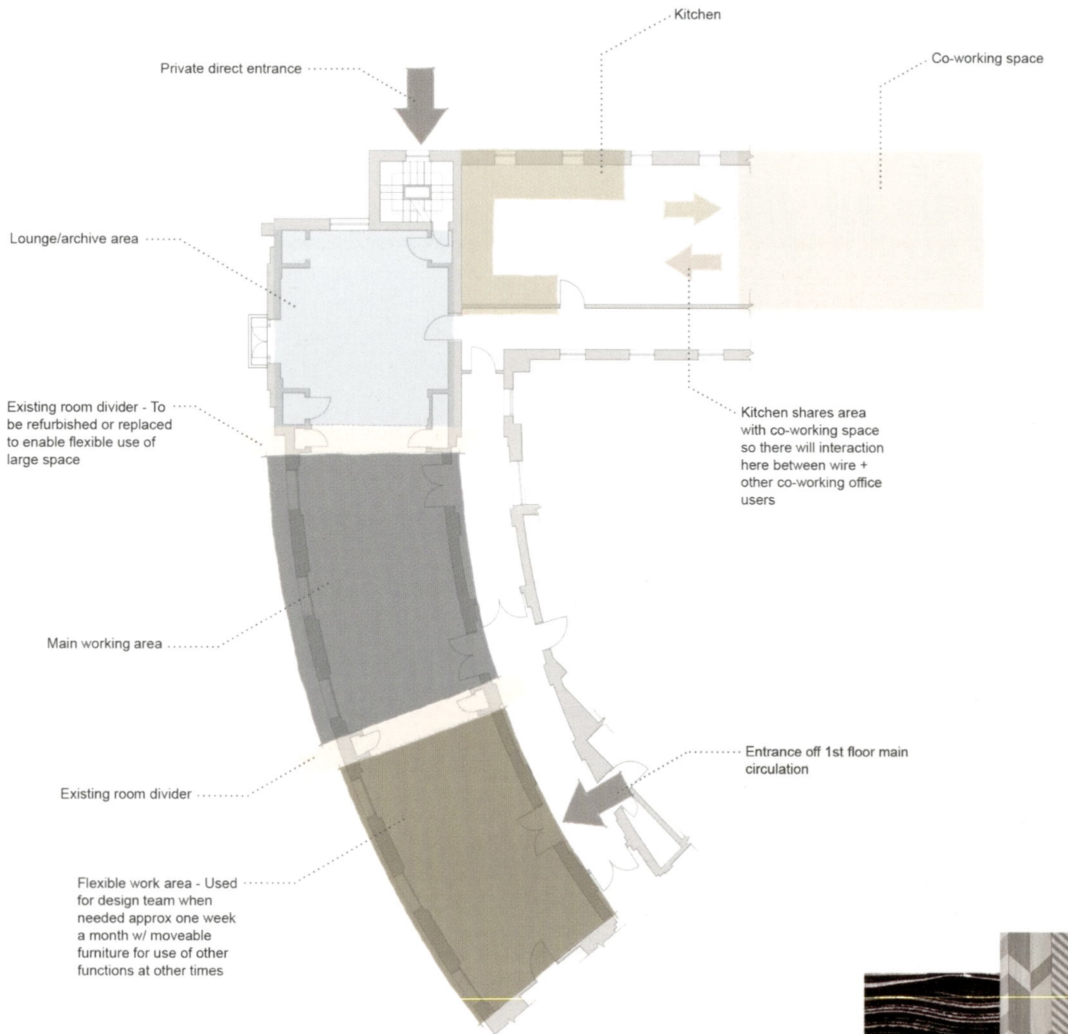

Lighting - Natural | Artificial

Storage - Can be removed and taken with | Individualised for person task

Journey - Curved space

Interaction - Private vs Open

Zoning

Jenny Olver

Interior Architecture & Design
BA(Hons)

Studio Studio 2:
The state of independence
Stoke Newington Town Hall

The Brief

The building we are redesigning is Stoke Newington Town Hall. We were tasked with designing one space for the community and one for a business called The Wire. Here is some of the area research of Stoke Newington town hall and its surrounding locations.

우리가 재설계하고 있는 건물은 스토크 뉴잉턴 타운 홀이다. 우리는 지역사회를 위한 공간과 WIRE 라는 사업을 위한 공간을 설계하는 임무를 맡았다. 여기 스토크 뉴잉턴 타운홀과 그 주변 지역의 지역연구가 있다.

Drawing of Stoke Newington Town Hall, Identify the location being next to the park and churches.

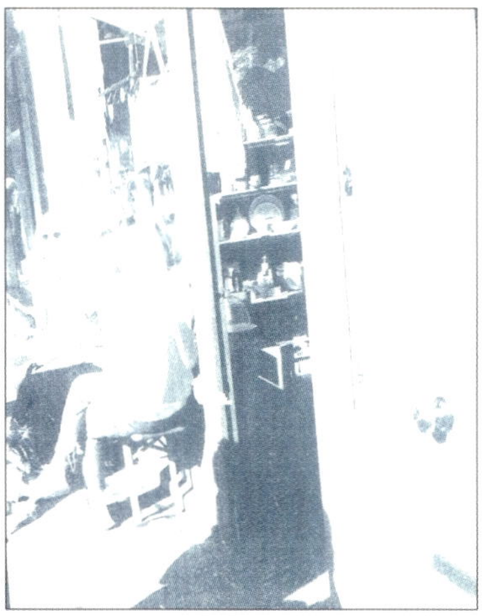

The difference between Church street and the high street is quite dramatic, church street is made up of bespoke artisan shops Vs the common high street shops. This gives Stoke Newington Town Hall a good footfall of people that could be using it daily. This map shows the local schools within the immediate area surrounding the town hall, due to the amount in such a small perimeter. I believe this is a good way to bring the community together, through the children.

Church street와 high street의 차이점은 매우 극적이다. Church street는 맞춤식 장인 상점들과 일반 상점들로 이루어져 있다. 이러한 상점들은 스토크 뉴잉턴 시청에 매일 그것을 사용할 수 있는 상당수의 고객을 제공한다. 이 지도는 읍내를 둘러싼 인접 지역 내의 지역 학교들을 보여 주고 있고 이것이 아이들을 통해 지역사회를 하나로 만드는 좋은 방법이라고 믿는다.

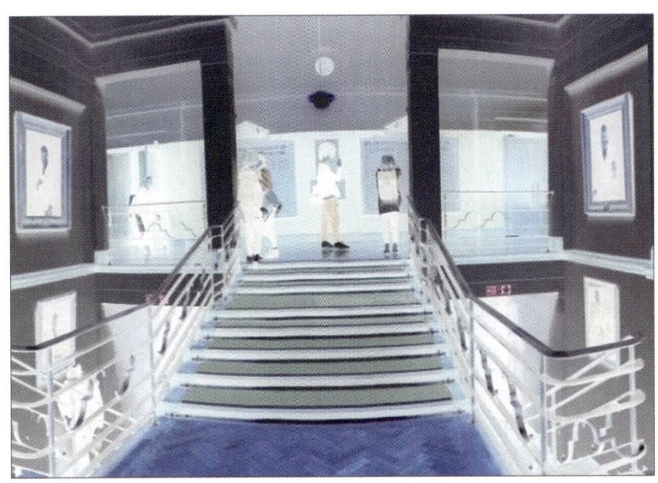

Way Finding

I used my transition space as the way finding around the building as I believe without people being able to find the building and feel comfortable around it then it wont attract people to enjoy the interiors. Using the benches around the exterior of the building and the park will encourage people to take some time out of their day and sit within nature. The community can get their book from the library and have somewhere to sit and enjoy the environment.

길 찾기

사람들이 건물을 찾을 수 없고 건물 주위에서 편안함을 느낄 수 없는 상황에서 내 전환 공간을 건물 주변을 찾는 방법으로 사용했다. 건물 외관과 공원의 벤치를 이용하면 사람들이 하루 중 시간을 내서 자연 속에 앉도록 할 수 있다. 지역사회는 도서관에서 책을 구할 수 있고, 앉아서 환경을 즐길 수 있는 장소를 가질 수 있다.

This is how the frontage of the building with look when the railing has been removed. It gives a completely different vibe of it and invites people to have a stroll through or sit and read their books within the grounds of the town hall. The softer appearance along with the new landscape can make this area an extension to the local park and encourage people to use the facilities more.

난간을 걷어냈을 때의 모습을 담은 건물의 정면은 이렇게 되어 있다. 그것은 전혀 다른 느낌을 주며, 사람들이 마을회관 구내에서 거닐거나 앉아서 책을 읽도록 초대한다. 새로운 풍경과 함께 더 부드러운 외관은 이 지역을 지역 공원으로 확장시키고 사람들이 이 시설을 더 많이 이용하도록 장려할 수 있다.

SOLO

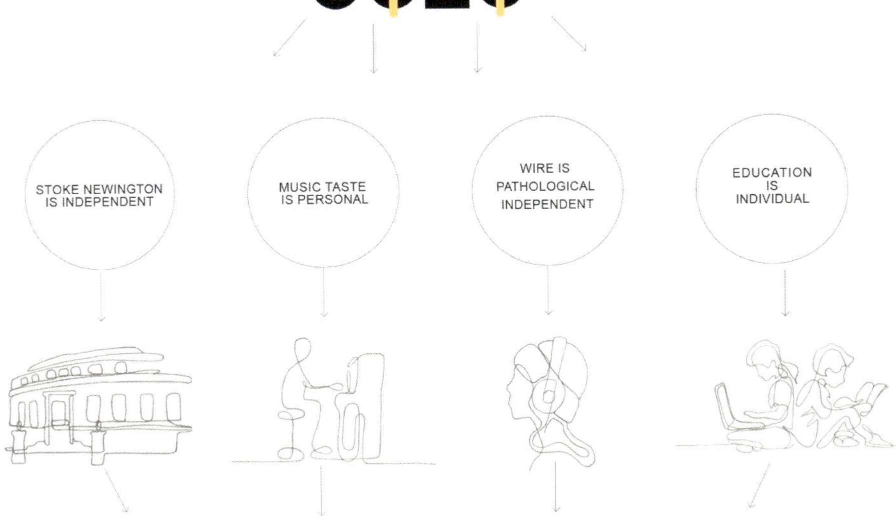

Totem which will be located around the exterior of the building.

This is the logo in-situ at the front of the building where the existing name was shown.

The reason the new name for the town hall is solo is because of the independent and individuals that will be apart of the new community within it. Solo also relates to music which is shown in the dictionary definition.

SOLO

adjective

1. For or done by one person alone; unaccompanied.

noun

2.a piece of vocal or instrumental music or a dance, or a part or passage in one, for one performer.

SOLO

This is a diagram to show the idea behind people coming together and uniting within the new solo building.

The text font is Europhonic- Regular

The yellow lines breaking into the circle are the individual people that will be using the town hall, becoming one as a community in the middle.

Visual Learning

About them:
- Remembers what they read and write.
- Enjoys visual projects and presentations.
- Can remember diagrams, charts, maps well.
- Understand information best when they SEE it.

Spaces that benefit their learning:
- Having pictures and visual simulating information to read.

- They like to have a small distance between themselves and the visual information.

- Standard classrooms are set up in a visual learning way, with a board and all facing the visual information.

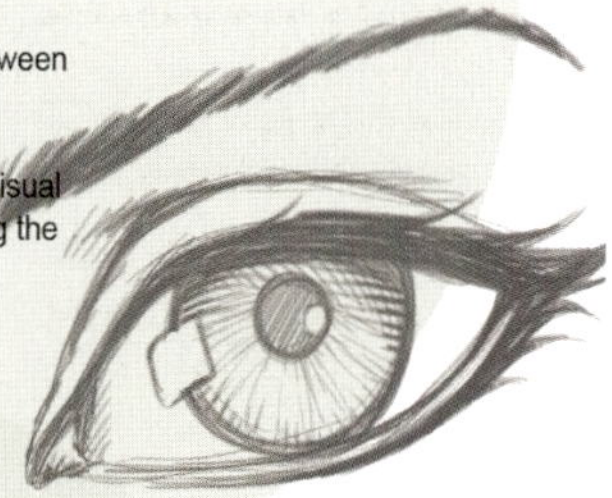

Noise - Silent

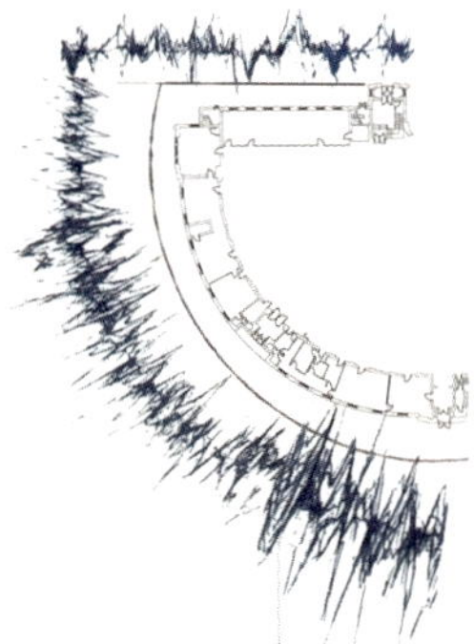

The building is located on a busy road, this affects the front of the building. Noise is something that can distract learners and effect them from learning, therefore is really important to identify which room is best suited to the noise levels. The back of the building is very quite due to being next to a dead end with only a park behind it.

Light - Dark

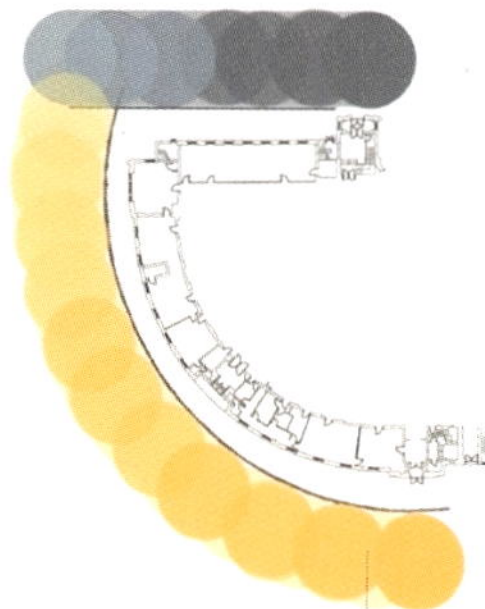

This diagram shows the performance of the space from the natural environment, the sun hits the front of the building through the whole day until about 4 pm where is moves towards the back of the building. I will be choosing where each space goes dependant on the needs and requirements of the activities and if they need light or a dark environment.

Focused Environment

Working within a space where there is a short view point of the information means that there can be any amount of people within the space.

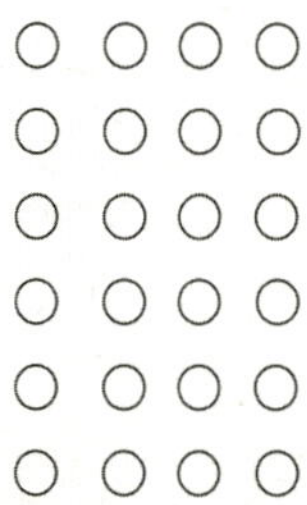

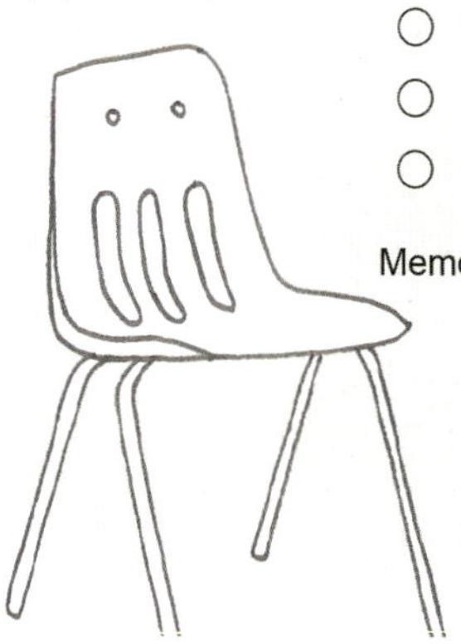

Memory

In this space on the ground floor, mild conversations would be had.

Quite Environment

In this space on the ground floor, it would be very quite space with little talking.

OR

Personal

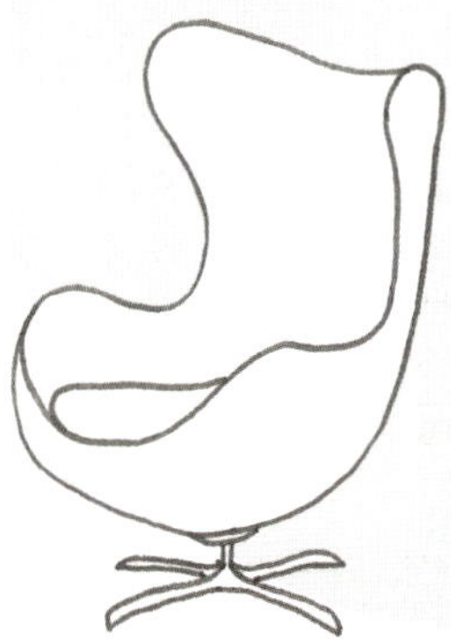

Work best with themselves listening to voice recordings or reading aloud to themselves. However can work with a small group, talking over information by hearing your own voice or a teachers.

For the community space I wanted to make a space where all types of learners could feel comfortable and be able to maximise their learning depending on the different learning environments needed.

Auditory Learning

About them:
- They learn by hearing and listening.
- Understand and remember things they have heard.
- You store information by the way it sounds
- You often learn by reading out loud

Spaces that benefit their learning:
- Having a space which can be separated into a single user space and a small group is the best space.

- Having a quite space is important for them to be able to hear the information they are trying to learn.

- Having headphones in the space would really benefit their learning.

Tactile Learning

About them:
• Learn by touching and doing.
• They understand and remember things through physical movement.
• They tend to learn better when some type of physical activity is involved.
• You need to be active and take frequent breaks
• They often speak with your hands and with gestures, and they may have difficulty sitting still.

Spaces that benefit their learning:
• Having interactive activities where touching, building, moving, or drawing is involved is the best space for tactile learners. Learning tacitly is best done alone where you can get stuck into it.

• Movement is a key part in their functioning brain, rock in a chair are a great way to help with learning.

• Tapping a pencil, shake your foot, or hold on to something while learning is a normal thing for tactile people so having a fidget element will push the potential for these student.

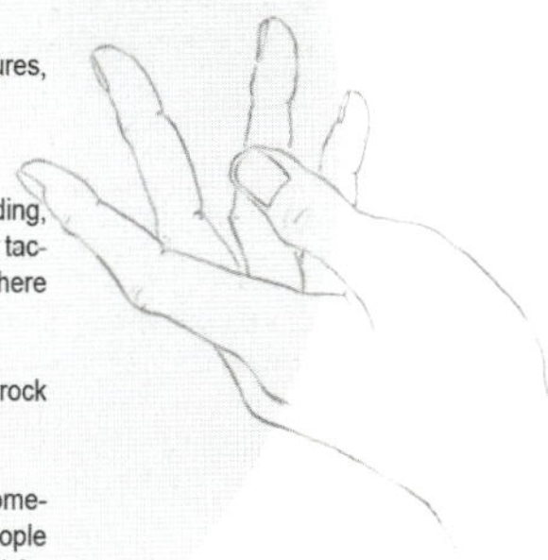

Moving Objects

This space is best for a tactile learning having space which is loud and busy with lots of activities to do on their own but surrounded by lots of people.

Loud Environment

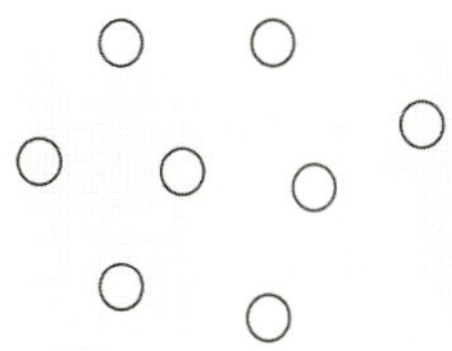

Tasks

In this space on the ground floor, it would be loud with lots of talking and noises from movement.

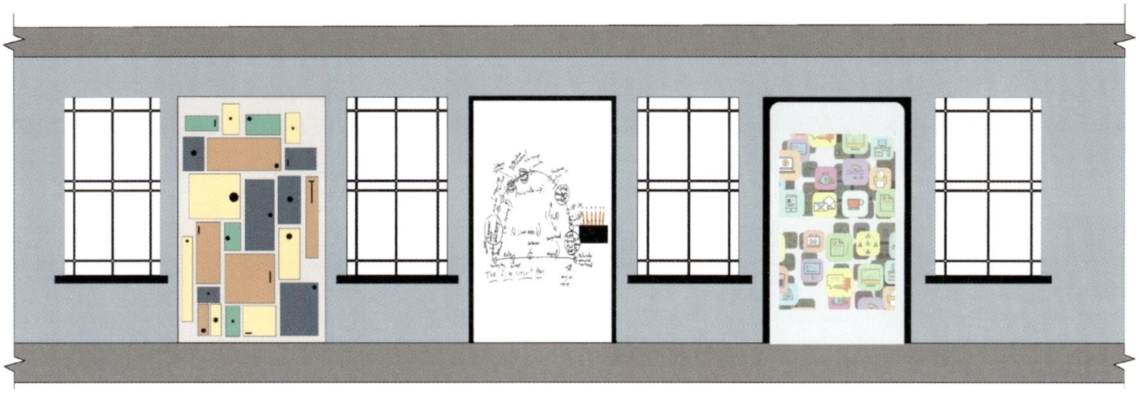

The Building

The buildings natural central point is at the back of the building, the heart of the space with connections to all areas of the building and is the most logically places to have the reception. Here is a diagram of the current receptions circulation path vs the new one if the reception moved to the large square room. The old reception only had two access points where having the new reception in square room allows for 4 access points which allows the circulation of the building to be a lot better.

Due to having children as my main focus for using the space, I believe it is best to show the signage for everyone to see and not limit children by the height the signage is at. Having the signage on the floor will allow them to interact with it and follow it.

건물의 자연 중심점은 건물의 모든 구역과 연결되어 있는 공간의 심장인 건물 뒤쪽에 있으며, 가장 논리적으로 리셉션을 할 수 있는 곳이다. 여기 리셉션이 큰 네모난 방으로 옮겨졌을 때 현재 리셉션 순환 경로 대 새로운 리셉션 순환 경로의 도표가 있다. 구형 접수창구는 네모난 방에 새로운 접수창구를 설치하면 건물의 순환이 훨씬 더 좋아질 수 있는 네 군데의 출입구를 가질 수 있었다.

주된 공간 사용은 아이들을 위한 것임으로, 아이들을 포함한 모든 사람들이 볼 수 있는 표지판을 보여주고, 표지판이 있는 높이에 따라 아이들을 제한하지 않는 것이 최선이라고 믿는다. 바닥에 팻말을 다는 것은 그들이 그것과 상호작용을 하고 그것을 따를 수 있게 해줄 것이다.

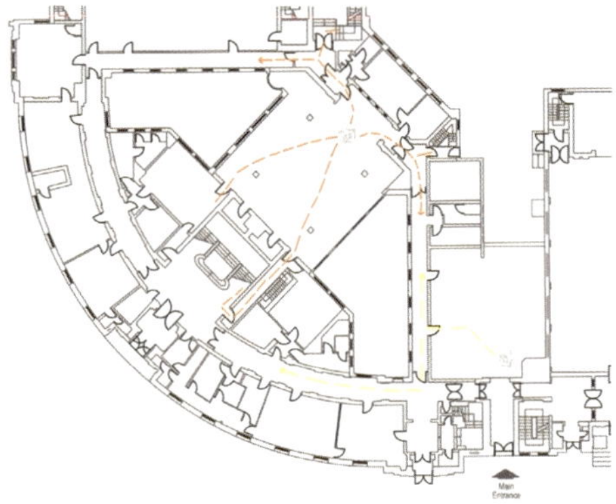

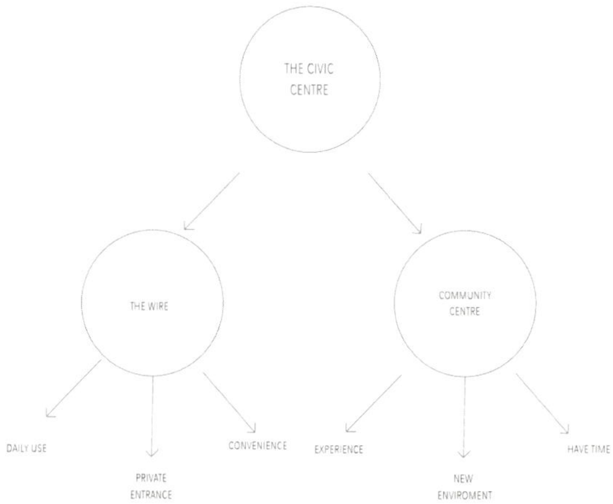

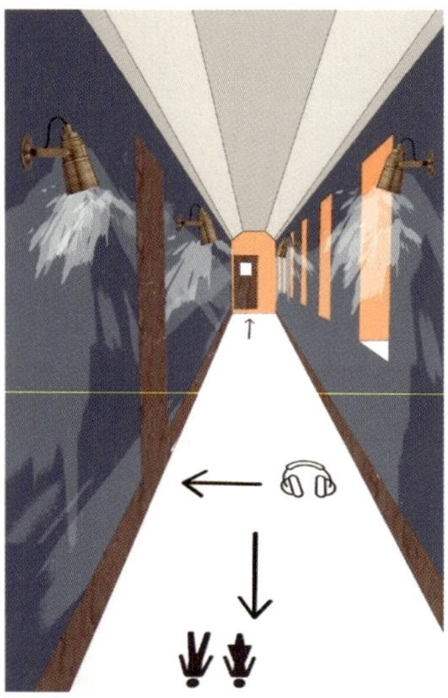

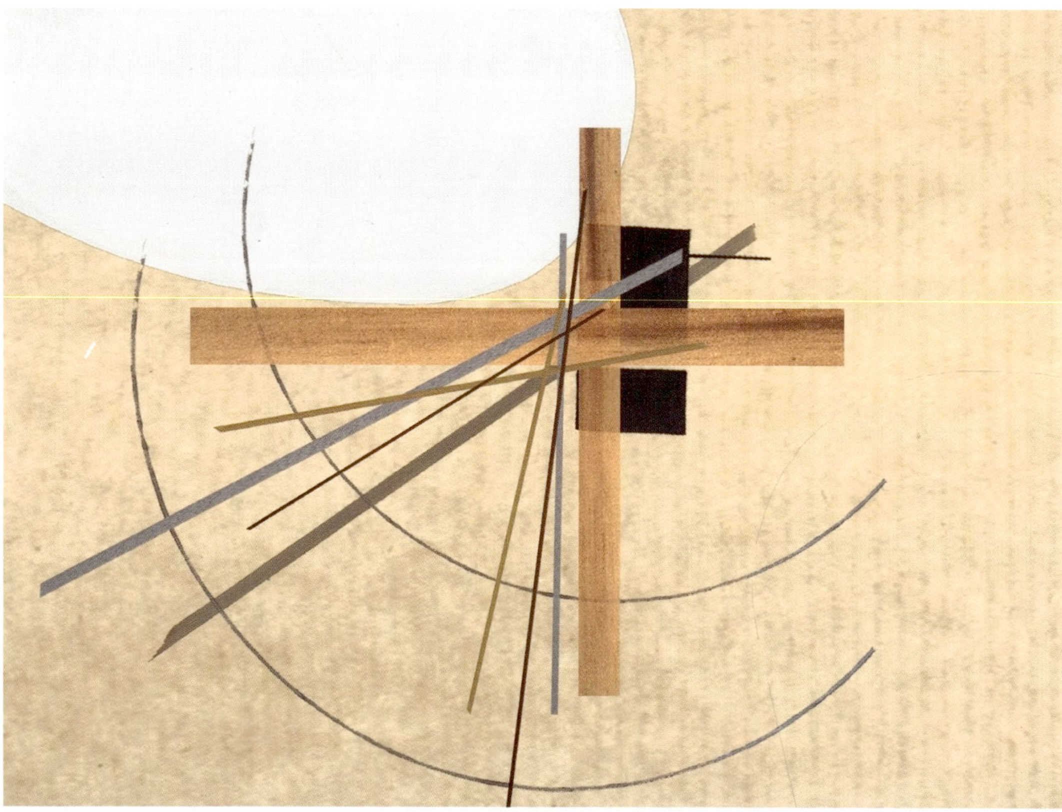

The Wire In Art Deco

The Wire office and Stoke Newington Town Hall have different external looks. However, the two have what each other need to create an ideal office space. Stroke Newington town hall is lacking life inside of it and needs order inside. Whereas the wire is missing space and the community around it of which the Town Hall has. The wire suits a more rural material aspect suiting the Art Deco style. This piece of art depicts the combination of the Stoke Newington Town Hall and The Wire as a Brand and company.

WIRE 오피스와 스토크 뉴잉턴 타운홀은 외관이 다르다. 그러나 두 곳은 서로 이상적인 사무실 공간을 만들기 위해 필요한 것을 가지고 있다. 스트로크 뉴잉턴 마을회관은 그 안에 생명이 부족하고 내부에 질서가 필요하다. 반면에 WIRE에는 공간이 부족하고 마을 회관이 있는 주변 지역도 빠져 있다. WIRE는 아르데코 스타일에 어울리는 시골 소재에 더 잘 어울린다. 이 작품은 스토크 뉴잉턴 타운 홀과 WIRE를 브랜드와 회사로서 결합한 것을 묘사하고 있다.

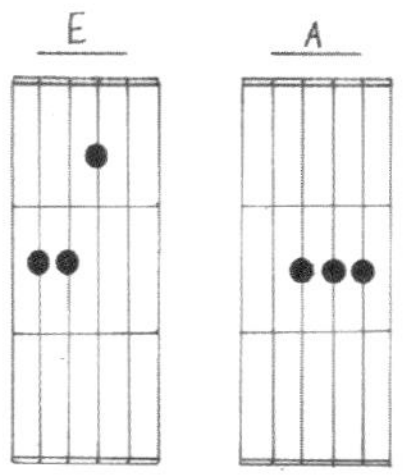

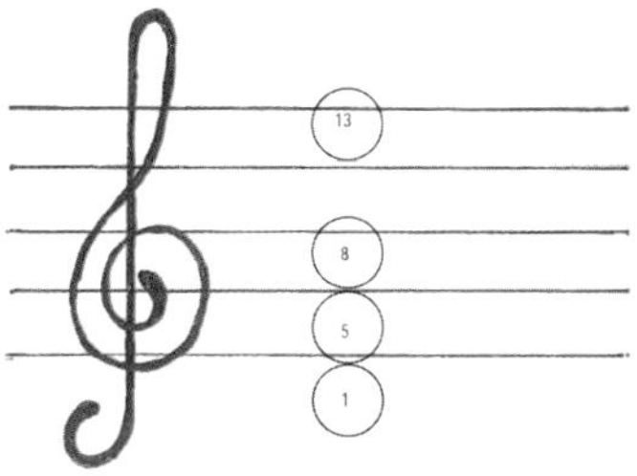

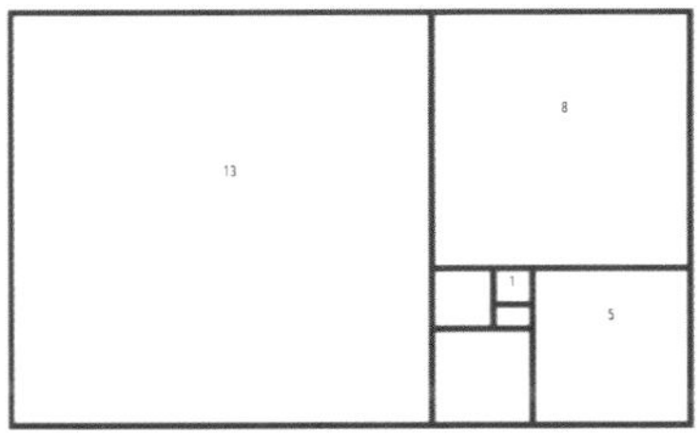

The golden section can be found within music as well as the multiple other areas this equation has been applied to. Fibonacci Sequence can be shown on the left of how music notes are made so that they are pleasing. The link I have made is that this can also be done with the exterior of Stroke Newington town hall.

황금 부분은 음악 내에서뿐만 아니라 이 방정식이 적용된 여러 다른 영역에서도 찾을 수 있다. 피보나치 순서는 음악 음이 어떻게 만들어지는지 왼쪽에 보여져서 기분이 좋아질 수 있다. 내가 만든 연결고리는 스트로크 뉴잉턴 타운 홀의 외관으로도 할 수 있다는 것이다.

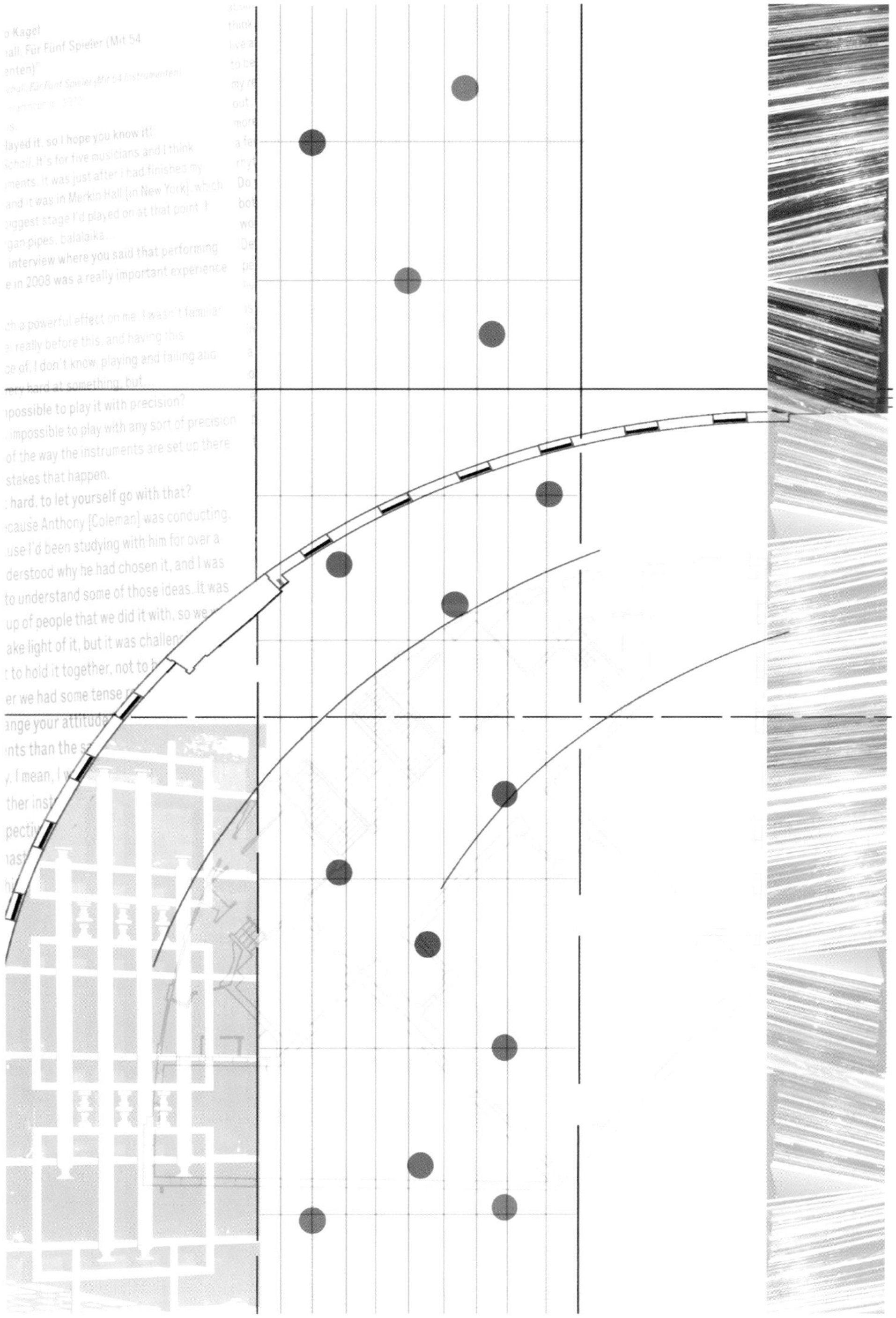

The Wire

This is the current Wire office, it has no natural sunlight making it feel quite small inside. It is organised mess with a place and function for everything within the office. There is a long desk which is all joint for everyone to be able to communicate, The wire has been an in-dependant music magazine since 1892.
I chose this type of artwork to represent

The Wire as it resembles by initial visualisation of the company. After, doing research into the company it became apparent that their visual branding leaves room for the imagination as it is branded with simplicity and contrast. However, this looks to be a clever strategy as they cover an array of music genres of which all carry a unique image.
On first thought I believed a punk rock type feel would suite The Wire office work place, however, upon meeting I realised that their website perfectly represented them and their relaxed nature. I now think a more natural, stripped back approach would be far better suited to the office space, allowing them to fill the office with their individual stamps.

The Wire have a well considered magazine which is quite poetic, when reading one of their article they have a set layout formula that follows through out all their issues. They have tab like this with a little comment which is written with elegant words to highlight the topic within the writing.

이곳은 현재의 WIRE 사무실이다. 그곳은 자연광이 없기 때문에 내부가 아주 작게 느껴지고 사무실 내의 모든 것을 위한 장소와 기능들로 조직되어 있다. 모두가 의사소통을 할 수 있는 긴 책상이 있다. WIRE은 1892년부터 독자적인 음악잡지였다.

회사의 초기 시각화로 인해 WIRE가 닮았다는 것을 나타내기 위해 이런 종류의 작품을 선택했다. 그 회사에 대한 조사를 한 후, 그들의 시각적 브랜딩은 단순함과 대조적으로 낙인이 찍히면서 상상의 여지를 남기는 것이 명백해졌다. 하지만, 이것은 그들이 모두 독특한 이미지를 가지고 있는 일련의 음악 장르를 다루기 때문에 영리한 전략으로 보인다.

처음에 펑크 록 타입의 느낌이 WIRE 사무실과 어울릴 것이라고 믿었지만, 만나자마자 나는 그들의 웹사이트가 그들과 그들의 편안한 성격을 완벽하게 나타낸다는 것을 깨달았다. 나는 이제 좀 더 자연스럽고 벗겨진 뒷면 접근법이 사무실 공간에 훨씬 더 적합할 것이라고 생각하는데, 이것은 그들이 사무실을 개별 우표로 채울 수 있게 해준다.

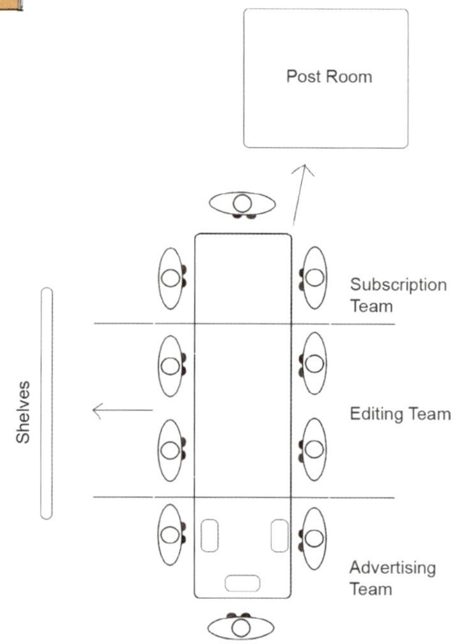

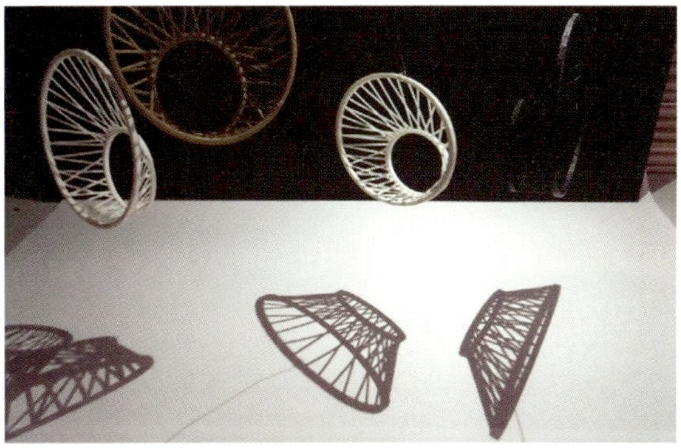

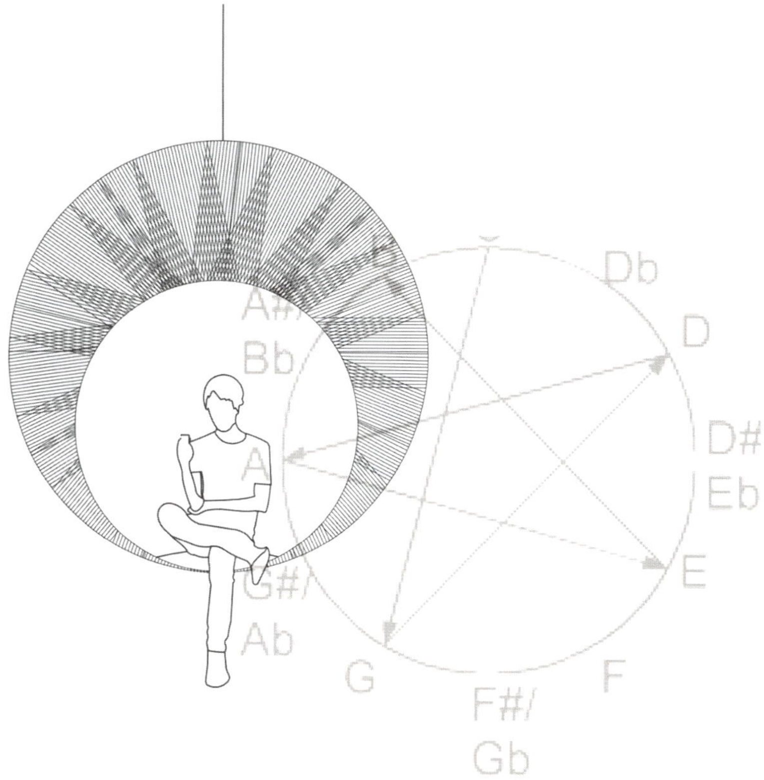

Pathological Piece of Funiture

The piece of furniture above is taken from the idea of a one dimensional object, with no depth. This is inspired by the tuning of instruments. All-fifths tuning refers to the set of tunings in which each interval between consecutive open strings is a perfect fifth. I will be using the lines as a weave for a soft finish on this piece of furniture.

This piece of furniture was inspired by a guitar and its 3 dimensional shape. I wanted a space where people could share a sit, be social and have a break from the office. Once I knew I wanted a seat that could host 4-5 people a bench was the ideal shape. I began to look at the features the guitar had and how the structure could become a bench. This creating 'The Ensemble'

가구 병리학적 조각 위의 가구 조각은 깊이가 없는 1차원 물체의 아이디어에서 나온 것이다. 이것은 악기들의 튜닝에서 영감을 얻었다. 전체 5분의 1 튜닝은 연속적으로 열려 있는 문자열 사이의 각 간격이 완벽한 5분의 1인 튜닝 세트를 말한다. 가구를 부드럽게 마무리하기 위해 그 선을 엮어 쓸 것이다.

이 가구는 기타와 3차원 형상에서 영감을 받았다. 사람들이 함께 앉아 있고, 사교적이고, 사무실에서 쉴 수 있는 공간을 원했다. 4-5명의 사람들을 수용할 수 있는 좌석을 원한다는 것을 알았다. 벤치는 이상적인 형태였다. 기타가 가지고 있는 특징과 구조가 어떻게 벤치가 될 수 있는지 살펴보기 시작했다. 이것이 '앙상블'을 만든다.

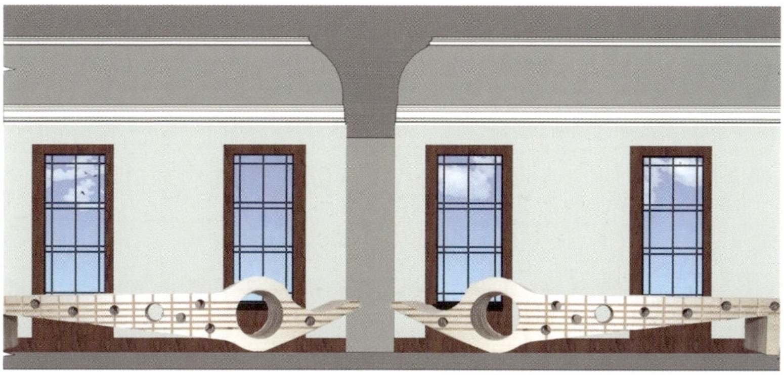

Paula Antonia Hinderer

Interior Architecture & Design
Studio Studio 2: The state of independence
Stoke Newington Town Hall

INDIVIDUAL BRIEF, & SITE PLAN

CLIENT

The London Borough of Hackney council has recently refurbished the Assembly rooms of the Stoke Newington Town Hall in Clissold, but they did not develop a new concept for the building. Now they are asking for concepts as how to use, connect and design the rest of the building.

해크니 시의회의 런던 보루는 최근 클리졸드에 있는 스토크 뉴잉턴 타운 홀의 회의실을 새로 단장했지만, 그들은 이 건물에 대한 새로운 개념을 개발하지 못했다. 이제 그들은 건물의 나머지 부분을 어떻게 사용하고 연결하고 설계할 것인지에 대한 개념을 요구하고 있다.

RE-DESIGN

Re-design five connected spaces in the building, for the reinvention of the Civic Centre, the music magazine the Wire and a second company. This includes community space and workspace. The needs of the community are to use an official place where they can come together as a community for their education and enjoyment. The Wire requires an archive, a gallery, a radio station, and a small performance space. The brief asks to translate the identity of the Wire magazine together with their sense of 'pathological independence' through the Town Hall. Furthermore 'the state of the independence' has to be developed within the building.

Civic Centre, Music Magazine The Wire 및 두 번째 회사의 재개를 위해 건물 내에 연결된 다섯 개의 공간을 재설계해야한다. 여기에는 커뮤니티 공간과 업무 공간이 포함된다. 공동체의 필요성은 그들이 그들의 교육과 즐거움을 위해 공동체로 모일 수 있는 공식적인 장소를 사용하는 것이다. WIRE는 아카이브, 갤러리, 라디오 방송국, 그리고 작은 공연 공간을 필요로 한다. 그 브리핑은 와이어 잡지의 정체성을 타운 홀을 통해 그들의 '병적 독립' 의식과 함께 번역해 줄 것을 요청한다. 게다가 '독립의 상태' 는 건물 내에서 개발되어야 한다.

OUTCOME

The project develops a concept that will fit the requirements of the companies in cooperation with the community space and thereby the community itself. Five areas are designed in detail. The outcome of the design research and development includes the use of a pattern and table for interior and exterior. This will raise awareness and change the mindset of the users based on the idea of 'the state of the independence.'

이 프로젝트는 지역사회 공간과 그에 따라 지역사회 자체에 협력하여 기업의 요구조건에 맞는 개념을 개발한다. 5개 영역이 상세하게 설계되어 있다. 설계 연구 개발의 결과에는 내외부를 위한 패턴과 테이블의 사용이 포함된다. 이것은 '독립의 상태' 라는 생각에 바탕을 둔 사용자들의 인식을 높이고 사고방식을 변화시킬 것이다.

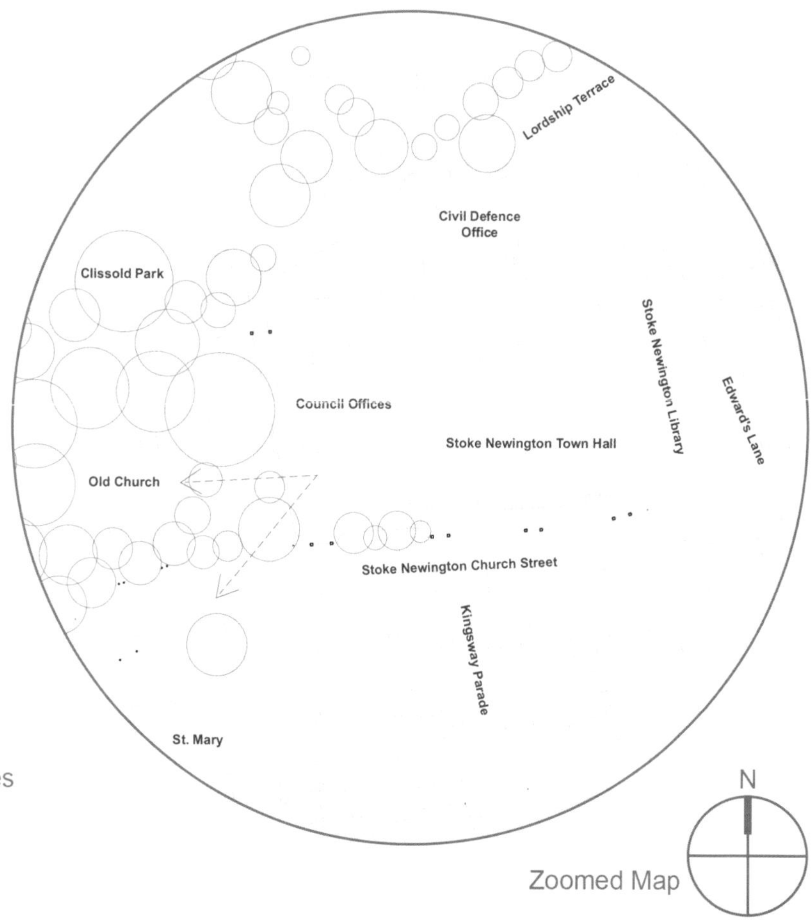

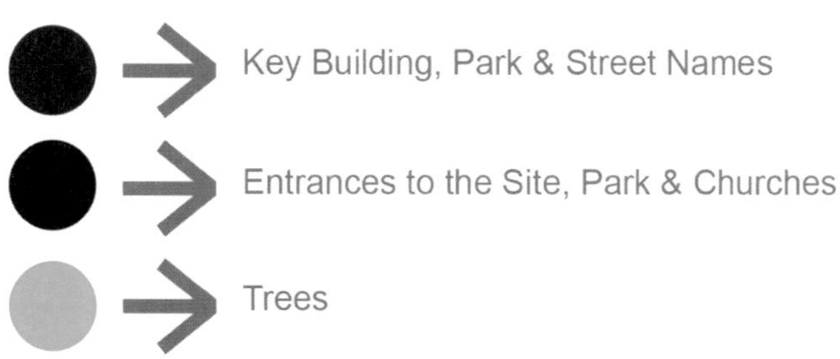

START - CREATE MIXTURE

Let Town have a voice (no hierarchy), companies and Civic Centre work together and organise successful events but also give everyone the freedom to be individual.

타운에 목소리(위계층 없음)를 가지도록 하고, 회사와 시민 센터는 함께 협력하여 성공적인 행사를 조직하되, 모든 사람들에게 개인적이 될 수 있는 자유를 준다.

Step - Lasting

Using a pattern of several parts coming together and interact will make the mixture last. In earlier days they tried to use art deco patterns for branding of the building, but failed in the development.

여러 부품이 서로 연결되어 상호작용하는 패턴을 사용하면 혼합물이 오래 지속된다. 초기에 그들은 그 건물의 브랜드화에 Art Deco 패턴을 사용하려 했지만 개발에는 실패했다.

Decision - Present

Clearly state and present what the pattern is and includes for Civic Centre therefore Community and Companies.

시민센터의 패턴이 무엇이며, 이에 따라 지역사회와 기업들을 위해 포함되는지 명확히 진술하고 제시한다.

Action - Grow

Expand the growth and development of pattern and how it works within the town.

패턴의 성장과 발전, 그리고 그것이 마을 내에서 작동하는 방식을 확장

End - Future

Everything is mixed, offices, residential, retail, restaurants, arts and culture = Successful pattern!

모든 것이 혼합되어 있으며, 사무실, 주거, 소매, 식당, 예술과 문화 = 성공적인 패턴!

INITIAL CONCEPT

WORKSPACES

Workspaces are the social interaction within the office. It is the area where the work is getting done. The centre of the office. For our client, the Wire, the centre is their long central desk. Interior architecture and design in workspaces today are not there anymore only to look nice. Everything has a special meaning and a specific reason why it is where it got placed. My concept will change the way thinking of the people. The Wire workspace reflects on their 'pathological independence.' Throughout the spaces meaningful transition points are build. The users can discover interior and exterior through the whole building and on their independent level. Therefore the user is given a voice. The whole concept creates space & community.

업무 공간은 사무실 내의 사회적 상호 작용이다. 그 곳은 그 일이 마무리되고 있는 지역이다. 사무실 중앙. 우리의 고객인 와이어에게 센터는 그들의 긴 중앙 책상이다. 오늘날 작업 공간의 실내 건축과 디자인은 더 이상 멋져 보이기 위해서만 있는 것이 아니다. 모든 것은 특별한 의미와 그것이 배치된 구체적인 이유가 있다. 내 개념은 사람들의 사고방식을 바꿀 것이다. WIRE 작업공간는 그들의 '병적 독립성' 을 반영한다. 공간 전체에 걸쳐 의미 있는 전환점이 구축된다. 사용자는 건물 전체와 독립적인 수준에서 내부와 외부를 발견할 수 있다. 따라서 사용자에게 발언권을 준다. 전체 개념은 공간과 공동체를 창조한다.

THE STATE OF THE INDEPENDENCE

The identity and the independence of the Town Hall and its users is reflected through meaningful transition points. Supported by created space, community and easy navigation. The concepts provide enough space for the Wire, the Event company and the Community to be independent. However it also offers plenty of room to interact and to be part of the whole gesture. This gives the user a voice. The design enables the building space to express itself - less is more. The Town Hall becomes the centre of the Town again. The created space & community is the big gesture of the Town Hall and the Town. The Town is unconventional and my design approach is the activist in the building.

타운홀과 그 이용자의 정체성과 독립성은 의미 있는 전환점을 통해 반영한다. 공간, 커뮤니티 및 손쉬운 내비게이션을 통해 지원된다. 이 개념은 WIRE, 이벤트 회사 및 커뮤니티가 독립할 수 있는 충분한 공간을 제공한다. 그러나 그것은 또한 상호작용을 할 수 있고 전체적인 제스처의 일부가 될 수 있는 많은 공간을 제공한다. 이것은 사용자에게 발언권을 준다. 그 디자인은 건물 공간이 자신을 표현할 수 있게 해준다. 작으면 작을수록 좋다. 시청은 다시 도시의 중심이 된다. 창조된 공간과 공동체는 시청과 마을의 큰 몸짓이다. 그 도시는 관습에 얽매이지 않고 나의 설계 방식은 그 건물의 활동가이다.

CIVIC CENTRE

A Civic Centre is where the community can come together. My Civic Centre concept in the Town Hall is all about community. The Town Hall is the Centre of the Town. Therefore a Civic Centre is created which is always accessible for the community to come together. A Civic Centre with a clear approach for its community. This can be provided through hosting weekly Civic events. This will be supported by the event company that found its home in the Town Hall. The weekly events are sometimes simpler, sometimes fancier. From flew-markets, bake sales, barbecues, to movie nights, Ted Talks, dance & music events and more. Events for old and young. The event timetable will change monthly. The event rooms create a meaningful transition for the community. The user will be given a voice. The whole concept creates space & community.

시민 센터는 지역사회가 함께 모일 수 있는 곳이다. 시청에 있는 나의 시민 센터 개념은 모두 지역사회에 관한 것이다. 시청은 그 도시의 중심이다. 그러므로 시민 센터는 지역사회가 항상 함께 모일 수 있도록 만들어진다. 지역사회를 위한 명확한 접근법을 가진 시민센터. 이것은 주간 시민 행사 개최를 통해 제공될 수 있다. 이것은 타운홀에 본사를 찾은 이벤트 회사에 의해 지원될 것이다. 주간 행사는 때때로 더 간단하고, 때로는 더 환상적이다. 플라이마켓부터 베이크 세일, 바비큐, 영화 밤, 테드 토크, 댄스 & 음악 행사 등. 노인과 젊은이를 위한 행사. 행사 시간표는 매달 바뀔 것이다. 행사장은 지역사회를 위한 의미 있는 전환을 만들어낸다. 사용자에게 발언권이 주어질 것이다. 전체 개념은 공간과 공동체를 창조한다.

Start
Big gesture research brought me to look deeper into patterns.

Step
Understanding how Art Deco and Tapper CD patterns are build up.

Decision
Realising that they are actually all the same --> build up out of simple lines.

STOKE NEWINGTON
TOWN HALL
& ASSEMBLY HALL
COMMEMORATING DEATH,
CELEBRATING BIRTH AND LIGHT
WIRE
Laurie Anderson
Completely in the present
SENYAWA
FUTURA RECORDS
STORM BUGS
Val Wilmer on
Cecil Taylor
Carl Stone on
Looney Tunes
Tansy Davies
Henry Threadgill
Derek Jarman
NON Worldwide
Proc Fiskal

TABLE DESIGN BUSINESS CASE

Materials & Layers

The base material is recycled paper. The paper surfaces can be coloured in any colour. The different layers can be ordered in individual materials or colours. Surfaces can always be changed and ordered differently again. Each costumer can select their wished and needed layer.

기본 재료는 재활용된 종이다. 종이 표면은 어떤 색으로도 채색될 수 있다. 다른 층은 개별 재료 또는 색상으로 주문할 수 있다. 표면은 항상 다르게 변경되고 다시 정렬될 수 있다. 각각의 고객은 원하는 레이어와 필요한 레이어를 선택할 수 있다.

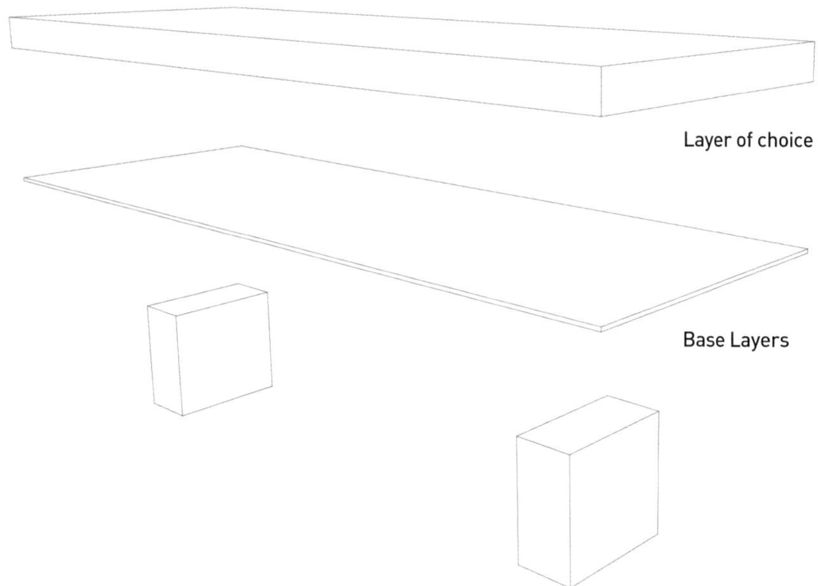

Layers of choice

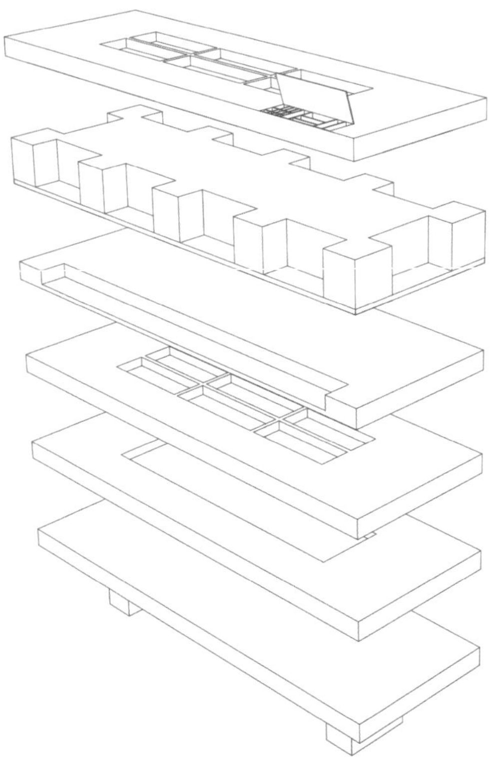

WIRE
WIRE
Larry Heard
From paper to wood.
to wool.

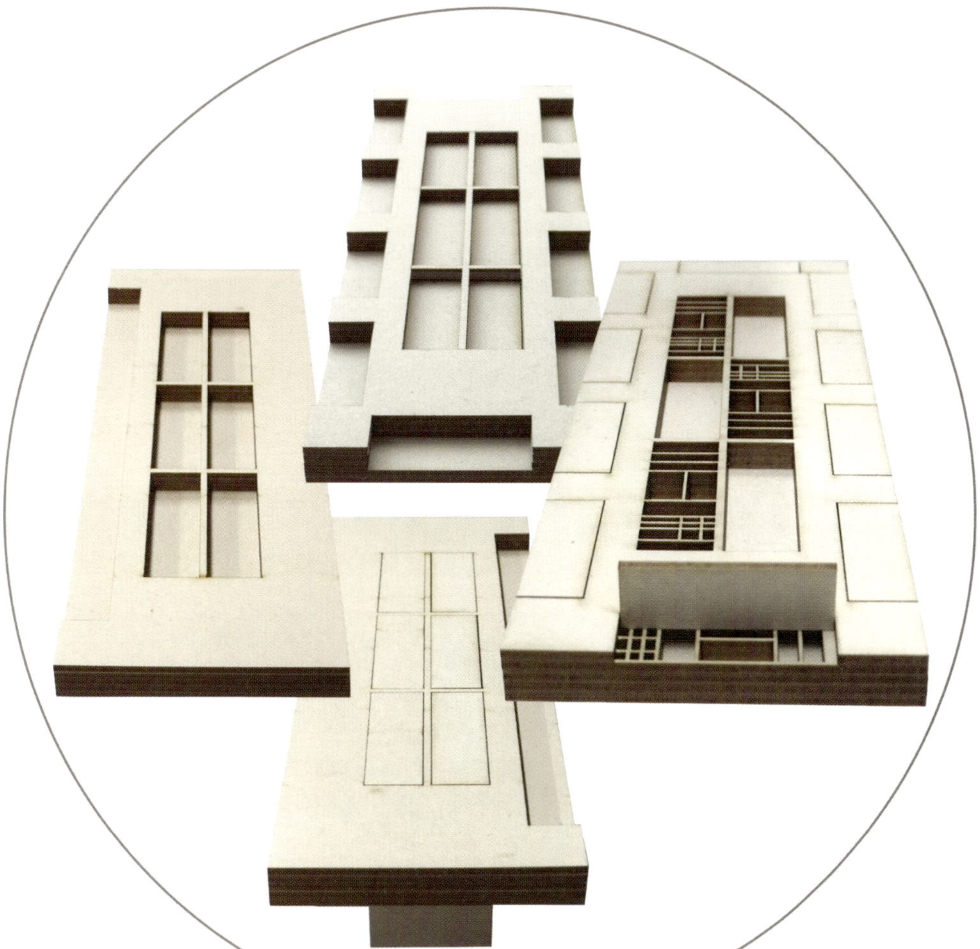

MATERIAL, LIGHT, & INTERIOR EXPLORATION

Start

Build / rebuild sustainable interior & exterior, that can expand, inspire, & be recycled. Sustainable concept will be included into community events. Same materials for all levels, interior & exterior to support concept --> no hierarchy.

확장, 영감을 주고 재활용할 수 있는 지속 가능한 내부 및 외부 구축/재구축. 지속 가능한 개념은 지역 사회 행사에 포함될 것이다. 모든 레벨에 동일한 재료, 개념을 지원하는 내부 및 외부 -> 계층 없음

Decision

Material choices include choosing certified sources, certified logs, certified mills, continuing with certified factories, & certified merchants & joiners. Everything what can be constructed on the site, will be constructed on site.

재료 선택에는 인증된 소스, 인증된 로그, 인증된 공장, 인증된 상인과 가입자의 선택이 포함된다. 현장에 건설될 수 있는 모든 것은 현장에 건설될 것이다.

Action

Demolished materials are reused. E.g. fence will be melted & the iron is used for fixings. A big part is to reduce our ecological footprint by lowering the total amount of waste we produce. Reused Paper will be the main material.

퇴화된 재료는 재사용된다. 예를 들어 울타리가 녹고 고정용 철이 사용된다. 큰 부분은 우리가 생산하는 총 폐기물 양을 줄임으로써 우리의 생태학적 발자취를 줄이는 것이다. 재사용된 종이가 주요 재료가 될 것이다.

Development

The first choice was to use pure materials like concrete. The change & development was that these are too heavy & in consequence the materials will be used only as layers to make the construction lighter.

첫 번째 선택은 콘크리트와 같은 순수한 물질을 사용하는 것이었다. 변경과 개발은 이것들이 너무 무겁고, 그 결과 재료들은 단지 레이어로만 사용되어 건설이 더 가벼워질 것이라는 것이었다.

Result

In consequence the sustainability concept of reducing, reusing, and recycle, will be reflected onto the community. The paper material is supported by many other sustainable materials and all together create the grand gesture.

결과적으로 감소, 재사용 및 재활용이라는 지속가능성 개념은 지역사회에 반영될 것이다. 종이 재료는 다른 많은 지속 가능한 재료들에 의해 지지되고 모두 함께 굉장한 제스처를 만들어낸다.

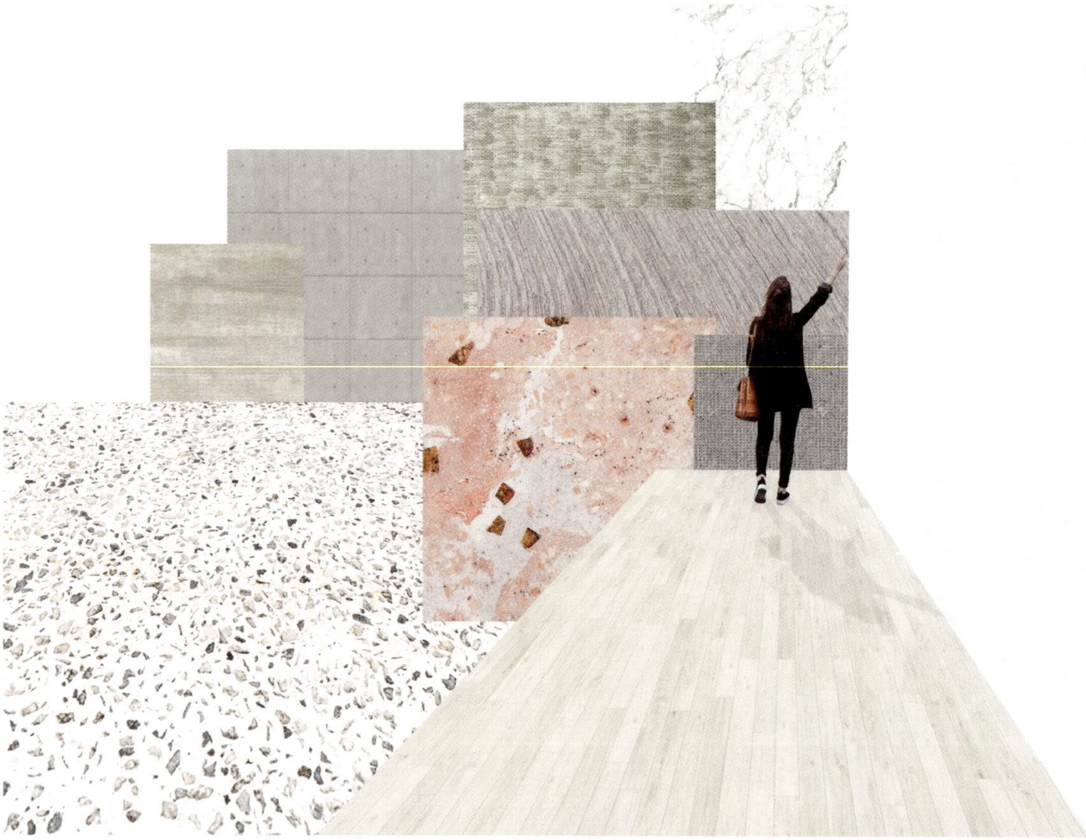

TABLE EXTERIOR DESIGN OUTPUT REALISED THROUGH MODELLING

Staircase Walls Materials

The Materials on the walls change the experience when walking on the stairs. The user will experience both interior and exterior design. Also the user will realize that materials first used on the floor suddenly appear on the walls of the staircase.

벽의 재료는 계단을 걸을 때 경험을 바꾼다. 사용자는 내부 및 외부 설계를 모두 경험할 것이다. 또한 사용자는 바닥에 처음 사용된 재료가 갑자기 계단 벽에 나타난다는 것을 깨닫게 될 것이다.

REALISATION, & VISUALISATION OF THE TABLE TERRACES

Concept
The interaction between the wire and the community is achieved through the meaningful transition points. The main one are the table terraces.

Old and New
The tables are build at the rear of the building. The connection of the tables to the building represent the interaction of the community and the wire - the old and the new coming together and creating new.

Noise, Light, Green
Very quiet at the rear of the building. Beautiful natural light. Directly next to the park and street. Perfect view over whole town.

Gesture
The multifunctional table terraces are the grand gesture. Bringing the table function from interior to exterior creates meaningful transition points and big interactions. The coming together is the grand gesture.

개념.
WIRE와 커뮤니티 간의 상호 작용은 의미 있는 전환점을 통해 달성된다. 주된 것은 식탁 테라스다.

오래된 것과 새로운 것
탁자는 건물 뒤쪽에 세워져 있다. 표와 건물의 연결은 공동체와 WIRE의 상호작용을 나타낸다. 오래된 것과 새로운 것이 함께 모여 새로운 것을 창조한다.

소음, 빛, 녹색
건물 뒤쪽의 조용함. 아름다운 자연광. 공원과 길 바로 옆. 도시 전체를 한 눈에 볼 수 있는 조망

제스쳐
다기능 탁상 테라스가 웅장한 몸짓이다. 테이블 기능을 내부에서 외부로 가져오는 것은 의미 있는 전환점과 큰 상호작용을 만들어낸다. 함께 하는 것이 굉장한 몸짓이다.

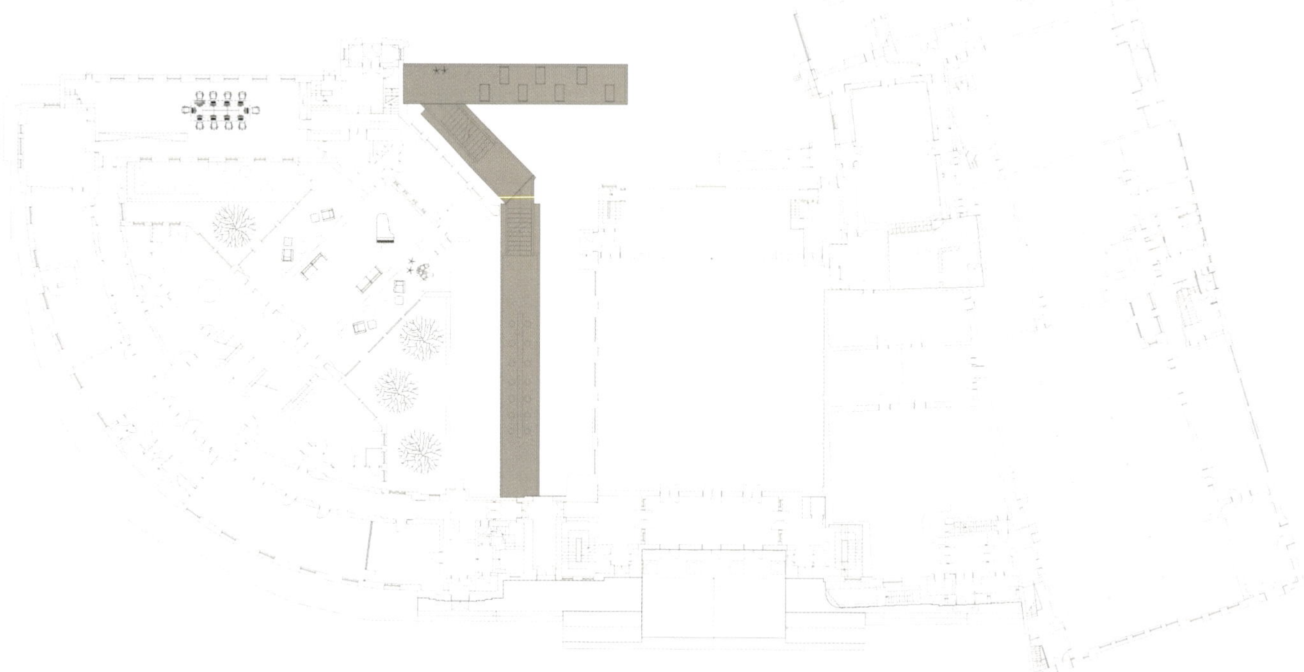

ROCK
FESTIVAL

REALISATION, & VISUALISATION OF THE WIRE

Concept
The Wire space reflects on their identity. It offers enough storage and space to be independent. It is reflected by meaningful transition points.

Gesture
The outcome of the grand gesture is the use of the precedents study less is more to enable gestures space to be expressed. The use of patterns & tables are the gesture & raise awareness of independence.

Privacy
Private rooms on the ground floor, own entrance, doors to close. However also enough opportunities for interaction and transition. The user is given a voice.

Noise, Light, Green
Very quiet at the back of the building. Beautiful natural light, a lot of windows. Directly next to the park, perfect view.

개념.
WIRE 공간은 그들의 정체성을 반영한다. 독립할 수 있는 충분한 저장 공간과 공간을 제공한다. 의미 있는 전환점에 의해 반영된다.

제스쳐
위대한 제스처의 결과는 제스처 공간을 표현하기 위해 선례 연구를 적게 사용하는 것이다. 패턴과 테이블을 사용하는 것은 독립에 대한 인식과 바로 몸짓이다.

사생활
1층에 있는 개인 방, 자기 출입구, 문 닫을 곳. 그러나 상호 작용과 전환을 위한 충분한 기회도 있다. 사용자에게 발언권이 주어진다.

소음, 빛, 녹색
건물 뒤쪽에 조용함. 아름다운 자연광. 많은 창문들. 공원 바로 옆. 완벽한 경치

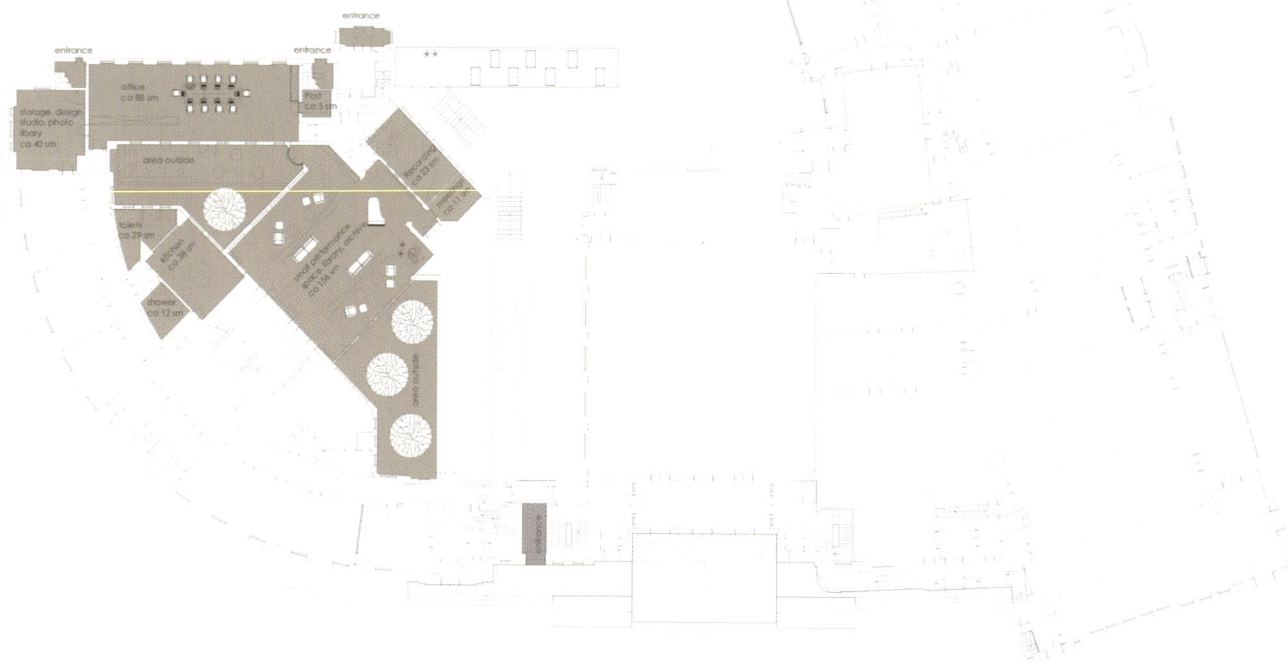

REALISATION, & VISUALISATION OF THE COMMUNITY SPACE

Concept
The community is given a voice and an official space where they can come together as a community.

Reinvention
Community space is upstairs to express no hierarchy. Offers a journey through the civic space - their community space.

Noise, Light, Green
Very quiet at the front of the building. Beautiful natural light, a lot of windows. Directly next to the park and street, perfect view over the town - communities town. High ceilings and Art Deco details.

Gesture
The daily hosted events and in consequence the interaction of the community are the grand gestures of the civic centre. Furthermore the gestures expands from making the town hall the centre of the town again.

개념.
공동체에 발언권이 주어지고 그들이 공동체로 모일 수 있는 공식 공간이 주어진다.

재창안
공동체 공간은 계층을 표현하지 않기 위해 위층에 있다. 시민 공간, 즉 그들의 지역사회 공간을 통과하는 여정을 제공한다.

소음, 빛, 녹색
건물 앞쪽은 매우 조용하다. 아름다운 자연광, 많은 창문들, 공원과 거리 바로 옆, 마을을 내려다 볼 수 있는 완벽한 경치 – 커뮤니티 타운. 높은 천장과 Art Deco 디테일

제스처
매일 개최되는 행사들과 그 결과 지역사회의 상호작용이 시민센터의 웅장한 몸짓이다. 게다가 그 몸짓은 마을 회관을 다시 마을의 중심지로 만드는 것에서 확장된다.

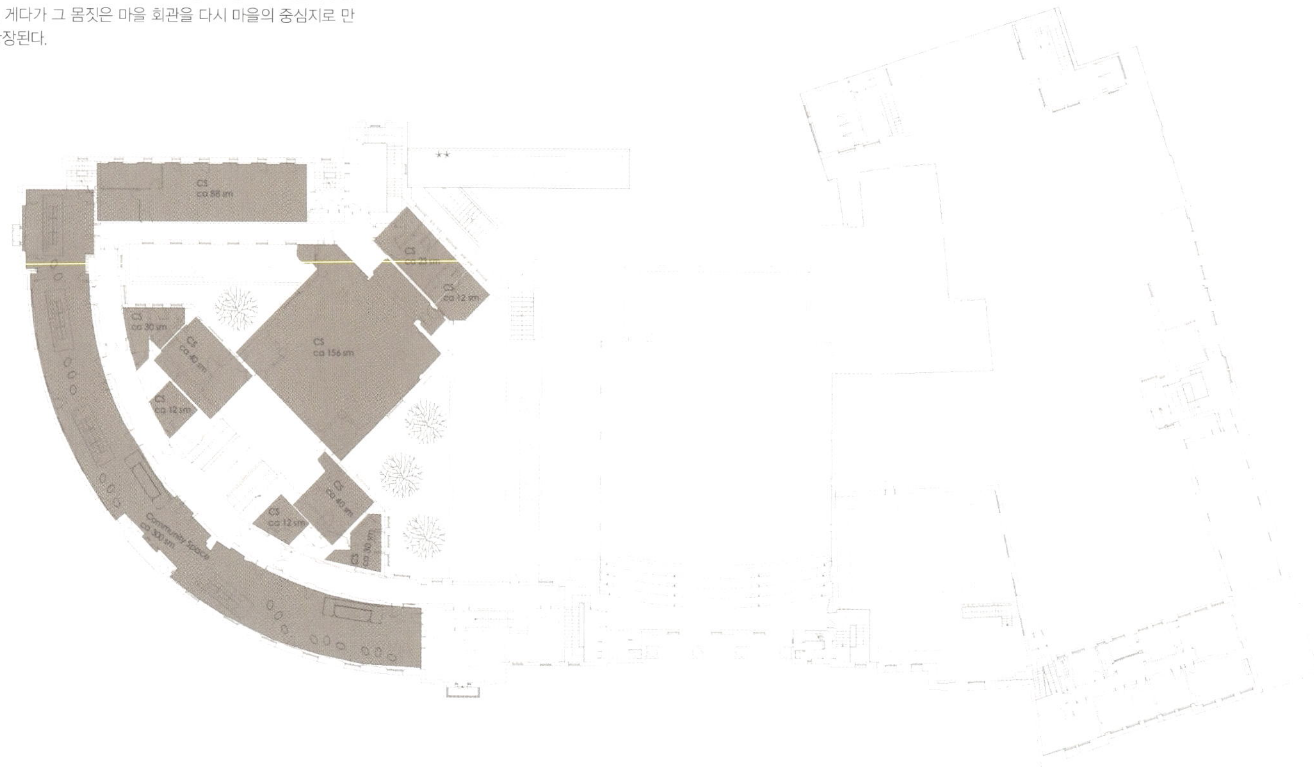

William Turner

Interior Architecture & Design
BA(Hons)

Studio Studio 2:
The state of independence
Stoke Newington Town Hall

The site - Stoke Newington Town Hall

Located on Stoke Newington Church Street, the Town Hall building is at the heart of the community. With the nearest overground station being ten to fifteen minutes away by foot, the area has managed to escape the investment of large chains and wealthy investors. As a result, Stoke Newington Church Street is a beautiful commercial street of independently owned shops and businesses - a real rarity in such a central London Location. The North side of the site borders historic Clissold park, renowned for beauty as well as Clissold House. At the weekend, crowds from all over North London come to make use of the wide range of leisure facilities at the park. The site is sandwiched between two incredibly busy local lei--sure destinations - The Church Street and Clissold Park. As it stands, the underused Town Hall acts almost as a blockade between the two, stopping the flow of circulation between the two and dividing the area. The high iron railings create in imposing presence on the Church Street and act as a deterrent to potential visitors. As a result, this section of the Church street is much quieter than further down towards the High Street.

스토크 뉴잉턴 교회 거리에 위치한 타운홀 건물은 지역사회의 중심에 있다. 가장 가까운 지상역이 걸어서 10분에서 15분 거리에 있는 이 지역은 대형 체인과 부유한 투자자들의 투자에서 가까스로 벗어났다. 그 결과 스토크 뉴잉턴 교회는 독립적으로 소유하는 상점들과 사업체들로 이루어진 아름다운 상업 거리로서, 런던 중심지에서는 정말 드문 곳이다. 이 유적지의 북쪽은 크리솔드 하우스 뿐만 아니라 아름다움으로 유명한 역사적인 클리솔드 공원과 경계를 이루고 있다. 주말에는 북런던 각지에서 모여든 인파가 공원의 다양한 레저 시설을 이용하기 위해 온다. 그 장소는 믿을 수 없을 정도로 바쁜 두 지역 레저 여행지 사이사이에 끼어 있다. 바로 Church street와 클리스톨드 공원이다. 서 있는 그대로, 활용도가 낮은 타운홀은 거의 둘 사이의 봉쇄로 작용하여, 둘 사이의 순환 흐름을 막고 지역을 나누는 역할을 한다. 높은 철제 난간은 Church street에 우뚝 솟아 있으며 잠재적인 방문객들을 억제하는 역할을 한다. 그 결과 교회 거리의 이 구간은 High street쪽으로 더 내려갈 때보다 훨씬 조용하다.

The Barometer Concept

Built in 1935, the building was designed to represent the community and population of the area at that time. Directly facing the new St Mary's Church, a clear statement of the imposing role of the Church of England is made. Now that the Church plays a much less dominant role in the community, I begin to speculate about what people gained from church in 1935. Aspects such as spiritual wellbeing and guidance are also available in from other places, so I carried out a survey to find out where local residents are going for Guidance, a sense of community, space for reflection, and a sense of wellbeing.

1935년에 지어진 이 건물은 당시 이 지역의 공동체와 인구를 대표하도록 설계되었다. 새로운 성모 교회를 정면으로 마주보고 있는 것은 영국 교회의 당당한 역할에 대한 명확한 진술이 제시되어 있다. 이제 교회가 사회에서 훨씬 덜 지배적인 역할을 하고 있기 때문에, 나는 1935년에 사람들이 교회로부터 무엇을 얻었는지 추측하기 시작한다. 영적 행복이나 지도 같은 측면도 다른 곳에서도 이용할 수 있기 때문에 현지 주민들이 안내를 위해 어디로 가고 있는지, 공동체 의식, 반성을 위한 공간, 행복감 등을 알아내기 위해 조사를 실시했다.

Through my survey I found that people are going to a hugely diverse range of places to acquire the same goals as people once used to gain from going to church. I also found that many of those places were highly exclusive to those with more financial means. I think it is a great shame that the community has lost the sense of togetherness that Church once provided.

조사를 통해 사람들이 한때 교회에 가서 얻었던 것과 같은 목표를 얻기 위해 엄청나게 다양한 장소에 가고 있다는 것을 알게 되었다. 그러한 장소들 중 많은 곳이 더 많은 재정적인 수단을 가진 사람들에게 매우 배타적이라는 것을 발견했다. 교회가 한때 제공했던 유대감을 지역사회가 잃어버린 것은 큰 수치라고 생각한다.

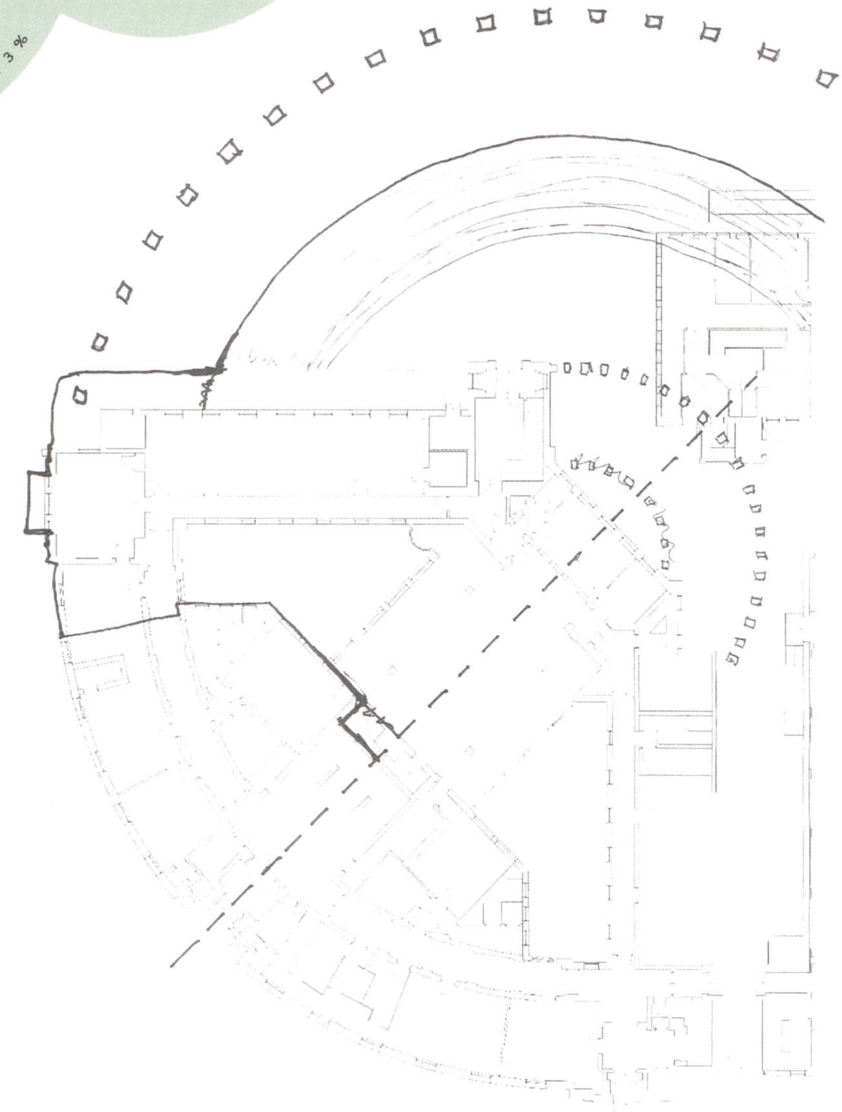

Bearing in mind how the building was originally designed to assimilate the community's connection to the Church, I begin to think about how the building can be adapted to cater for the hugely diverse local demographic of today's Stoke Newignton. Through exploring rhythms and patterns in the building, I begin to develop the possibility of continuing the curve of the building, and opening the building up to the park.

그 건물이 원래 어떻게 교회와의 연결을 동화시키도록 설계되었는지를 염두에 두고, 그 건물이 오늘날의 스토크 뉴잉턴의 매우 다양한 지역 인구를 충족시키기 위해 어떻게 개조될 수 있는지에 대해 생각하기 시작했다. 건물의 리듬과 패턴을 탐구함으로써, 나는 건물의 곡선을 구부리고 공원으로 건물을 개방할 수 있는 가능성을 개발하기 시작한다.

The Barometer

Something which reflects changes in circumstances or opinions,
The Barometer seems to be a fitting name for Stoke Newignton's new cylindrical community hub.
A space that ties together the rapidly evolving local demographic. A completely inclusive space.
A space for anyone to come and feel a part of the community.

상황이나 의견의 변화를 반영하는 것.
바로미터는 스토크 뉴잉턴의 새로운 원통형 커뮤니티 허브에 적합한 이름인 것 같다.
빠르게 발전하고 있는 지역 인구통계학을 하나로 묶는 공간. 완전히 포괄적인 공간.
누구나 와서 공동체의 한 부분을 느낄 수 있는 공간.

Shift in focus

With the presence of the building mainly felt on the Church Street, my intervention aims to continue the curve of the original building, and shift the focus of towards Clissold park, at the rear of the building. By creating a circular plan, visitors are invited to enter the building from any angle, and explore the buildings' layers, prompting interaction amongst one another.

그 건물의 존재가 주로 교회 거리에서 느껴졌기 때문에, 원래의 건물의 곡선을 계속하여 건물의 뒤쪽에 있는 클리솔드 공원으로 초점을 옮기는 것을 목표로 한다. 원형 계획을 수립함으로써 방문객들은 어떤 각도에서든 건물 안으로 들어가 건물 층을 탐색하여 상호작용을 유도한다.

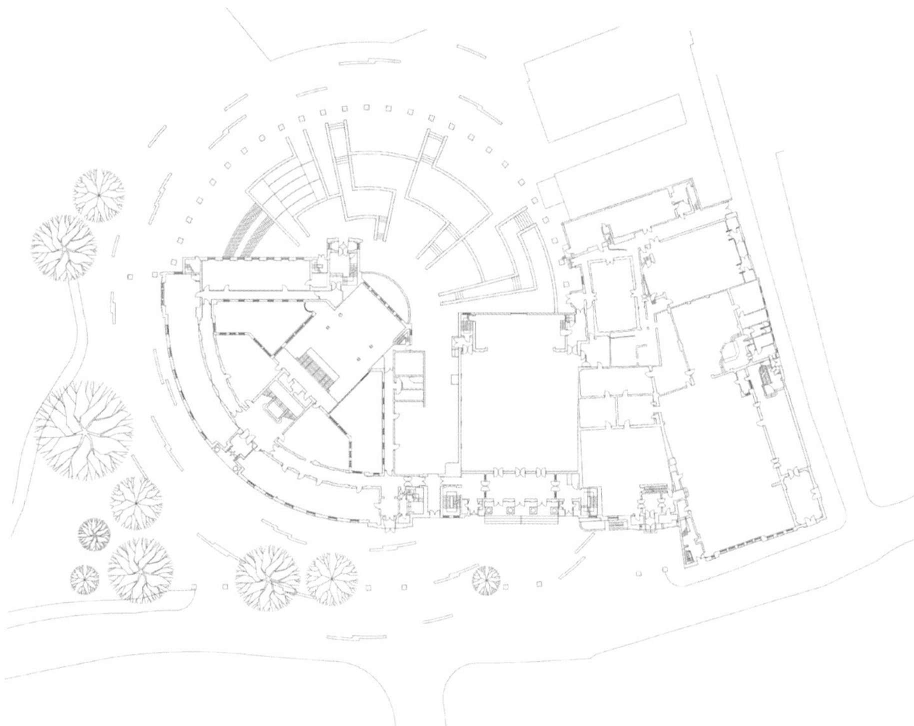

Exposing the Council Chamber

The existing building was built at a time where great social divides were at the core of society. The building's structure is evidence that these social divides once existed. The Council Chamber was once hidden from the normal citizen, a place which was usually only seen by the Council members. In order to reflect todays society, my intervention literally breaks down the concrete partitions and brings ordinary people into the space. The diverse community are now invited to come and look up at the beautiful council chamber which was once the centre of the community, and look around them at the people which now make up the heart of Stoke Newington.

기존 건물은 사회적으로 큰 분열이 있었던 시기에 지어졌다. 그 건물의 구조는 이러한 사회적 분열이 한 때 존재했다는 증거다. 평의회는 한때 보통 평의회 의원들만이 볼 수 있는 장소인 일반 시민으로부터 숨겨져 있었다. 오늘날의 사회를 반영하기 위해, 나의 개입은 말 그대로 콘크리트 칸막이를 무너뜨리고 평범한 사람들을 공간으로 데려온다. 이제 다양한 지역사회가 초대되어 한때 지역사회의 중심이었던 아름다운 의회 회의실을 올려다보고, 이제 스토크 뉴잉턴의 심장부를 구성하고 있는 사람들을 둘러본다.

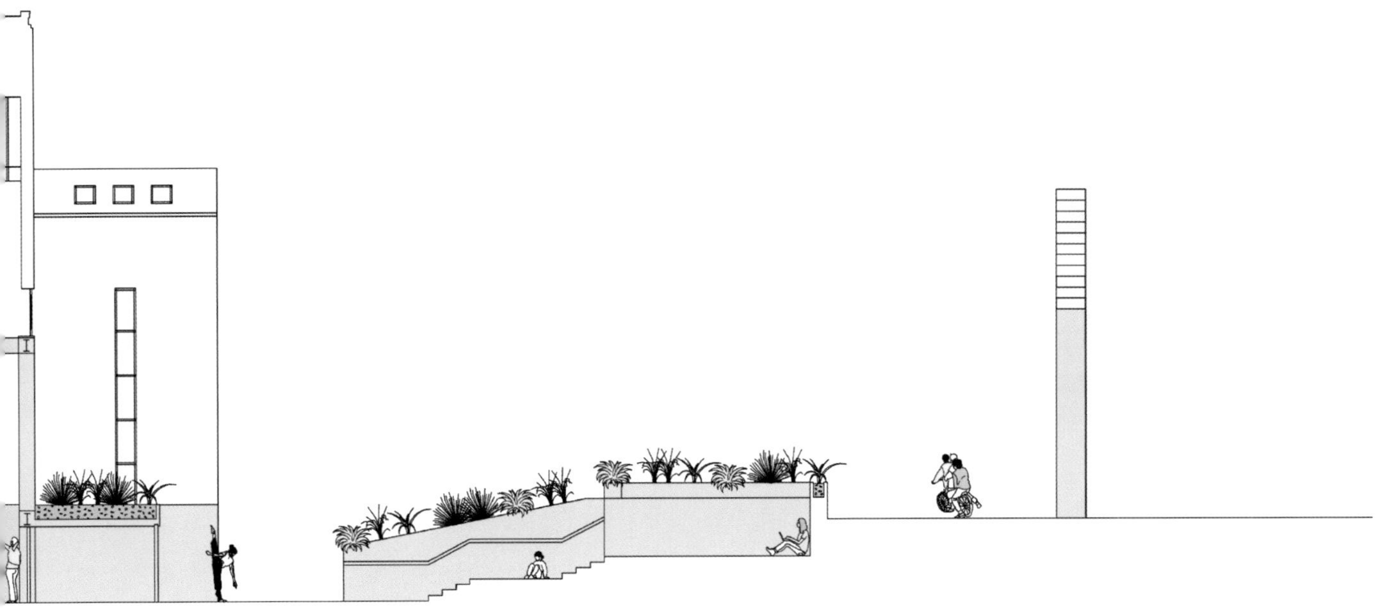

"Their use of gold leaf and gold dust was not mere extravagance. Its reflective properties were put to use as a source of illumination"

- Junichiro Tanizaki, In Praise of Shadows

Due to its shape and positioning, the building's interior is dark. In addition, the original council chamber was designed to avoid the afternoon light, saving the council members from the hot glare of the summer sun. Rather than adapt the building to become lighter, I have decided to embrace the original design intent and indulge in the beauty of darkness. The brass lines from the paneling will meet at the floor and become a functional way-finding strategy around the building.

Designed to mirror the original art deco staircase, the new staircase (opposite) has brass rails . The brass is brushes (not polished), and the banisters are vertical rather than horizontal in order to pay homage to the original designer, but still distinguishing old from new.

그 건물의 모양과 배치 때문에, 그 건물의 내부는 어둡다. 게다가 원래 평의회는 오후의 불빛을 피하도록 설계되어 평의원들이 여름 햇살의 뜨거운 눈부심으로부터 구했다. 건물을 더 가벼워지도록 개조하기보다는 원래의 설계의도를 받아들여 어둠의 아름다움에 빠져들기로 했다. 패널에서 나온 황동 라인은 바닥에서 만나 건물 주변의 기능적인 길찾기 전략이 될 것이다. 원래의 art-deco 계단을 거울로 만들도록 설계된 이 새로운 계단에는 황동 난간이 있다. 황동은 연마하지 않으며, 배니스터는 원래 디자이너에게 경의를 표하기 위해 수평이 아닌 수직이면서도 여전히 새 것과 구별된다.

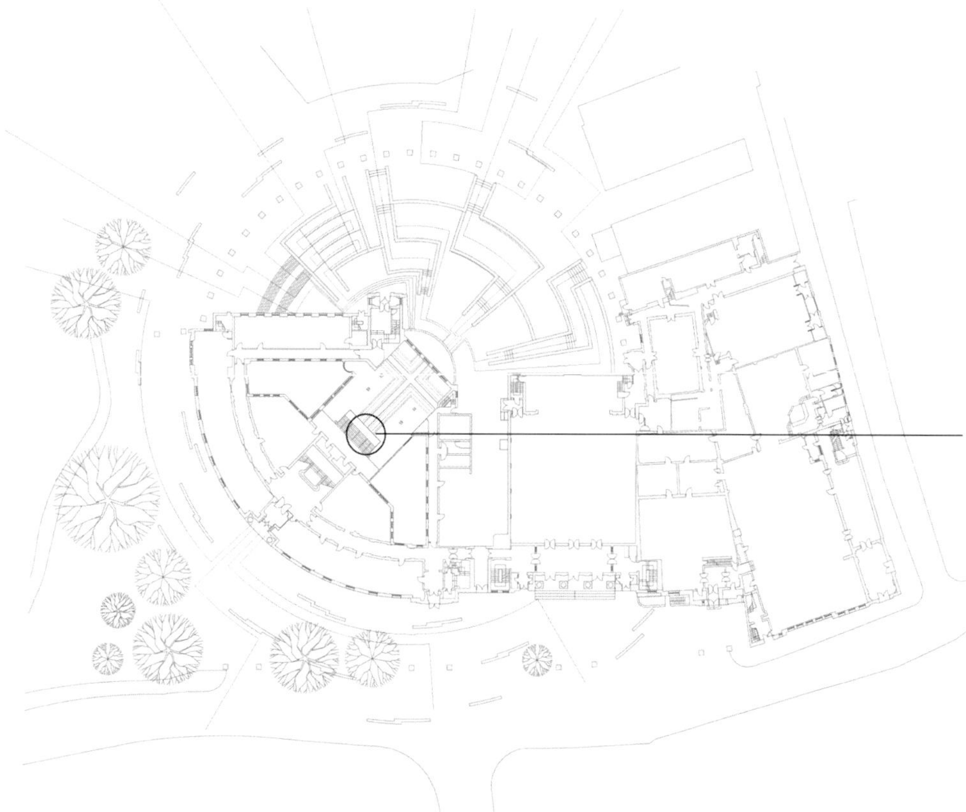

Location of stairwell depicted opposite

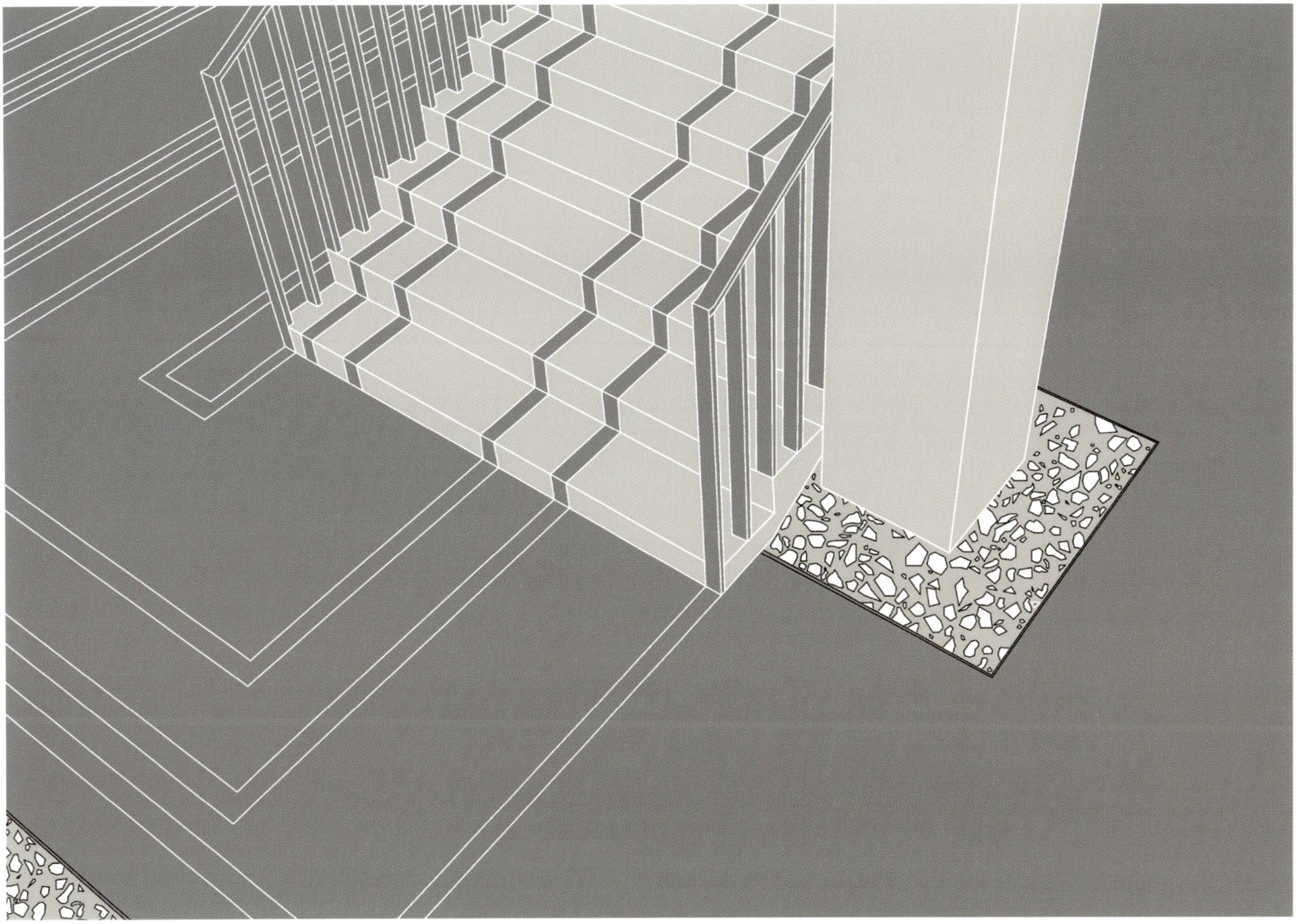

Taking cues from the 1930s terrazzo in the original building, I specified a dark terrazzo floor, with a different variation of terrazzo to border the room and the columns. A small mosaic tile will divide the two variations of terrazzo.

원래 건물에서 1930년대의 테라쪼를 단서로, 어두운 테라쪼 바닥을 지정했는데, 그 방과 기둥에 경계를 두기 위해 테라초와는 다른 변형이 있었다. 작은 모자이크 타일은 테라쪼의 두 가지 변형을 나눌 것이다.

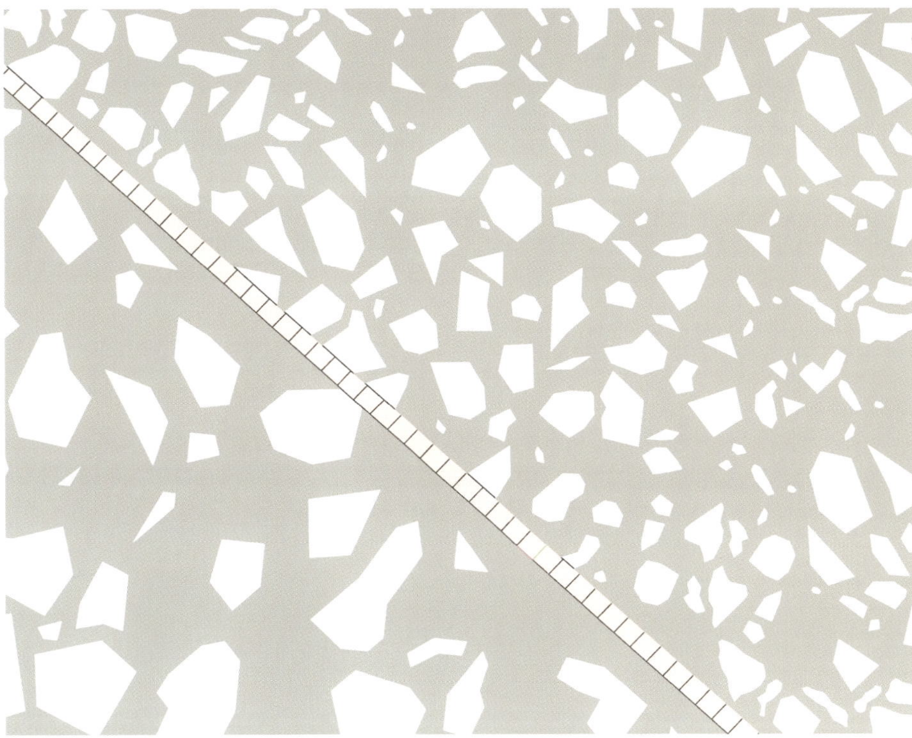

At the centre of the community

From beautiful interior way-finding mechanism, this feature will continue to explore the local community. The brass line will seamlessly link the indoors with the outdoors and reach out to local community spaces, linking the Barometer to local schools such as Stoke Newignton School, Grazebook Primary , Sir Thomas Abney School and Northwold Primary. This will encourage children and parents to use the Barometer as a space to exercise before school, or a place to relax in the late afternoon. It will encourage school trips where children will come from across North London, instilling a firm sense of community in children from a young age.

아름다운 내부 길찾기 메커니즘에서, 이 특징은 지역사회를 계속 탐구할 것이다. 황동선은 실내와 실외를 원활하게 연결하고 지역사회 공간에 도달할 것이며, 바로미터와 스토크 뉴잉턴 학교, 그레이즈북 소학교, 토마스 아비니 경, 노스월드 소학교와 같은 지역 학교들을 연결할 것이다. 이렇게 되면 어린이와 학부모들은 학교에 가기 전에 운동할 수 있는 공간, 혹은 늦은 오후에 휴식을 취할 수 있는 공간으로 바로미터를 사용하도록 장려할 것이다. 그것은 어린 시절부터 아이들에게 확고한 공동체 의식을 심어주면서, 북런던을 가로질러 아이들이 오는 수학여행을 장려할 것이다.

Stoke Newington Church Street

The brass lines will also reach out to Hackney Family and Community centre and Schonfield Square Care home , breaching the gap between ages amongst the community.

황동선은 또한 해크니 패밀리 앤 커뮤니티 센터와 스촌필드 스퀘어 케어 하우스에 연결하여 지역사회 사이의 나이 차이를 뚫을 것이다.

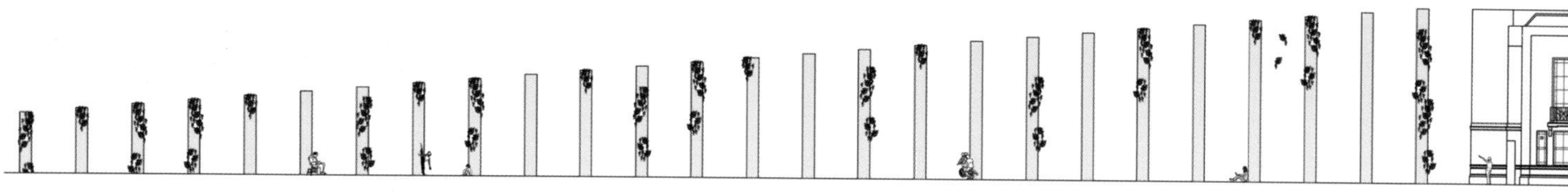

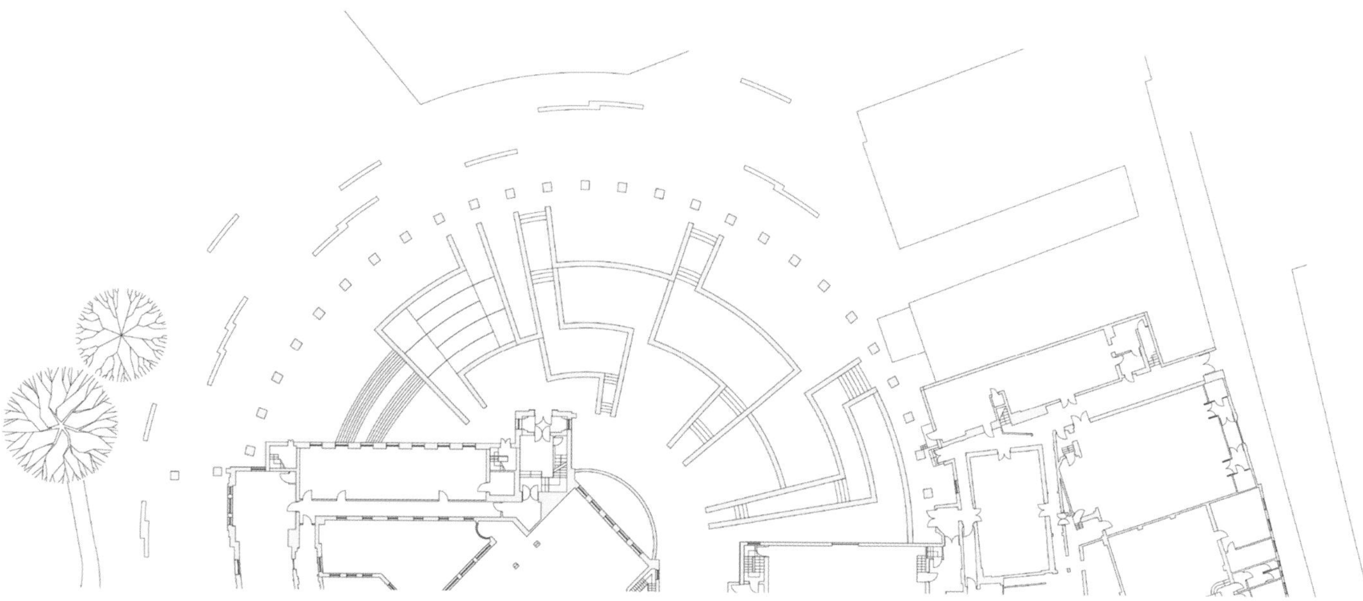

A stunning level of care and attention to detail has been given to the building's facade. This will remain untouched. However, where the previous design left the back of the building in a less considered manner, my intervention aims to bring life to this previously overlooked part of the building, My addition to the elevation pays homage to Truelove's original design, continuing the rhythm of the windows in the form of vertical columns; and continuing the fraction of a circle which the original design contains. Rather than maintaining the whole height, the columns gradually shorten, creating an elegant sweeping effect that will be visible from the park, without being too imposing .

그 건물의 정면에는 놀랄 만큼 세심한 주의가 요구된다. 이것은 손대지 않고 남아 있을 것이다. 그러나 이전의 설계가 건물의 뒷면을 덜 고려된 상태로 남겨둔 경우, 나의 디자인은 이전에 간과되었던 이 부분에 생명을 불어넣는 것을 목표로 하고 있다. 보완작업은 Truelove의 원래 설계에 경의를 표하고, 수직 기둥의 형태로 창문의 리듬을 계속하며, 원래 디자인에 포함된 원의 소수부분 역시 계속할 것이다. 전체 높이를 유지하는 대신 기둥은 점차 짧아져 너무 늠름하지 않고 공원에서 볼 수 있는 우아한 쓸림 효과를 만들어 낸다.

Front elevation

Osmosis with the Church Street

The original building has an incredibly imposing iron fence at its border with the church street. This acted as a physical divide between the building and the Church Street, but it also intimidates potential visitors and pushes them away from the site. In order to make the site permeable with the bustling Church Street, I propose removing the fence, and repaving the whole street in order to create fluidity between the spaces. Similarly to the road system on Exhibition Road, South Kensington, pedestrians and motorists will be forced to co exist and become more aware of one another.

The new paving will take the form of the concentric circles laid out by the original building design, and include sections of grass and other greenery. Partial planted barriers will deter motorists from the front of the building, but also tempt curious pedestrians into the space.

The beautiful gate posts will remain as a visual reminder of the old system, and signify that the community is open to everyone.

원래 건물은 교회 거리와의 경계에 믿을 수 없을 정도로 당당한 철제 울타리를 가지고 있다. 이것은 건물과 교회 거리 사이의 물리적 구분의 역할을 했지만, 그것은 또한 잠재적인 방문객들을 위협하고 그들을 그 장소에서 멀어지게 한다. 분주한 교회 거리에 그 부지를 스며들게 하기 위해, 나는 울타리를 없애고, 공간 사이에 유동성을 만들기 위해 거리 전체를 다시 꾸밀 것을 제안한다. 사우스 켄싱턴, 사우스 켄싱턴의 도로 시스템과 유사하게 보행자와 운전자들은 서로에 대해 더 잘 알고 있어야 할 것이다.

새로운 포장재는 원래의 건물 설계에 의해 펼쳐지는 동심원 형태를 취할 것이며, 풀과 다른 녹색 부분을 포함할 것이다. 부분적으로 심은 방벽은 운전자들이 건물 앞부분을 막는 동시에 호기심 많은 보행자들을 공간으로 유인할 것이다.

아름다운 관문, 기둥은 옛 제도를 시각적으로 상기시키는 것으로 남을 것이며, 공동체가 모두에게 열려 있음을 의미한다.

View from Clissold Park

Zora Krakan

Interior Architecture & Design
BA(Hons)

Studio Studio 2:
The state of independence
Stoke Newington Town Hall

SITE SURVEY

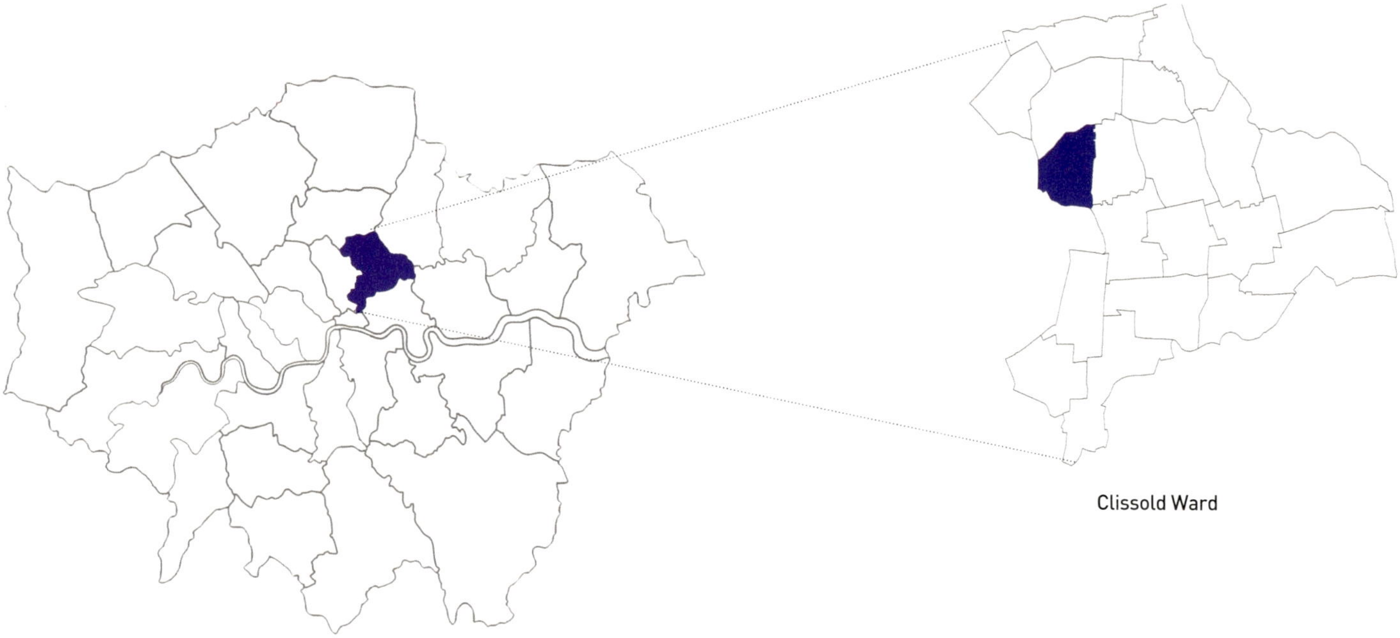

Clissold Ward

London Borough of Hackney

View of the Stoke Newington Building Complex

LOCATION

The Complex is located in Clissold ward, in the north-western part of Borough of Hackney (London, UK).

Fronting Stoke Newington Church Street, the Complex is bound by Edwards Lane to the east, an electricity substation to the north-east, Lordship Terrace to the north, St Mary's Church to the west, and contains six principal elements, namely: the Town Hall and municipal offices; the Assembly Hall; the Former Reference Library; Stoke Newington Library; Library ancillary rooms and the Rear (The Civil Defence Building, Data Centre and bunker).

콤플렉스는 해크니(영국 런던)의 보루(Borough of Hackney)의 북서쪽에 있는 클리솔드 병동에 위치해 있고 앞쪽 스토크 뉴잉턴 교회 거리, 동쪽으로는 에드워드 레인, 북동쪽으로는 영주 테라스, 서쪽으로는 세인트 메리 교회와 연결되어 있으며, 6개의 주요 요소를 포함하고 있다. 즉, 시청과 시립 사무소, 국회의사당, 이전 참조 도서관, 스토크 뉴잉턴 도서관,: 도서관 보조실 및 후면(민방위 건물, 데이터 센터 및 벙커).

CYANOTYPE WORKSHOP

'Photography is an unthinkable part of our practice, it is a means of viewing architecture. Today, architectural history is not transmitted by built works but rather by their photographic documentation. Our perception and sensibilities as designers have been formed above all by photography.' (B. Tschumi)

사진술은 우리의 관행에서 생각할 수 없는 부분이다. 그것은 건축을 보는 수단이다. 오늘날, 건축사는 지어진 작품이 아니라 그들의 사진 문서에 의해 전달된다. 디자이너로서의 우리의 인식과 감성은 무엇보다도 사진에 의해 형성되어 왔다

Through the photographic survey and the cyanotype workshop, an almost 'historic atmosphere' was created and these actions aided in getting a different perspective on spatial understanding of the building and 'putting experience on paper': the way light enters a room, movement within space, materials, details that might have been overseen, etc.

(베르나르 츄미의 사진) 조사와 카노타이드 워크샵을 통해 거의 '역사적 분위기' 가 조성되었으며, 이러한 행동은 건물에 대한 공간적 이해와 '종이 위에 경험' 즉 빛이 어떻게 방에 들어오는 방식, 공간 내부의 움직임, 재료, 관리를 통한 세부사항등에 대한 다른 관점을 얻는 데 도움이 되었다.

THE CLIENT

The WIRE, founded in London, in 1982, is a music and performance magazine that supports and reflects the work of grass roots musicians and performance artists across the world. They describe themselves as 'patho logically independent' and it is essential that the WIRE occupy their 'own space' in order to keep their original values and philosophies true as much through the space that they occupy as in the copy that they write.

1982년 런던에 창간된 WIRE는 전 세계의 인디 뮤지션과 공연예술가의 작품을 지지하고 반영하는 음악 및 공연잡지다. 그들은 스스로를 '병적으로 독립적' 이라고 묘사하고 있으며, 그들이 쓰는 사본에서처럼 그들이 차지하고 있는 공간을 통해 그들의 원래 가치와 철학을 진실하게 유지하기 위해서는 WIRE가 그들의 '소유적 공간' 을 차지하는 것이 필수적이다.

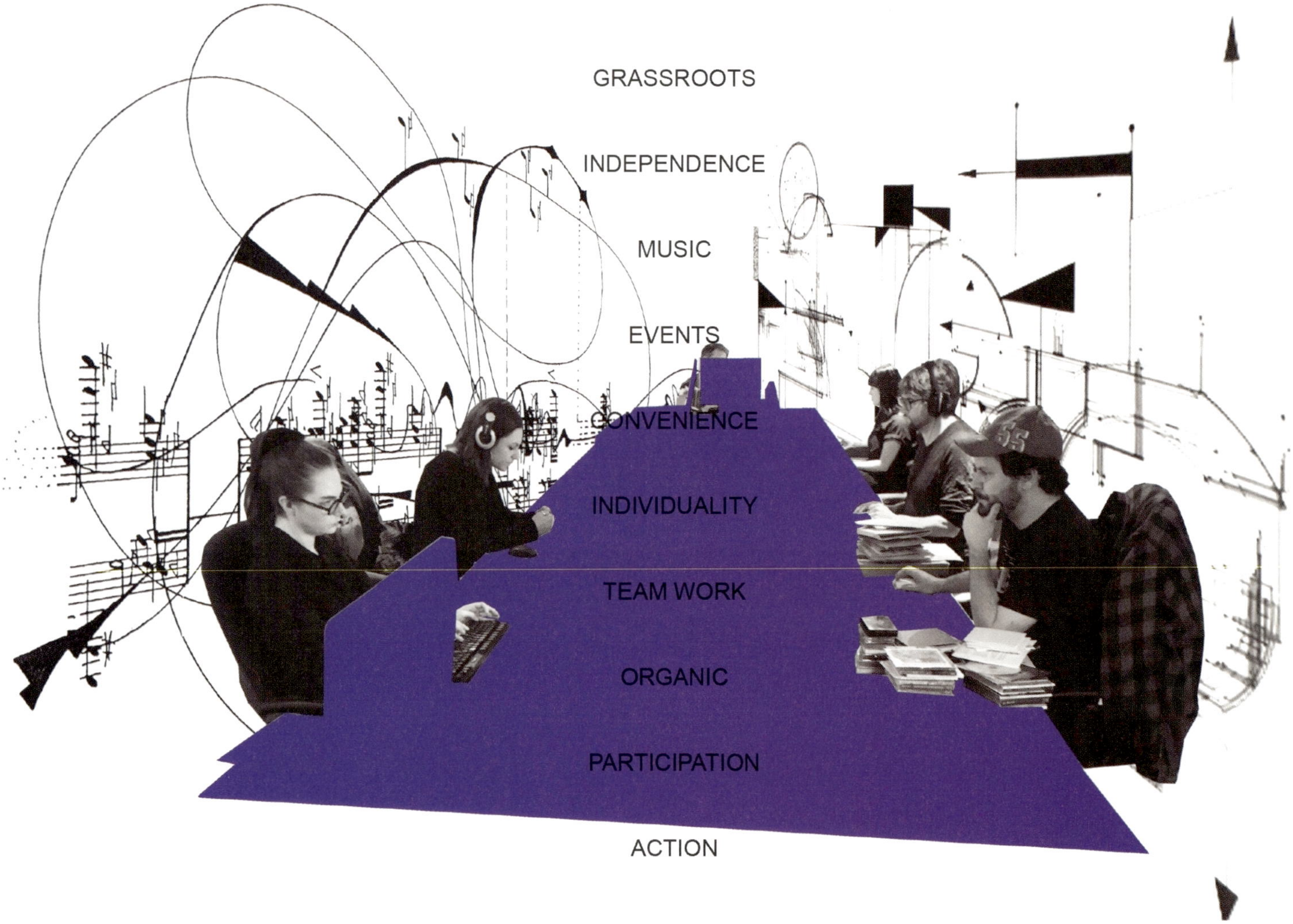

CONCEPT DEVELOPMENT SYMBIOTIC RELATIONSHIPS

The first diagram shows Tony Herrington's statements about the Wire Magazine which immediately lead towards concepts that can be observed in nature between two organisms (symbionts) through symbiotic relationships. Therefore, a parallel to architecture can be drawn through this biological analogy. Platon already described in 'Politeia' the city as a 'macro - anthropos', comparing the human body to the 'polis' - the city.
'No organism is an island - each one has a relationship to other organisms' - directly or indirectly.
These relationships, when transfered to Architecture and Design, can be mutualistic, commensual or parasitic, depending on the influence symbionts have on each others structures, mate rials, forms and spatial organisations. These relationships, when transfered to Architecture and Design, can be mutualistic, commensual or parasitic,depending on the influence symbionts have on each others structures, materials, forms and spatial organisations.
The following types of symbiotic relationship between the Studio's Client, programme and building are possible and will be examined through Venn diagrams:

첫 번째 도표는 WIRE 매거진에 대한 토니 헤링턴의 진술을 보여준다. WIRE 매거진은 공생관계를 통해 자연에서 관찰할 수 있는 개념으로 즉시 유도한다. 그러므로 이 생물학적 유추를 통해 구조와 유사하게 그려질 수 있다. 플라톤은 이미 '폴리티아'에서 인간의 몸을 '폴리스'인 도시에 비유하며 '매크로 – 인류'라고 묘사했다. '어떤 유기체도 섬이 아니다. 각각은 다른 유기체들과 관계가 있다.' 직간접적으로 말이다. 이러한 관계는, 건축과 설계로 이행될 때, 공생물이 서로의 구조, 재료, 형태 및 공간 조직에 미치는 영향에 따라, 상호주의적이거나, 공통적이거나, 기생적일 수 있다. 이러한 관계는 건축과 설계로 이행될 때 상호주의적이거나, 공통적이거나, 기생적일 수 있다. 공생물은 서로에 미치는 영향에 따라 구조, 재료, 형태 및 공간적 조직에 영향을 미친다. 스튜디오의 고객, 프로그램 및 건물 간의 다음과 같은 유형의 공생 관계가 가능하며 벤 도표를 통해 검토될 것이다.
공생물은 서로에 미치는 영향에 따라 구조, 재료, 형태 및 공간적 조직에 영향을 미친다. 스튜디오의 고객, 프로그램 및 건물 간의 다음과 같은 유형의 공생 관계가 가능하며 벤 도표를 통해 검토될 것이다.

MUTUALISM: a type of a symbiotic association, where both partners benefit from the relationship.
A reciprocal exchange of nutrients exists even though the extent to which each symbiont benefits may vary.

COMMENSALISM: refers to a relationship where one of the symbionts benefits in terms of nutritional or protective benefit, and the other is not harmed nor helped. Therefore, commensalism is a relationship in which one species benefits, whereas neutral impact exists on the other.

PARASITISM: an organism that grows, feeds and is sheltered by its host while contributing nothing to the host's survival.But it also understands its host, it knows how to use it to its full potential without killing it.

상호론: 공생적 연합의 일종으로, 두 파트너 모두 그 관계로부터 이익을 얻는다. 비록 각각의 공생력이 얻을 수 있는 정도의 차이는 있지만 영양소의 상호 교환은 존재한다.

공생론: 공생 중 하나가 영양상 또는 보호상의 이익으로 이익을 얻고, 다른 하나는 해를 입히거나 돕지 않는 관계를 말한다. 그러므로 공생주의는 한 종이 이익을 얻는 반면 다른 종에는 중립적인 영향이 존재하는 관계다.

기생론: 숙주의 생존에는 아무런 기여도 하지 않으면서 자라서 먹고 사는 유기체하지만 그것은 또한 숙주를 이해하고, 그것을 죽이지 않고 최대한의 잠재력을 발휘할 수 있는 방법을 알고 있다.

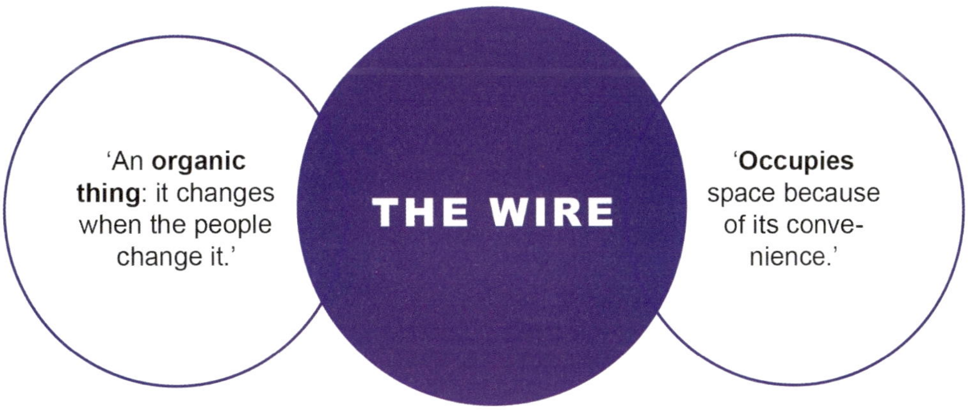

THE WIRE	INDEPENDENT
	GRASS ROOTS
	DEMOCRATIC
	ORGANIC

THE WIRE + SPACE	INFORMAL
	CONVENIENCE
	PRACTICALITY
	FROM SCRATCH

CONCEPT DEVELOPMENT SYMBIOTIC RELATIONSHIPS

THE WIRE (W) + TOWN HALL (TH)

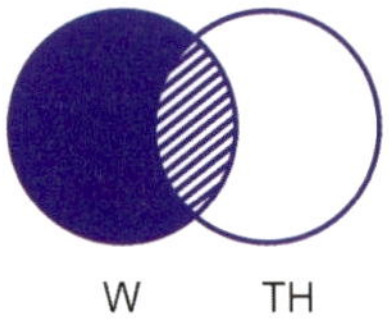

parasitism: The new Wire Office will change the buildings structure, materials, forms and spatial organisation through new interventions

기생론: 새로운 WIRE 오피스는 새로운 개입을 통해 건물 구조, 재료, 형태, 공간 구조를 바꿀 것이다.

THE WIRE (W) + CIVIC CENTRE (CC)

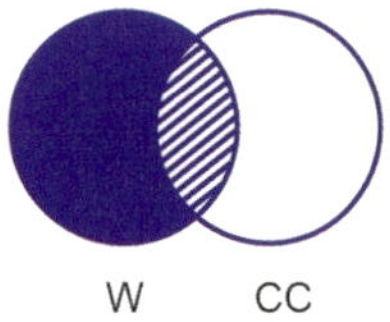

mutualism: The Wire Office and the Civic Centre will both benefit from each other through a new programme: the Event space - in which both can organise various events.

상호론: 와이어 오피스와 시민 센터는 새로운 프로그램인 이벤트 공간을 통해 서로 이득을 볼 것이다. 이 프로그램은 둘 다 다양한 이벤트를 준비할 수 있다.

THE CIVIC CENTRE

NEW RESTAURANT (R) + TOWN HALL (TH)

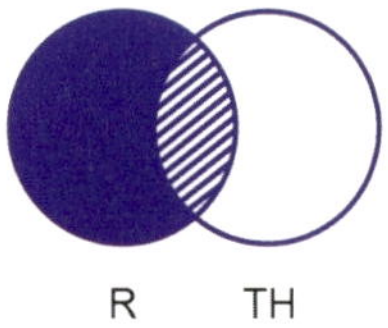

parasitism / mutualism: Even though the Restaurant will have a parasitic relationship with the building through new structural and material interventions, it will through its programme aid in generating income and draw the community into the Town Hall

기생론/상호론: 비록 레스토랑은 새로운 구조적, 재료적 개입을 통해 건물과 기생적인 관계를 갖게 될 것이지만, 그것은 소득 창출에 대한 프로그램 지원을 통해 지역사회를 시청으로 끌어들일 것이다.

NEW SQUARE (REAR) (S) + TOWN HALL (TH)

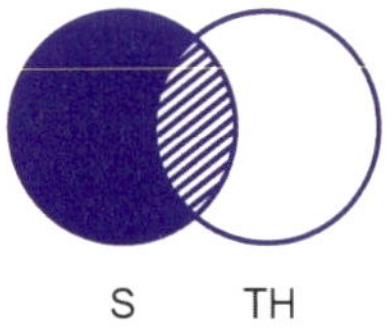

commensalism / mutualism: The new Square at the rear of the Complex will require structural and material interventions. However, it will not change the building, but the building will benefit from the new programme

공생론/상호론: 단지 뒤쪽에 있는 새 광장은 구조적 및 재료적 개입이 필요할 것이다. 하지만, 그것은 건물을 바꾸지 않을 것이지만, 그 건물은 새로운 프로그램의 혜택을 받을 것이다.

STUDY SPACE (S) + TOWN HALL (TH)

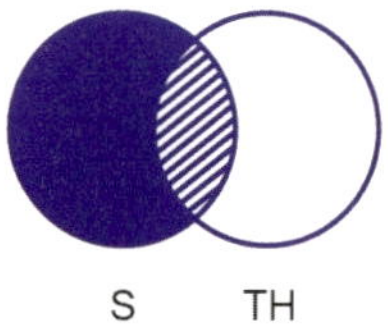

commensalism: The Study space on the first floor wire will require material changes, but this will not harm the building.

공생론: 1층 와이어의 스터디 공간은 재료 교환이 필요하지만 건물에 해를 끼치지는 않는다.

CONCEPT DEVELOPMENT EXTERIORITY - INTERIORITY

The 'Civicness' will start with the Exterior through the conceptualisation of the context: by 'turning the constraints of a context into the driving force behind the development of an architectural idea' (B. Tschumi). The result will be an activity spot - a public space at the Rear which will work its way through into the Interior through new architectural interventions such as a glazed facade on the ground floor, which will also serve as the main Restaurant entrance into the historic Town Hall building. The new square is ment to serve as an activator for bringing the community together, as well as other parts of the programme for the Civic Centre and connect the rear with the Clissold Park and front part of the Complex.

'시민성' 은 문맥의 개념화를 통해 외부로부터 시작할 것이다. 즉, '문맥의 제약조건을 건축사상의 발전의 원동력으로 변화시킴으로써' (베르나르 츄미). 그 결과는 활동 장소 – 1층에 유리를 칠한 면과 같은 새로운 건축적 개입을 통해 내부로 들어갈 수 있는 리어에 있는 공공 공간이 될 것이며, 이 공간은 역사적인 타운 홀 건물로 들어가는 주요 레스토랑 입구 역할을 할 것이다. 이 새로운 광장은 시민 센터를 위한 프로그램의 다른 부분뿐만 아니라 지역사회를 하나로 모으기 위한 활동가 역할을 하고 뒤쪽을 클리솔드 공원과 콤플렉스의 앞부분과 연결하기 위한 것이다.

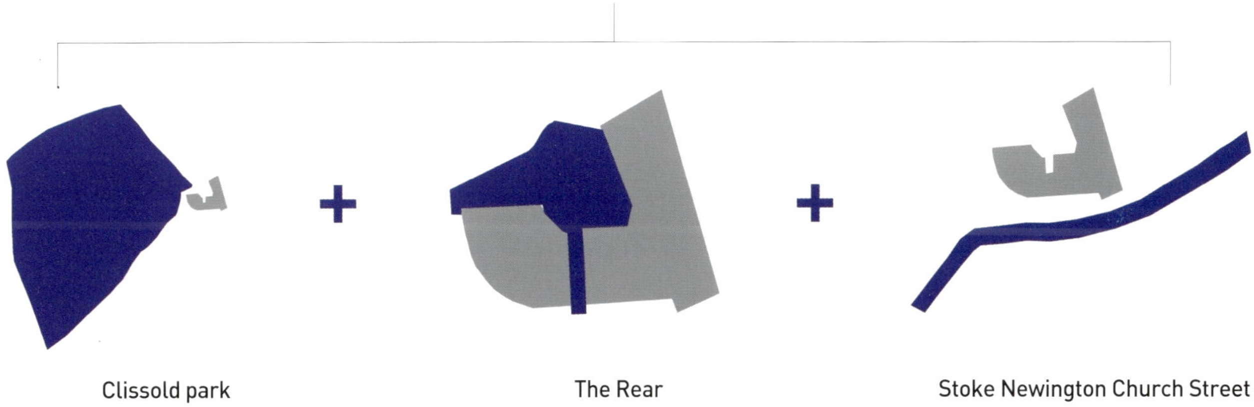

Demolition of the Civil Defence building, Data Centre building and pavement for the new public space

SPATIAL ADJACENCIES PROPOSAL

Ground Floor

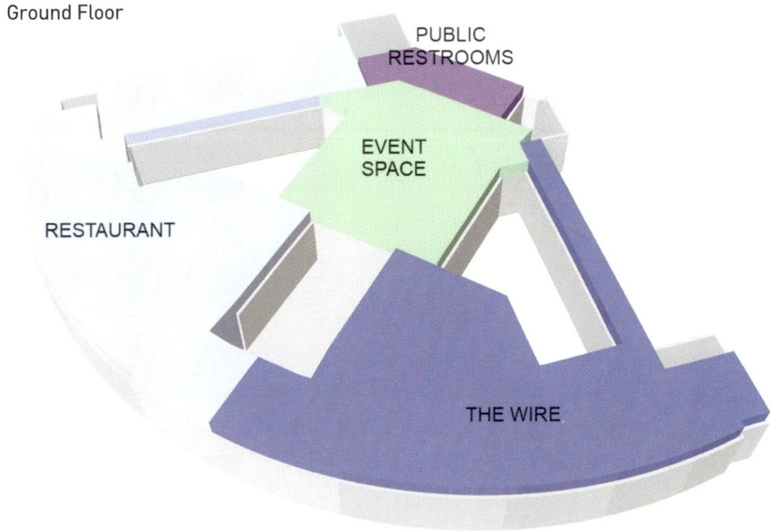

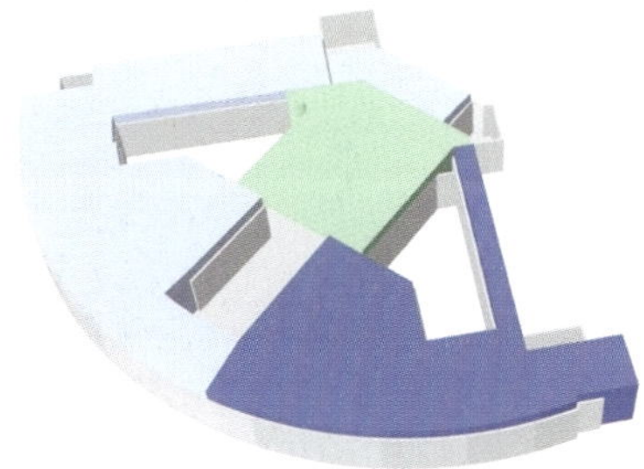

First Floor

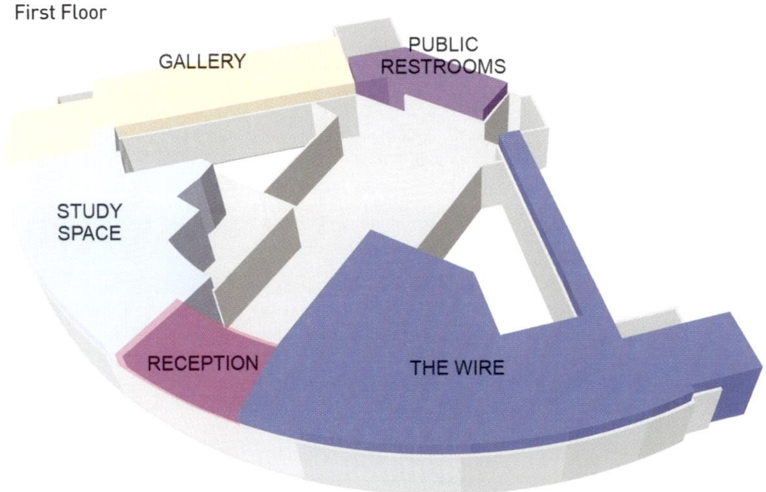

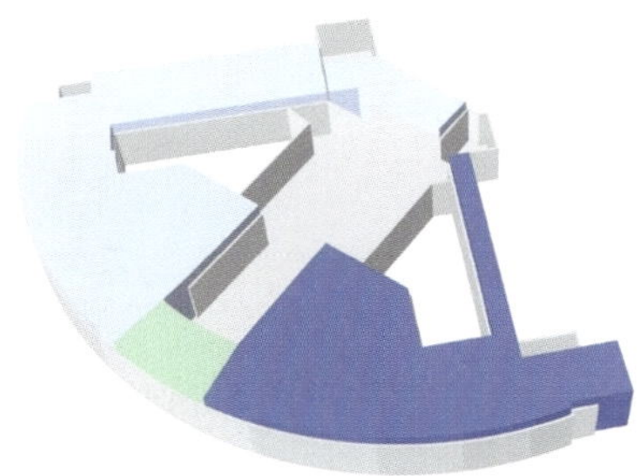

Public - Civic Centre

Private - The Wire

Public - used by both

The Wire Office, Ground Floor

Event Space, Ground Floor

The chairs can be stored within cupboards,hidden behind wooden acoustic panels to allow more space for other type of events such as small music performances or exhibitions.

의자는 찬장 안에 보관할 수 있으며, 작은 음악 공연이나 전시회와 같은 다른 종류의 행사를 위해 더 많은 공간을 확보할 수 있다.

Restaurant, Ground Floor

Study Space, First Floor

PATHOLOGICALLY INDEPENDENT PIECE OF FURNITURE

The piece of furniture is accessible from all sides, which was derived from the idea of the Wire's 'democratic' work ethic, as well as Tony Herrington's description of the Wire who "occupy the space because of convenience" and the Magazine being an "organic thing that changes when the people change it".
The independent piece of furniture is purposely made narrow for a fetus position, since the occupier wont be able to maintain the position for a long time and will eventually have to stretch. After this happens, the user's limbs exit the structure (host) and new symbiotic relationships are formed, which is also the main concept of the entire Studio project.
It is a fun and engaging conceptual furniture piece, which will spark curiosity amongst visitors of the new Wire office and Civic Centre within the spaces of the Stoke Newington Town Hall.

이 가구 조각은 WIRE의 '민주적' 직업 윤리에 대한 아이디어에서 비롯되었으며, 토니 헤링턴은 WIRE를 "편리 때문에 공간을 점령한다" 고 기술했고, 매거진은 "사람들이 그것을 바꾸면 변하는 유기적인 것" 이라고 묘사했다.독립된 가구 조각은 원래의 놓여진 위치를 오랫동안 유지할 수 없을 것이고 결국 스트레칭을 해야 하기 때문에 일부러 좁게 만들어진다. 이렇게 되면 사용자의 팔다리가 구조(호스트)를 벗어나 새로운 공생관계가 형성되는데, 이 역시 스튜디오 프로젝트 전체의 주요 개념이다. 이 작품은 스토크 뉴잉턴 타운 홀의 공간에 있는 새로운 WIRE 사무실과 시민 센터의 방문객들 사이에 호기심을 불러일으킬 재미있고 매력적인 컨셉의 모피가구 작품이다.

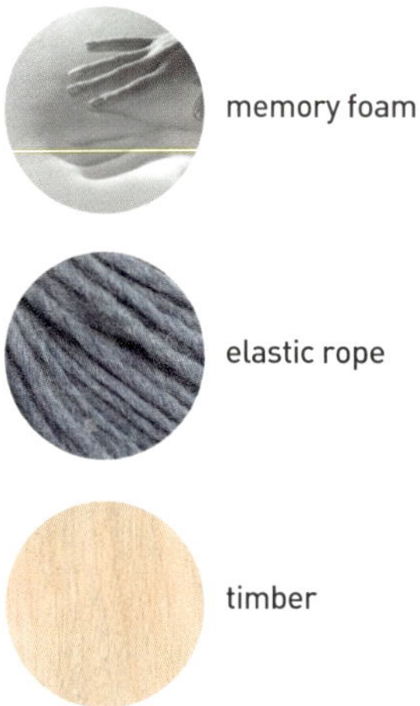

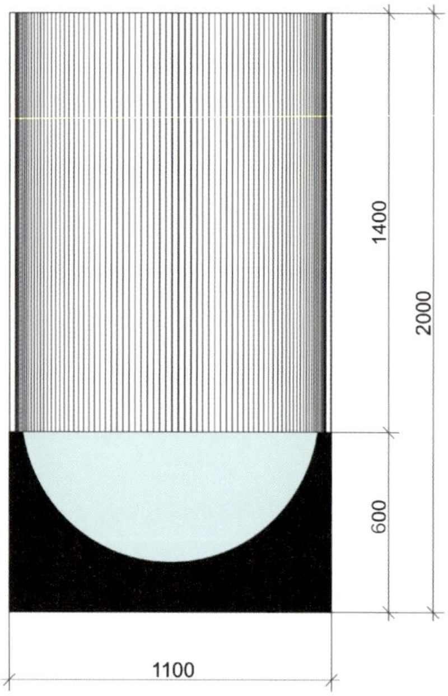

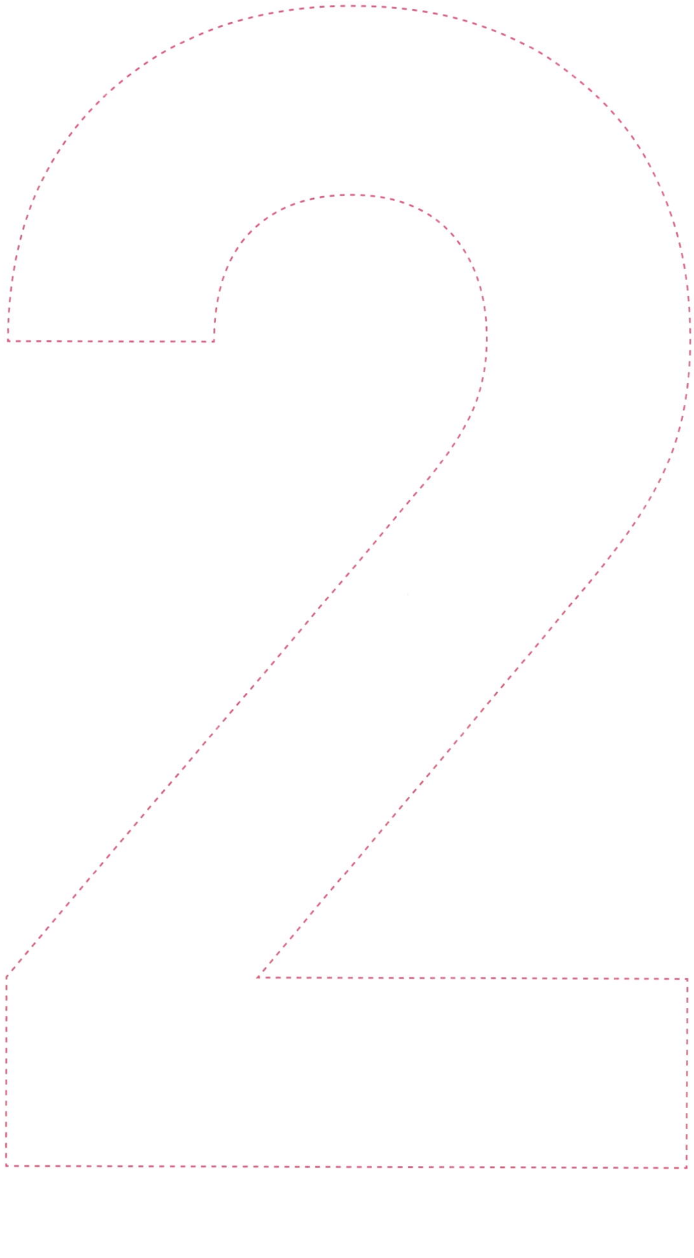

DONGYANG MIRAE University
interior & environmental design

DONGYANG MIRAE
UNIVERSITY

DONGYANG MIRAE UNIVERSITY

Department of Interior Environmental Design

2018 sophomore design studio

The 10th Hidden Space Project

REMEDITAION

동양 미래대학교 실내환경디자인과
2018 2학년 2학기 설계스튜디오 결과물

152-714
DONGYANG MIRAE University, Dept.of Interior Environmental Design
62-160, gocheok-dong, guro-gu, seoul, korea
phone + 82 2 2610 1962
fax + 82 2 2610 1960

http://interior.dongyang.ac.kr

SITE
서울특별시 구로구 중앙로6길 16(지번주소 고척동 78-2)

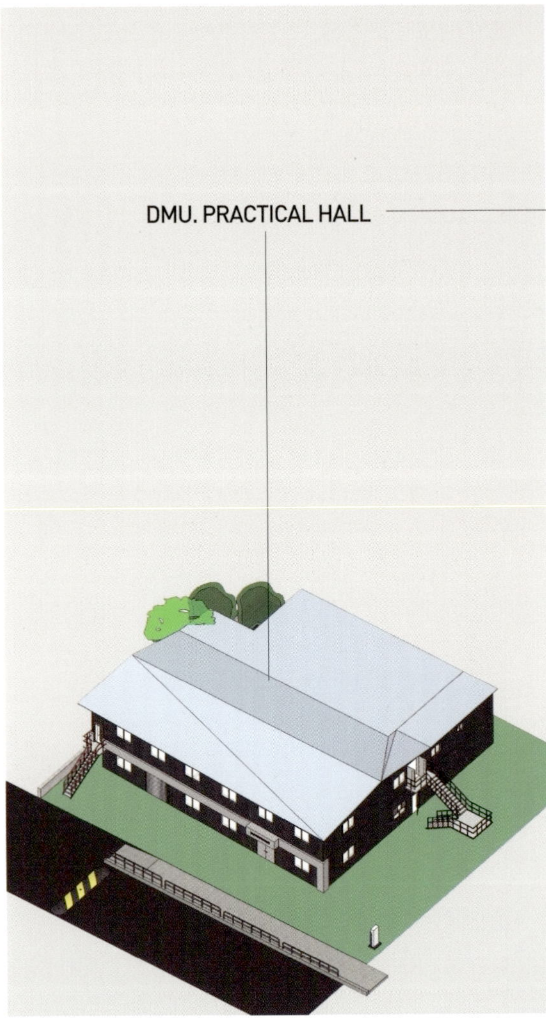

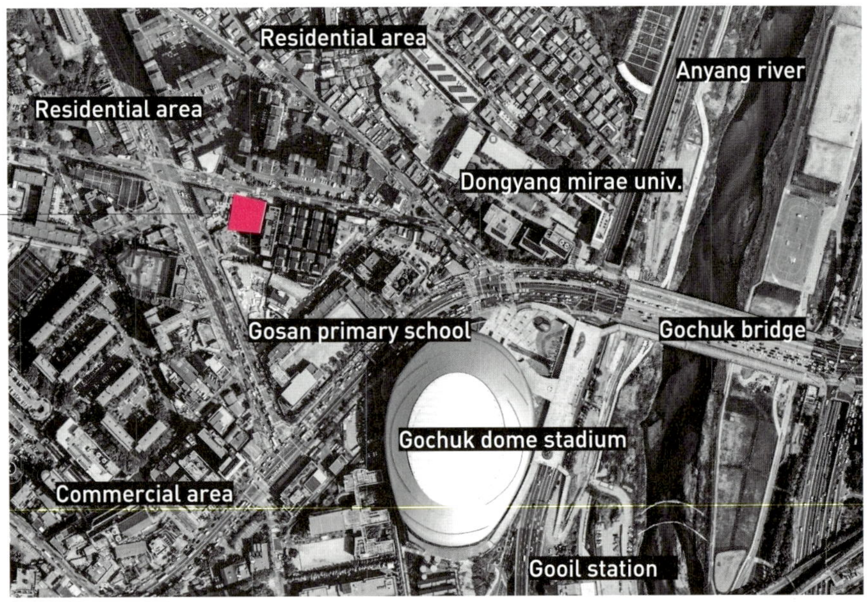

2학년 설계담당 **박영태 교수**
동양미래대학교 생활환경공학부 실내환경디자인과

responsible for sophomores | Prof. Park, Young-Tae
Department of Interior Environmental Design, DONGYANG MIRAE Univesity
robotomy@naver.com

히든 스페이스 프로젝트 열 번째 에피소드
REMEDIATION 再媒介

기술과 자본의 상호작용은 삶의 모든 콘텐츠들을 경계 없이 용해한다. 미디어를 매개로 시공간을 초월한 읽기와 쓰기를 포함한 모든 정보활동으로부터 주거와 일, 소통과 만남의 방식 전체가 변화하고 있다. 10번째 히든 스페이스 주제는 '공간을 통한 미디어의 재미디어화'를 생각해본다. 주로 지면을 통해 오프라인으로 활약해 온 미디어(매체)들의 변화가 어떤 식으로 변하고, 변하게 될지 상상해본다.

대중들의 라이프스타일에 밀접한 무언가를 어떤 방식으로 생산하고 전달해 줄지, 그들이 일하는 태도와 방식에 대해 깊이 고민해본다. 판매와 홍보를 포함한 콘텐츠의 지속가능 전략을, 향후 요구되는, 혹은 도래할 '일하는 공간'의 모습을 생각하며, '지각과 감각'을 주제어로 10개 매체를 대상으로 새로운 공간형식의 재매개(REMEDIATION)를 제안한다.

THE OTHER WORKING SPACE: MEDIA PLATFORM COMPANY / PUBLISHING HOUSE
'새로운 개념의 매체회사(출판사/잡지사)'

THE 3RD DIRECTIVITY / BETWEEN PERCEPTION & SENSE, BEYOND AFFECTION
Theme: 제 3의 지향성(指向性) / 지각과 감각 사이 / 정동을 넘어서기

1. 업무유형
- 매체, 지식산업, 정보를 가공, 편집하는 공간
- 관련업계의 지식과 정보를 생산, 유통하는 온 오프라인 잡지사
- 업무공간 이외 전시/이벤트 공간, 커뮤니티 공간
- 직원 10–15명

2. 프로그램 설정
2.1. 업무공간(출판사 공간에 필요한 일반적 업무공간)
2.2. 외부(지역, 학교, 주민 등)와의 커뮤니티 공간
2.3. 전시/이벤트 공간

3. 프로젝트 반영내용:
3.1. 지각과 감각에 대한 테마 〈 – 〉 재료와 질감에 대한 테마 〈 – 〉 인터렉션 개념 등을 공간의 기능과 프로그램과 연결하기.
3.2. 공간적 요소 : 지붕, 바닥, 벽, 계단, 보이드 vs 솔리드, 부분 vs 전체, 관계와 연결, 내부와 외부, 투명, 반투명, 가구(집기), 설비에 대한 테마 설정하기
3.3. 문화자본과의 연결 :
- 해당지역 사이트의 역사, 사회, 문화적 맥락과의 연결, 공간, 장소의 기억문제
- 안티테제의 설정 –> 자본주의, 첨단기술의 발전, 남성중심 가부장제, 권력과 위계중심 사회, 현대인들의 소외, 개인주의, 공동체주의, 지역주의, 파벌주의, 비/반도시성
- 개인이 가진 일하는 공간, 장소, 환경에 대한 경험과 기억의 호출, 부정적, 긍정적 경험, 체험...에 의한 서사, 스토리텔링

Module Summary (2018.09.11)

01 REQUIRED COMPULSORY SUBJECTS
Sophomore "Inteiror Design Studio III"

4 Of 18 Credit Modules in a Semester
- 6 Hours (tuesday) / 1 Week
- 1, or 2 Students / 1 Team
- Total 11 Projects

02 THEME
"REMEDIATION"

03 PROCESS & OUTPUT
- Research & Analysis (include FGI & survey, SWOT)
- Space Scenario making
- Sketch(analog & digital method) process
- Complex Space Program(zoning, planning....)
- Drawing
- 3D Simulation
- Scale Modeling
- Panel Presentation
- & Every Process(Photo & Text) Recording
  - for the Final Documentation

04 SCHEDULE
- 2018. 1st Week - Kick off & Orientation
- 2018.10.Last Week - Mid Term Critic Exhibition & Panel presentation
- 2018. 12. Last Week - Final Exhibition

05 KEY WORD
- REMEDIATION
- NEW LIFESTYLE
- CONTENTS
- PRODUCTION, CONSUMPTION, DISTRIBUTION
- INFORMATION
- PERCEPTION & SENSATION
- AFFECTION
- EXPERIENCE
- CULTURE CAPITAL
- ART, LABOR, EDUCATION, EVERYDAYNESS

MEDIA ISSUE 10

tumblbug

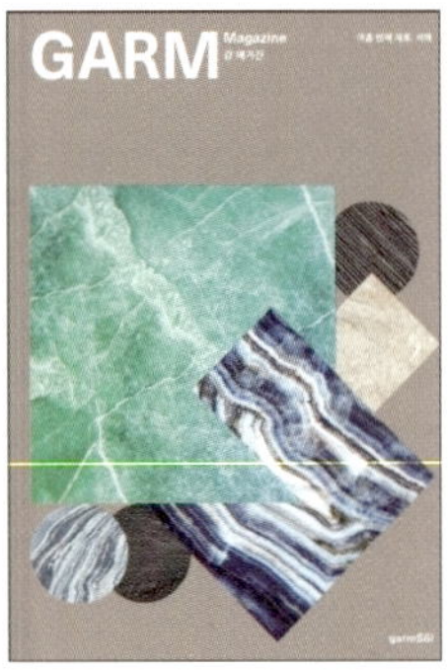

01 tumblbug - Oh Su Jin Park | Hye Jung "MOVING SIGHT"

02 magazine 'Pnorama' - Han Sang Mi | Jang Hye Lin "AT THAT MOMENT"

03 magazine 'Coffee' - Oh Jeong | Hyun Park Jeong Hyun "The Other Balance"

04 magazine 'JAZZ PEOPLE' - Lee Chae Rim | Ma Jeong Yeon "JAZZ PEOPLE"

05 magazine 'Decanter' - Hong Chun Myung | Choi Jae Wonn "SENSORIUM"

06 magazine 'RUNCITY' - Lee So Min | Eum Hyo Won "BREATH"

07 magazine ' MOM & ENFANT' - Cho Su Yeong | Park Se Jin "MOM & ENFANT magazine"

08 magazine 'GARM' - Choi Se Ryoung | Choi Hyo Jin "Ensemble of Space"

09 magazine 'THE SCENT' - Lee Yoon Seo | Kim Kyeong Gon "A sense of contrast"

10 magazine ' THE BEERPOST' - Choi Da Yeon | Kim Seo Ho "SCENT OF SCENT "

SPATIAL **REMEDITAION**; Contents A~Z

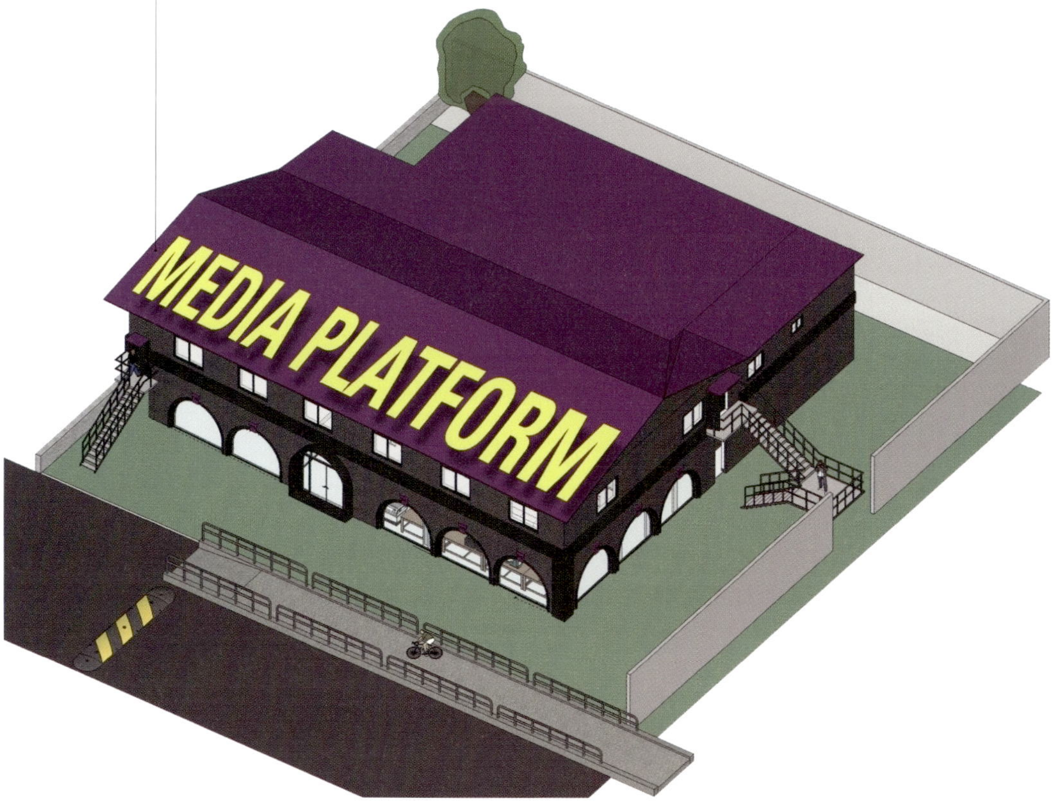

special lecture 01 - perception, sensation / Architect Park, Chng Hyn

special lecture 02 - spacial contents / Space Designer Lim, Tae Hee

Midterm Critique 2018, 11.06.

Kang, Hyung-seok (intune, CEO) / Lee, Chang-Han(JH Partners, CEO)

1st day_18, september ORIENTATION | The 7th Bldg. 5F AM 09:00~

Final Critique - Space Desiner Park, Jin Ho - 2018. 12.26.

Final Critique - Space Desiner Park, Jin Ho - 2018. 12.26.

오수진 I Oh, Su Jin 박혜정 I Park, Hye Jung

MOVING SIGHT tumblbug

영화적 시선의 개념을 적용한 텀블벅 업무공간 계획
Work space plan with application of concept of movie perspective.

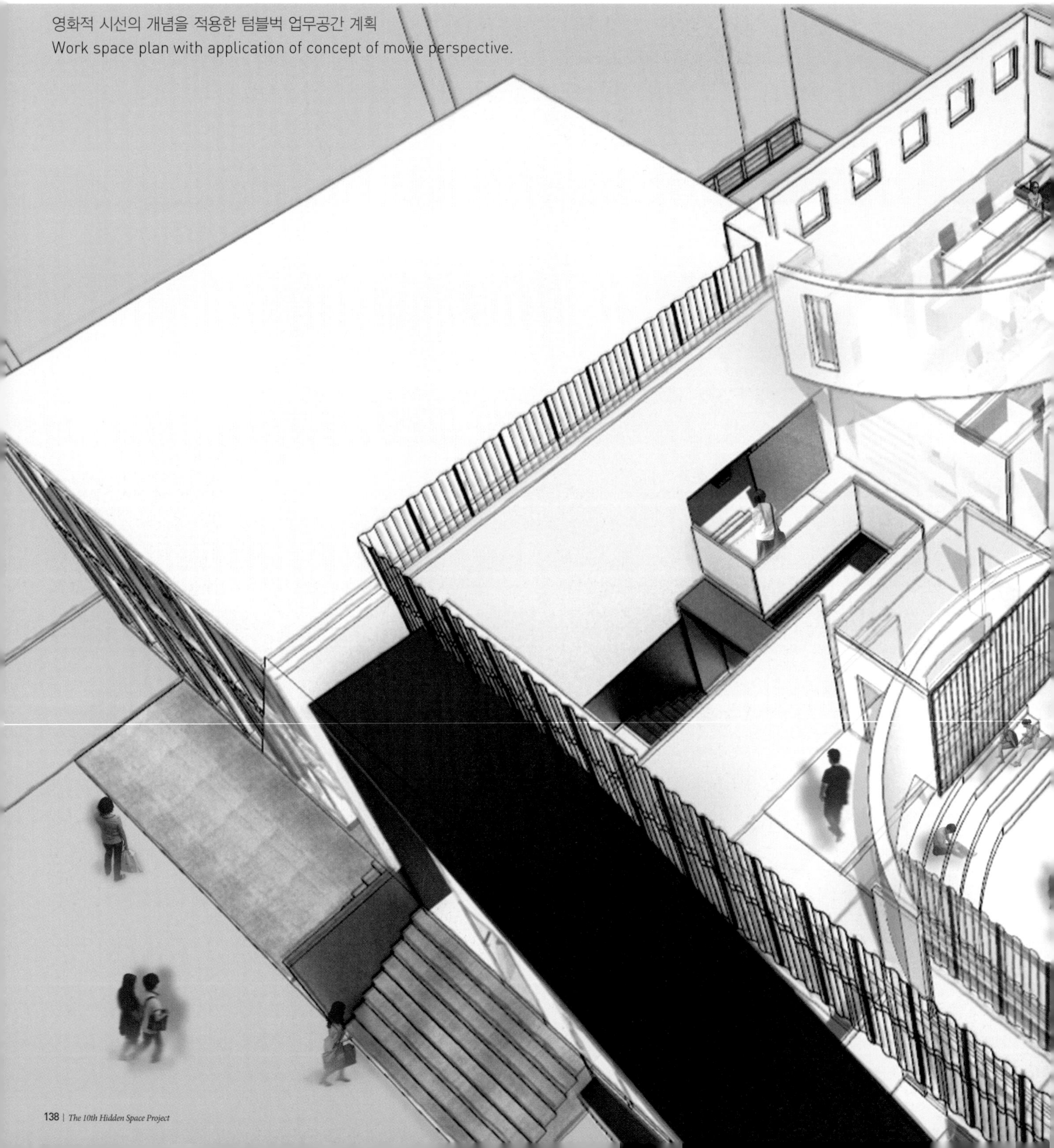

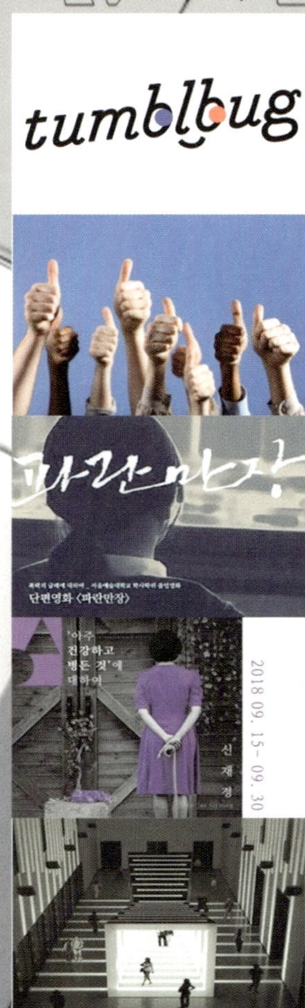

SITE

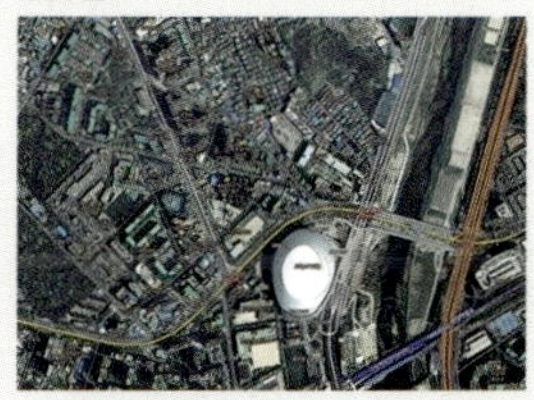

Location: 16, Jungang-ro 6-gil, Guro-gu, Seoul
Area: 670 m2

SITE ANALYSIS

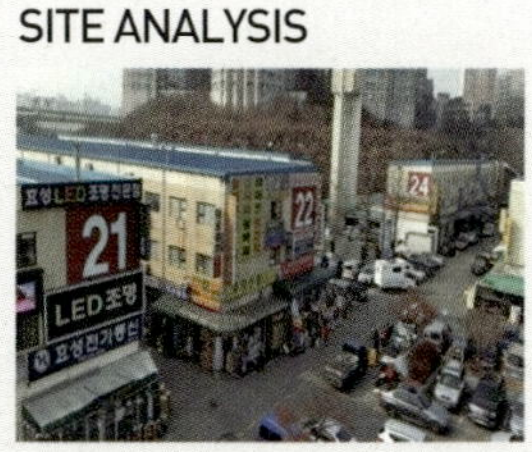

Gocheokdong, Gurogu where site is located is highly accessible for neighbors as there are elementary, middle, and high school, apartments, university, and baseball stadium around.

BACKGROUND

창작자와 후원자는 온라인 텀블벅 플랫폼을 통해 서로 끊임없이 교류하며 연결된다. 텀블벅의 직원 역시 창작자 및 후원자와 교류하게 된다. 이러한 특성은 텀블벅이 다루고 있는 다양한 장르 중 '영화' 매체와 유사하다. 영화는 종합예술로써 다양한 소통방법을 사용하지만, 그 기반은 '시각' 정보의 교환에 있다고 할 수 있다. 우리가 제안하는 'Moving sight: tumblbug' 은 텀블벅을 통해 이루어지는 사람들의 '교류' 와 영화 속에서 이루어지는 사람들의 다양한 '움직임' 을, 그것을 바라보는 카메라적 '시선' 으로 해석한 공간이다. 창작자들이 직접 이용할 수 있는 작업공간, 텀블벅 직원들과 원활히 소통할 수 있는 모임장소, 지역 주민 및 예술을 사랑하는 사람이라면 누구나 이용할 수 있는 도서관 공간을 통해 온라인에 머물던 사람들의 이야기가 오프라인으로 확장되는 복합업무공간을 제안한다.

Creator and sponsor are connected and interchange with each other through Tunblbug platform. Also, Tunblbug staff get chances of exchanging with creator and sponsor. Such feature is very similar to 'movie' one of the media various genres that Tunblbug deals with. As a comprehensive arts, movie utilizes different communication methods but its foundation is on exchange of 'visual' information.

'Moving sight-Tumblbug' offered by us is a space which a place for 'exchange' of different people, 'movement' of people in the movie, and its interpretation from the 'view' of camera watching them. It suggests complex working space for creators, meeting place for active communication with Tumblbug staff, and library for any person and locals liking arts, this space will expand communication of people from online to offline.

DESIGN PROCESS

MASS PROCESS
: 레벨차를 통한 틈,시선과 공간의 연결

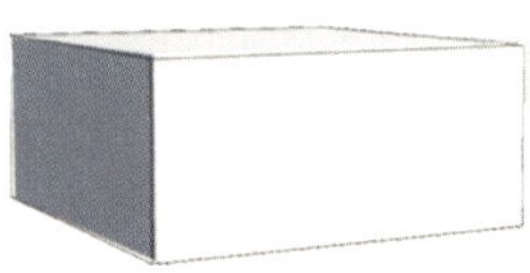

기존의 MASS

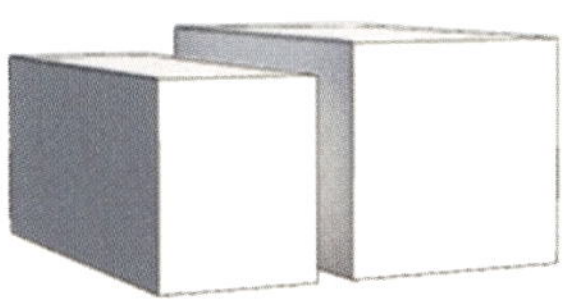

분절/ 공간의 용도에 따른 MASS 분리

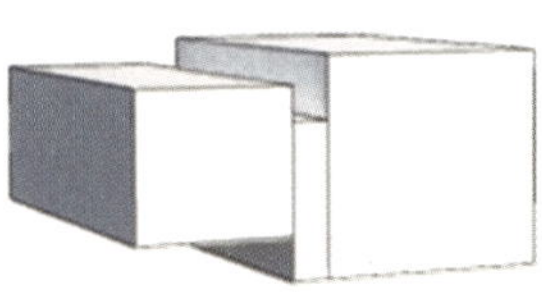

엇갈려 연결/ 틈을 통한 시선의 교류

개구부와 투명도 조절 : 시선의 종류

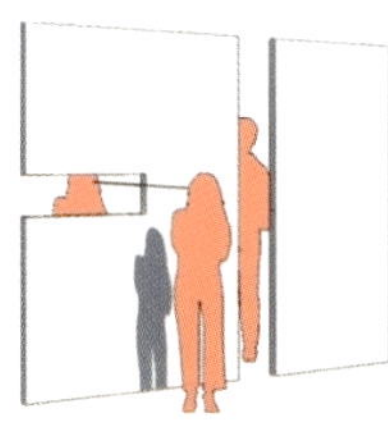

시선의 맞교환

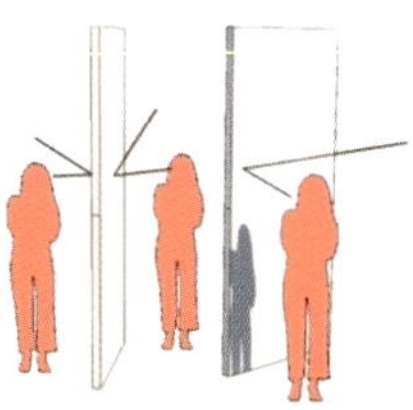

가려지는 시선

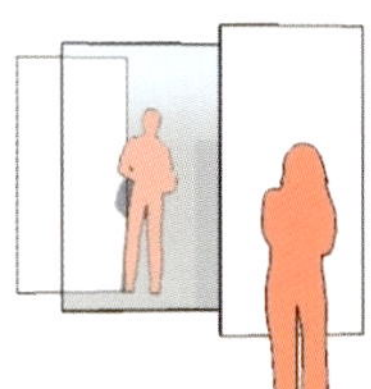

제 3자의 시선 – 카메라적 시선

KINETIC : 영화적 도형을 통한 공간 구획

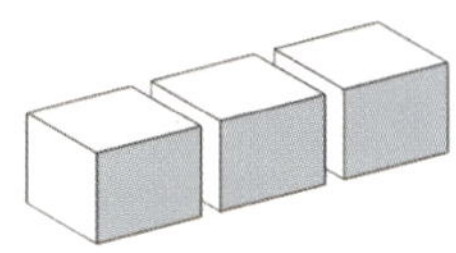

영화 시나리오 : BOX

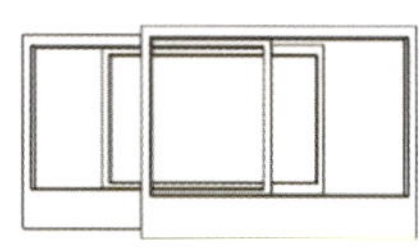

영화적 장면 : FRAME

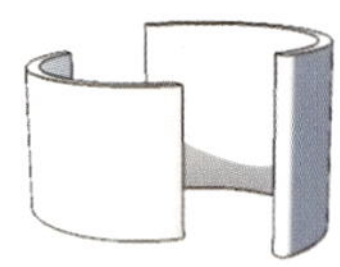

영화의 동작, 움직임 : 곡면 / REEL

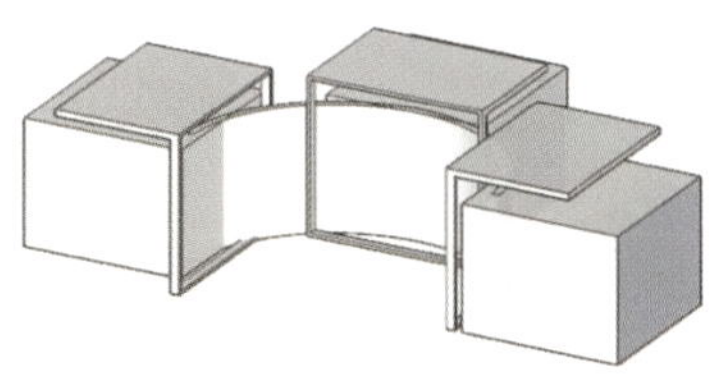

연결을 통한 공간구성

KEYWORD MAP

CONCEPT_1 시선의 교환 View exchange

우리는 텀블벅에서 이루어지는 사람들 간의 '교류' 와 영화 매체에서 이루어지는 '시각' 적 소통을, 카메라적인 '시선' 을 통해 바라봄으로써 서로의 시선을 통해 상호작용 할 수 있는 공간을 만들고자 하였다. 공간을 분할하는 벽체는 투명도의 조절을 통해 교류의 단계를 조절하며, 다양한 시선의 종류를 만들어낸다. 또한 매스(mass)의 분할을 통해 레벨차를 구성하며 운동감 있는 상호작용한다.

we tried to make a space for mutual communication through different views; from 'view' of camera looking 'exchange' of people and 'visual' communication in the media called movie. The wall separating spaces will control level of exchange by varying its transparency and creates various view points. What is more, it delivers dynamic interchange by separation of mass.

CONCEPT_2 영화적 운동감 Movie movement

영화 속 장면들은 분할(decoupage)과 재조합(montage)을 반복하며 이야기를 구성한다. 이를 영화의 물질적 구성요소인 릴과 프레임, 그리고 비물질적 구성요소인 장면(scene)과 내러티브(narrative)를 통해 교차시켜, 직선과 곡선이 교차하는 공간을 구성한다. 이를 통해 개별의 공간을 구성하고 레벨차를 만들며, 서로 연결될 여지를 제공한다.

Those scenes in the movie make a story by repeating separation (decoupage) and reunion (montage).
This forms a space where straight and curve lines cross each other by crossing material factors of reel and frame and immaterial factors of scene and narrative.

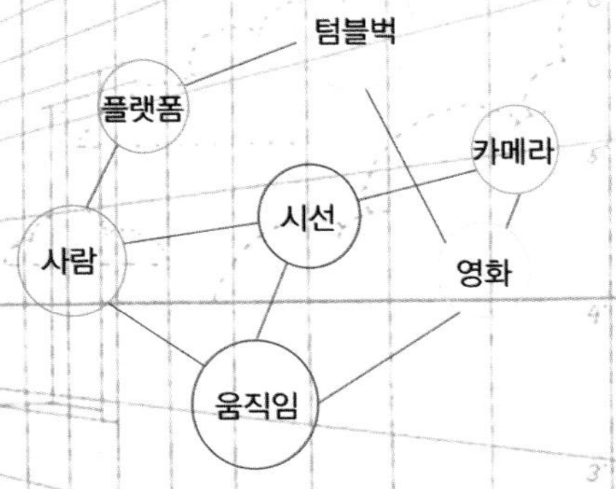

Keywords
: Platform, Moving, Human / Movie, Camera, View

SECTION - A

직원들 외의 외부 사람들도 이용할 수 있는 라이브러리, 영화 속 장면을 보는 듯한 프레임 공간을 표현한 필로티, 코워킹스페이스와 업무공간이 연결되는 썬큰가든 등의 접근성과 연결성이 좋은 텀블벅의 공간은 사용자의 시선의 교환를 꾀하고 쾌적한 공간을 제공한다.

Tumblbug space, providing good con\nection and accessibility such as library for staff and other people, pilotis showing frame space like a scene in movie, and sunken garden connecting co-working space and working space will be a wide space for exchanging different views.

PLAN

A side ▶

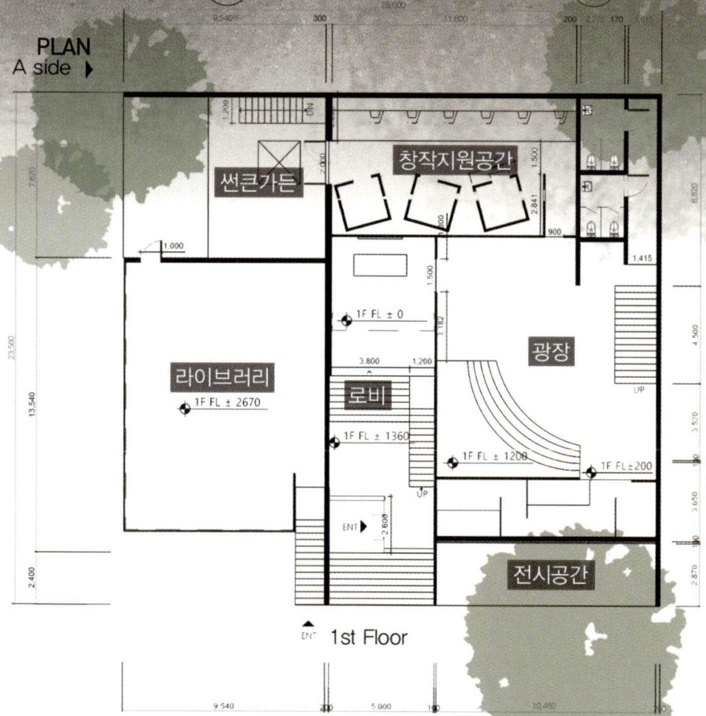

B side ▶

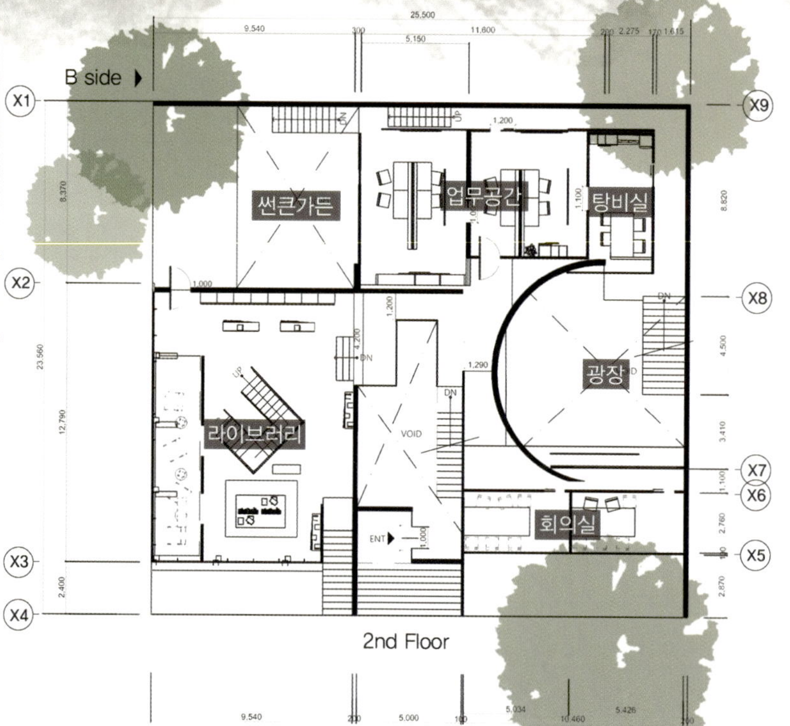

SECTION - B

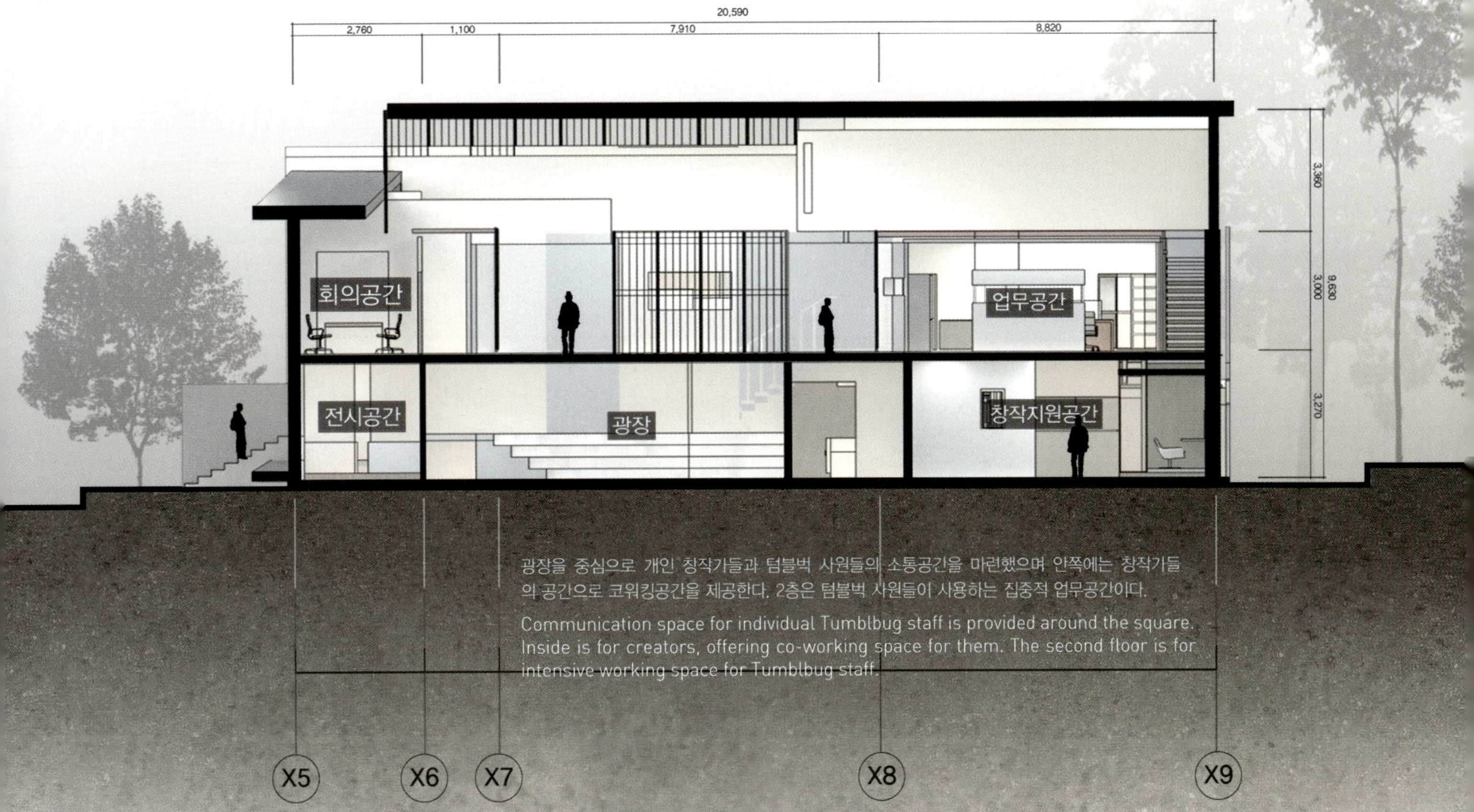

광장을 중심으로 개인 창작가들과 텀블벅 사원들의 소통공간을 마련했으며 안쪽에는 창작가들의 공간으로 코워킹공간을 제공한다. 2층은 텀블벅 사원들이 사용하는 집중적 업무공간이다.

Communication space for individual Tumblbug staff is provided around the square. Inside is for creators, offering co-working space for them. The second floor is for intensive working space for Tumblbug staff.

FACADE

전면에 큰 존재감을 가지는 프레임을 강조한 파사드이다. 프레임의 각도 살짝 기울여 기존의 것을 넘어서는 트렌디한 회사인 텀블벅의 특성을 표현하였다. 폴리카보네이트 재질의 유리를 사용하여서 가벼운 느낌을 주어 시선의 통함과 개방감을 강조했다.

The front side is facade showing a strong presence. Frame angle is little bit tinted to represent trendy features of Tumblbug. Polycarbonate glass is applied to provide light feeling, with a great view and openness.

PROGRAM

LIBRARY

1F : 동네주민, 개인창작가, 직원 모두를 위한 라이브러리이다. 업무공간의 출입문과 구별을 두어 자유로운 입장이 가능하며 소통의 공간이다.
2F : 개방된 라이브러리 공간으로 창 쪽에 빛이 들어오는 캐주얼 공간에서 책을 읽을 수있도록 디자인 하였다.

1F : This is a library for neighborhood, individual creator, and all staff. This working space is separated from the entrance to allow liberal communication and access.
2F: In a casual space with sun light coming in from open library space is designed for people to read books

OFFICE SPACE

1F : 개인창작가와 사내 직원들이 사용하는 개방된 원형 계단이 있는 커뮤니케이션 광장 공간이다. 안쪽에는 창작가들이 코워킹하는 폐쇄적인 공간이 있다.
2F : 사내 직원들의 업무공간이다. 중첩된 곡면과 BOX의 형태가 특징이다.

1F : This is communication square space with circular stairs for creators and staff. Inside, there is closed co-working space for creators.
2F : This is space for company staff. Its special feature is overlapped curve and box.

SUNKEN GARDEN

라이브러리와 업무공간의 통합을 꾀하는 썬큰 가든으로 업무를 보다가 지치면 리프레쉬 할 수 있는 공간이다. 중첩되는 프레임을 표현한 필로티공간과 이어지는 전이공간을 제공한다.

Sunken garden is a space for connecting library and working space where people can refresh themselves. This is a transfer space connecting to pilotis showing overlapped frame.

MOVING SIGHT / 움직이는 시선

라이브러리, 전시공간, 업무공간, 광장공간 등 기능에 따라 분리 된 공간들을 곡면의 벽체와 직선의 벽체가 이어줌으로써 시각적 리듬감을 선사한다. 이러한 리듬감 있는 벽체를 관통하는 투명, 반투명, 불투명의 벽들은 시선을 차단, 살짝 보여줌, 완전히 보여줌의 단계를 거쳐가며, '벽' 이라는 지각적 요소가 불규칙한 필터 효과를 통해 파편적 이미지를 공간화시킨다. 파편적 이미지로 구성된 공간은 인간의 움직임에 따라 변화하는 시점을 또 다른 움직임의 요소로 재구성한다.

Those separated spaces based on functions such as library, exhibition, working, and square spaces are connected by curved walls to offer visual rhythm. Those transparent, semi-transparent, and opaque walls going through wall frame show steps of blocking, little view, and full view. This perceptual factor 'wall' turns fragmental images into a space through irregular filtering effect. The space formed by fragmental images re-organizes views changing by movement into another movement factors.

3rd Floor Office

ISOMETRIC

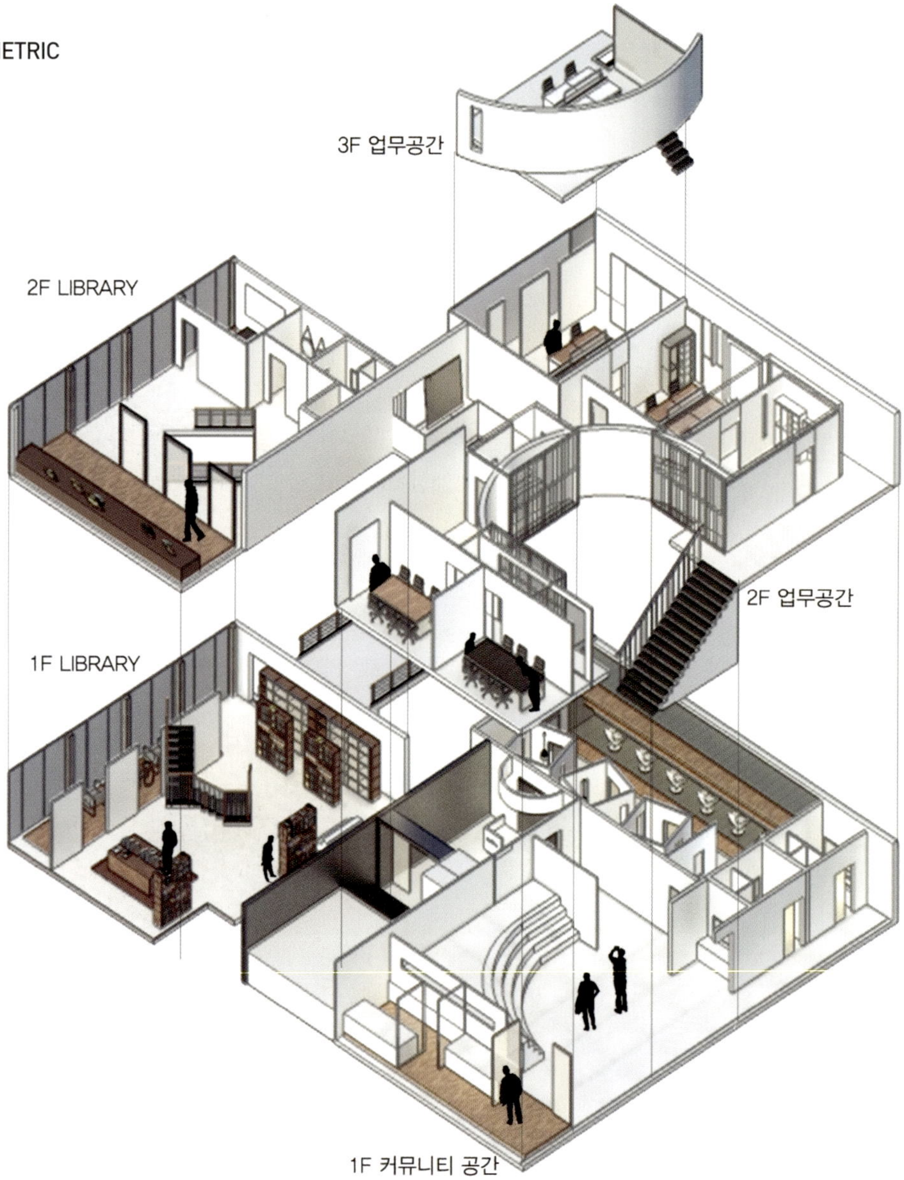

2nd Floor Aisle

2nd Floor flex & strait

1st Floor Co-Work

1st Floor Entrance

Composition of wall rhythm

높은 천장고 아래 곡면, 직선, 박스로 구성된 벽체의 높낮이 변화로 인해 리듬감이 구성된다.

Composition of wall rhythm: Changes in height of wall composed of high ceiling, lower curve, straight line, and box creates rhythm.

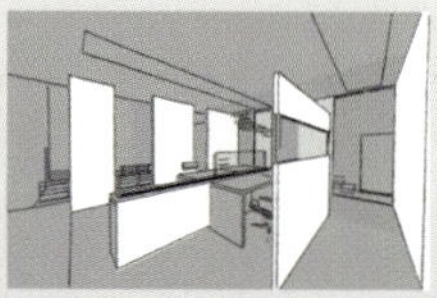
Irregular filter of wall

벽체의 투명도와 배치를 통해 불규칙한 장면을 유도하는 필터를 씌워, 시점을 다양화한다.

filter making irregular scene through transparency and arrangement of wall is place to make different views.

Space from fragmented images

리듬감과 다양한 시선으로 파편화 된 요소는 기능을 가진 공간으로써 재배치된다.

fragmented factors from rhythm and different views are re-organized into a functional space.

메인 광장공간 Main Squar

텀블벅의 주요 공간인 업무공간, 창작지원공간, 회의공간과 전시공간을 연결하는 중심 통로이자 광장 공간이다. 곡면의 호로 이루어진 벽체를 감싸는 직선의 박스와 프레임은 반투명 유리 재질로써 곡면의 유연함을 유지하며 열린 틈으로 반대편 공간까지 시선을 연결한다. 3층 높이 천정고의 공간을 감싸는 곡면 벽과 박스는 스페이스 인 스페이스의 개념으로 공간을 구성한다.

Tumblbug is a main path and square space connecting main places like working space, creation sponsor space, and meeting space. The straight box and frame covering curved wall is made up of semi-transparentglass material to maintain curl and flexibility, connecting view to the opposite space through open space. The curve wall and box covering the 3rd floor high ceiling form a space based on space-in-space concept.

1F

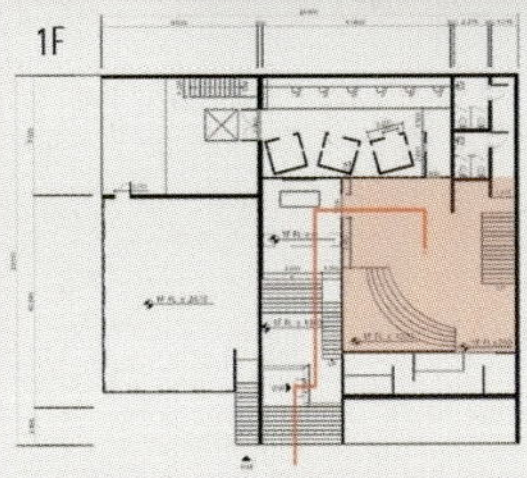

2F

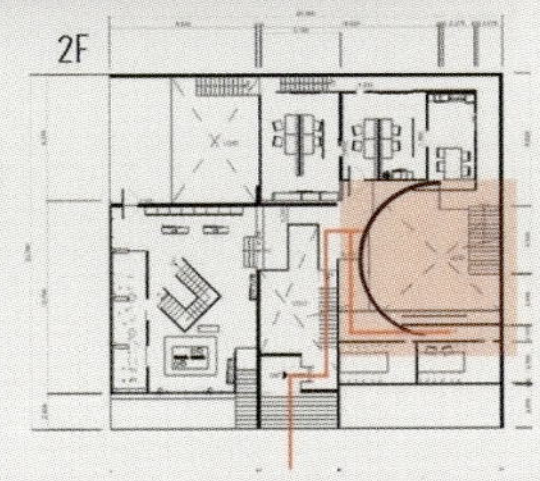

업무공간

스페이스 인 스페이스 형식으로 구성된 2개의 박스와 그 위에 얹어진 1개의 도형으로 구성된 공간이다. 내부는 벽체의 투명도 조절을 통해 직원들 간 프라이버시를 지켜주는 동시에 개구부의 연결로 연결되는 시선을 만들어낸다.

This space is made up of space-in-space form based 2 boxes and 1 shape on top of it. Inside wall makes connected view connecting entrance, at the same time, it protects privacy between staff by adjusting transparency.

2층 업무공간 – 창작지원팀

회의공간, 1층 광장공간, 3층 업무공간과 인접한 곳에 위치한 업무공간으로 다양한 부서와의 협업을 중요시하는 텀블벅의 핵심 업무공간이다.

This space is a working space located near meeting space, 1st floor square space, and 3rd floor working space, which is a core area for Tumblbug considering cooperation with other departments very important.

2층 업무공간 – 창작지원팀

곡면의 벽으로 둘러쌓인 텀블벅 3층 업무공간은 조금 더 집중을 요하는 업무를 위한 공간이다. 내부 계단을 통해 조금 더 은밀하게 들어가는 시퀀스를 만들어낸다.

3rd floor working space of Tumblbug covered by curve wall is a space for works requiring intensive job. It creates a sequence of entering the place through inside stairs in a more private way.

2F

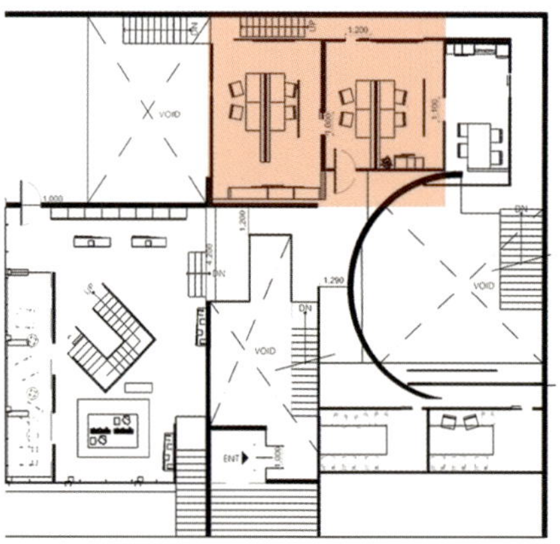

썬큰가든쪽을 바라보는 방향으로 설계 된 업무공간은, 세로 격자로 배열된 가벽을 통해 빛이 쪼개져 들어오며 외부의 장면을 부분적으로 차단해 파편화시키며 또다른 장면을 만들어낸다.

Working space designed towards sunken garden has sun light coming in splitted through fake wall arranged in vertical grid. This partially covers outside scene, making another scene.

창작 지원 공간

한 방향으로 개구부가 뚫린 박스공간은 서로 다른 공간에서 다른 업무를 하는 개별의 창작가들을 시선적으로 연결시켜주는 역할을 한다. 개별의 공간에서는 자유로운 회의 및 팀 작업을, 바 테이블에 마련된 공간에서는 개별의 업무를 선별적으로 할 수 있다. 1층 광장공간을 거쳐 이동할 수 있다.

Box space with one-way opening visually connects individual creators working for a different job in different places. In individual space, one can freely do team work and in a space with bar table, people can selectively do one's own job. One can move to this space through square space on the 1st floor.

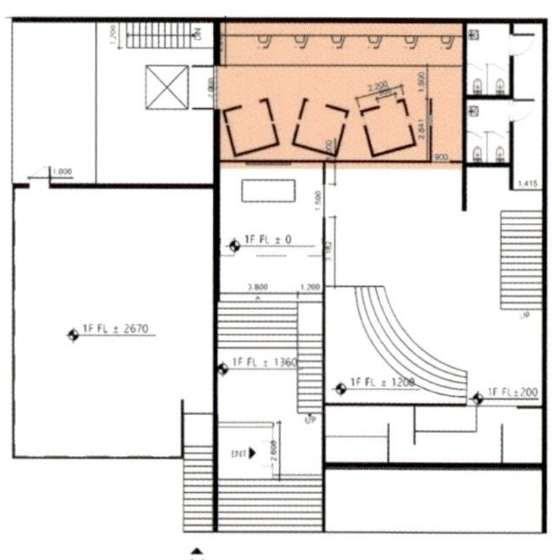

창작자, 텀블벅 직원 및 외부인들의 출입이 자유로운 개방된 공간이다. 업무공간, 썬큰가든 그리고 외부로 연결되어 있어 동선과 접근이 원활하다. 외부에서 진입할 수 있는 출입구와 중앙 통로로 진입하여 로비로 들어와 진입할 수 있는 사내직원의 출입구의 분리로 동선이 나뉘어져 있다.

This is a open space with free access for creators, Tumblbug staff, and other people. The pathway is separated from working space, sunken garden, and the entrance of staff, which can be entered from main pathway to the lobby.

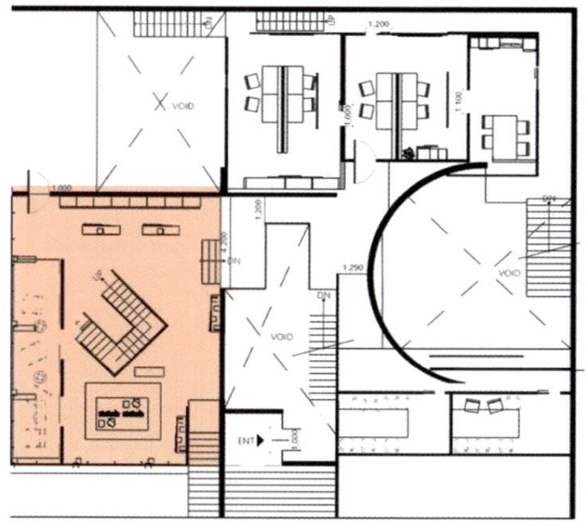

LIB 1F

라이브러리 1층

본래의 업무공간과는 다르게 누구나 이용할 수 있는 공간으로 서적이 구비되어 있고, 이를 누구나 읽을 수 있다. 꺾여 올라가는 계단은 투명, 불투명, 반투명의 벽이 감싸고 있으며 상부로 부터 조명을 받으면, 움직임에 따라 밝기가 변환되는 모습이 보인다.

The ordinary working space is open for all people and has books for any person. The whirl stair is covered by transparent, semi-transparent, and opaque wall and if shine by the light from the top, its brightness changes depending on the movement.

LIBRARY CASUAL SPACE

반투명 폴리카보네이트 벽체를 통해 빛이 투과되어 밝고 쾌적한 분위기의 공간.

Bright and fresh space with light coming in through semi-transparent polycarbonate wall.

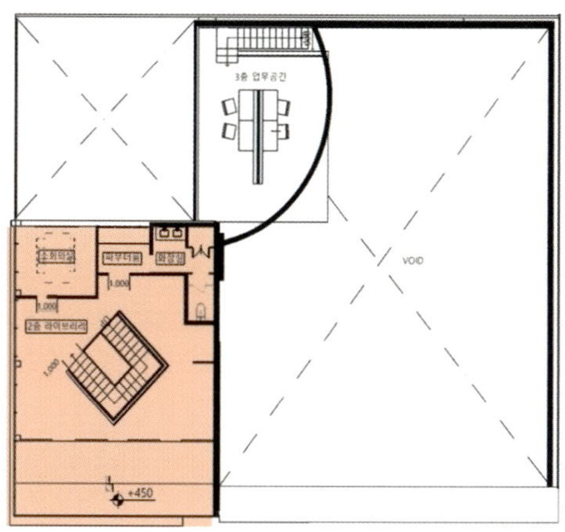

LIB 2F

라이브러리 2층

1층 공간에서 책을 골라와 자유롭게 읽을 수 있는 공간으로 구성되어 있다. 불투명 벽이 2층 천장고부터 1층 바닥까지 관통되어 연결성을 띈다.

This is a free space where people can choose a book from the 1st floor and come for reading. The opaque wall is connected from the 2nd floor ceiling to the 1st floor.

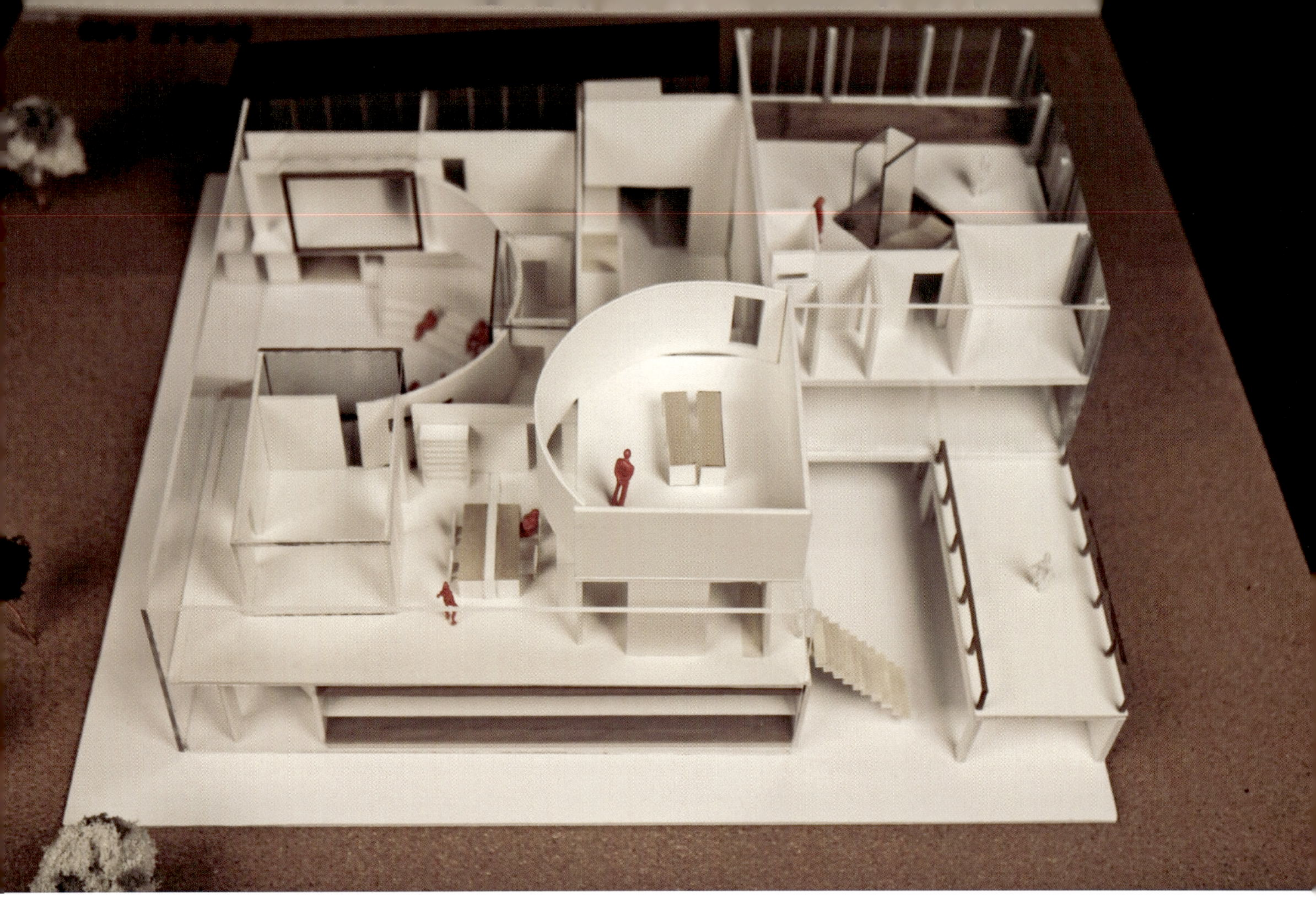

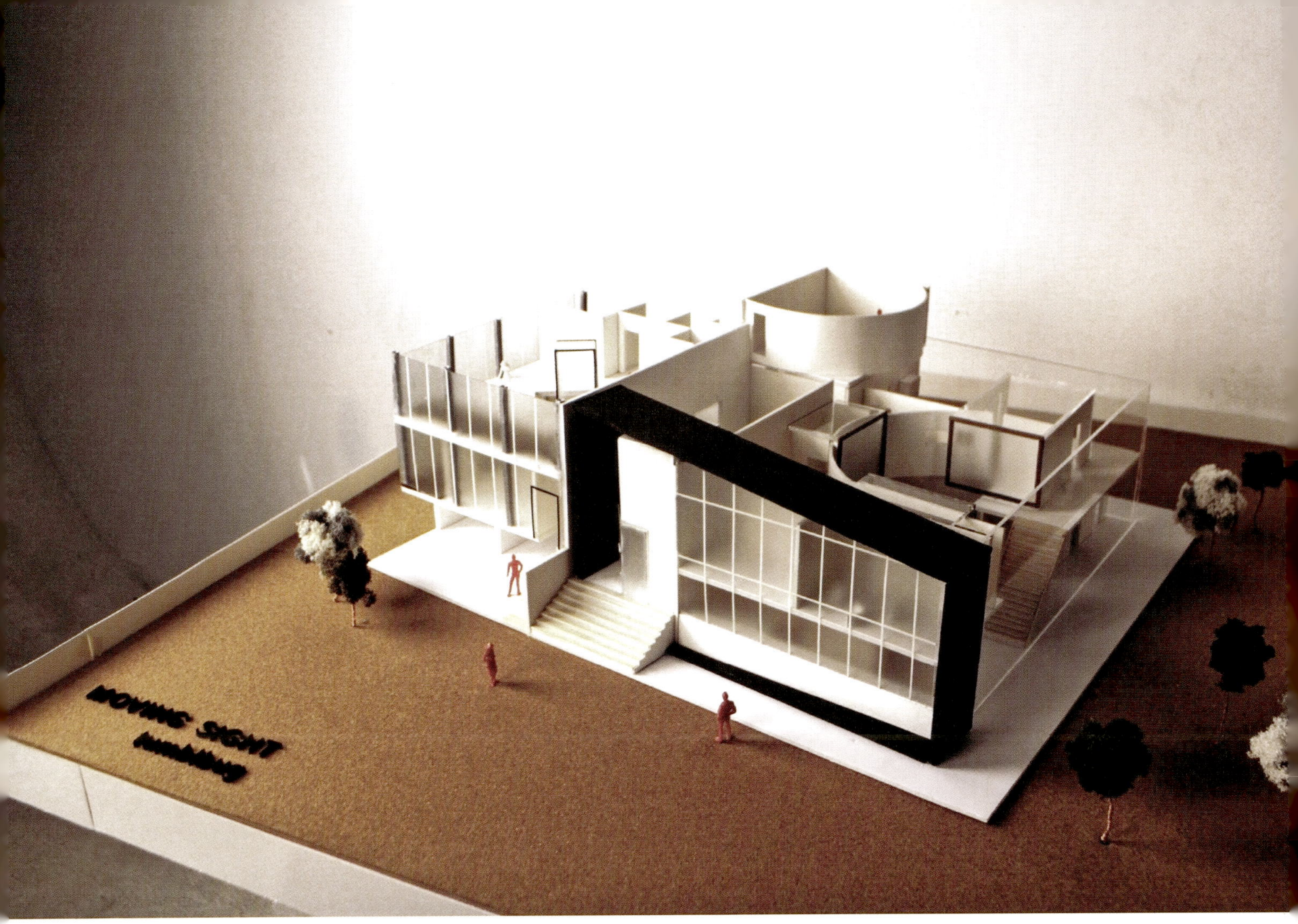
MOVING SIGHT

한상미 I Han,Sang Mi 장혜린 I Jang, Hye Lin

AT THAT MOMENT

'매거진 파노라마' 를 무의식과 의식의 사이의 개념으로 공간화한 업무공간

SITE

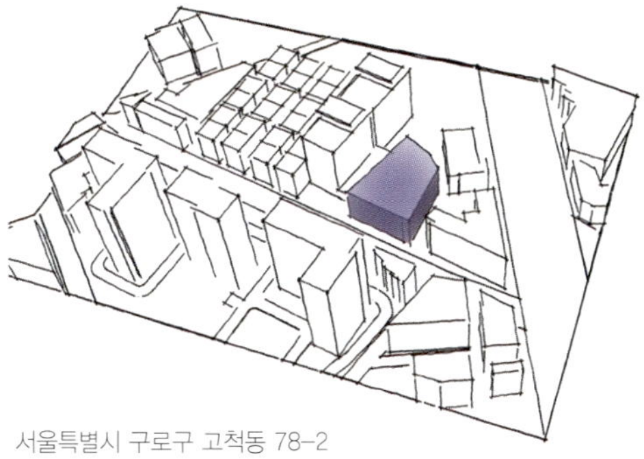

서울특별시 구로구 고척동 78-2

78-2 Gocheok-dong, Guro-gu, Seoul

SITE ANALYSIS

인근에 대형 병원과 소방서, 대학교 등이 자리잡고 있어 지리적 편리성을 갖고 있으며, 서울 주변 지역인 부천, 인천, 광명, 시흥을 연결하는 대중교통편이 많아 지리적 접근성도 용이한 특징이 나타난다. 또한 고척돔과 대형쇼핑센터도 인근에 위치하고 있어 문화적 접근성도 편리하다.

There are large hospital, fire station and university nearby, so it has geographical convenience. Also, this area is easy to access as many transportation systems connect this area with Bucheon, Incheon, Gwangmyeong and Siheung which are neibouring to Seoul. Furthermore, Gocheok dome and large shopping center are close to this area, making it culturally accessible.

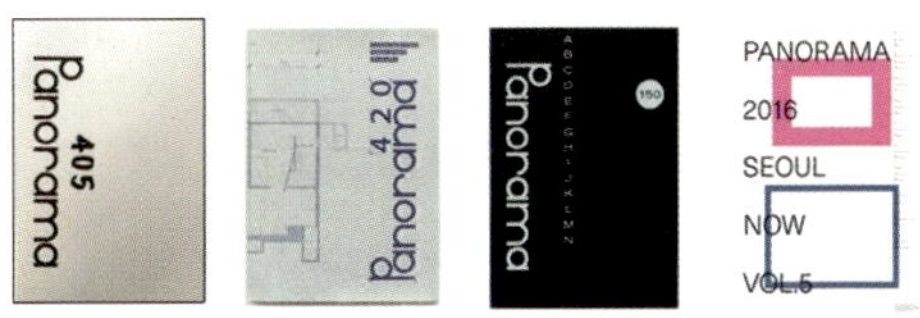

버스 노선 위의 건축에 대한 매거진

건축을 쉽게 접할수 있도록 버스와 같은 일상 속에서 이야기를 풀어나가는 스토리텔링 형식으로 구성되어 있는 잡지이다. 누구나 읽을수 있는 잡지를 지향하며, 축약되고 왜곡되는 수많은 컨텐츠의 바다에서 건축물들에 대한 시선들을 차분히 담아냈다.

This is a magazine composed of story which is happening in daily life such as 'bus', so that the people can readily learn about 'architecture'. This aims to be a magazine which everyone can read, and shows its own perspectives about the buildings in the ocean of the contents which are abandoned and distorted.

무의식과 의식의 사이

일상 (버스) 속에서 무의식에 스쳐지나가는 건축들을 '매거진 파노라마' 라는 잡지를 통해 의식하게 된다. 누구나 서울에 좋아하는 건축물이 하나쯤은 있었으면 좋겠다는 초기 설립자의 지향 즉, 건축의 접근의 어려움을 없애고자 하는 '매거진 파노라마' 의 지향성의 연장선이다. 의식과 무의식에 따라 명확히 구분되기도, 애매모호하게 합쳐지기도 하는 공간구분으로 감각과 지각의 변화를 표현한다.

Between the Unconsciousness and the Consciousness

Readers realise the buildings which they usually pass through unconsciously, through 'Magazine Panorama'. Early founder aimed that everyone would have one's favourite buildings and 'Magazine Panorama' tried to eliminate the barrier to access architecture to follow his aim. This concept is an extension of the directivity of the magazine. This expresses the changes of the sense and perception by spaces which are distictly dividing or indefinitely combining according to the consciouness and unconsciousness.

CONCEPT PROCESS

사용자는 어두고 좁은 복도를 무의식적으로 따라 이동하다 자연스럽게 넓고 밝은 의식의 공간에 도달한다. -> 무의식과 의식의 애매모호한 경계를 표현함.
좁고 어두운 복도는 의식과 무의식의 연결통로의 역할을 하고 미로같은 공간배치로 무의식과 의식의 개념의 복잡함을 나타낸다.

Users unconsciously walk along the narrow and dark corridor to reach a wide and bright space of consciousness.
--> This shows inapparent boundary of unconsciousness and consciousness.
The narrow and dark corridor not only plays a role as a link of unconsciousness and consciousness, but also implies a complexity of a conception of unconsciousness and consciousness through labyrinthian space arrangement.

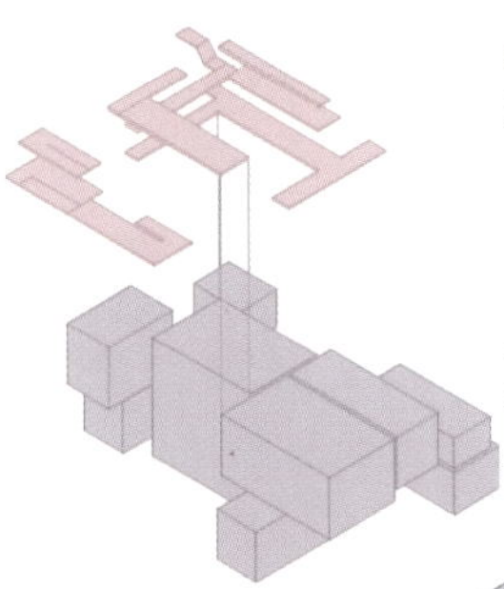

무의식의 공간 (복도)
Space of unconsciousness (Hallway)

의식의 공간 (역할공간)
Space of consciousness (Space for activity)

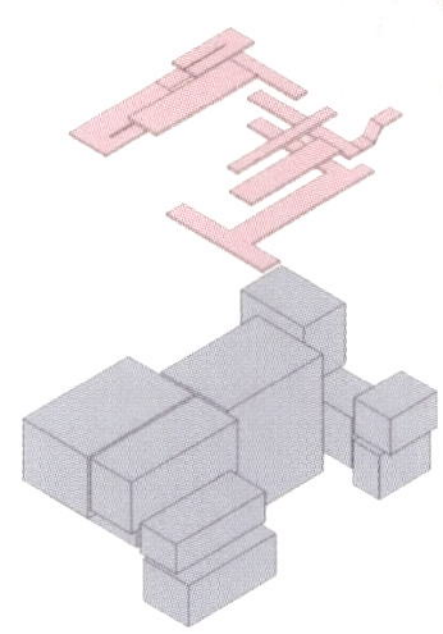

무의식의 공간 (복도)
Space of unconsciousness (Hallway)

의식의 공간 (역할공간)
Space of consciousness (Space for activity)

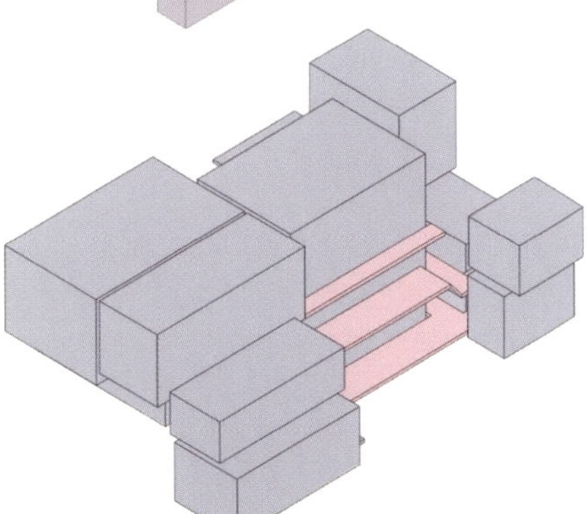

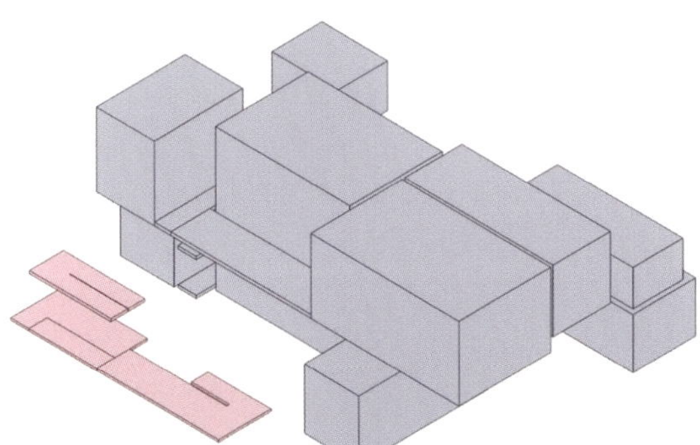

MIND MAP

매거진 파노라마의 핵심 키워드인 버스, 파노라마, 건축을 통해 의식과 무의식, 시간, 연속, 스쳐감의 키워드를 최종적으로 도출하였다.

The keywords consciousness, unconsciousness, time, continuity and perception are derived from the core keyword of 'Magazine Panorama', which are bus, panorama and architecture.

EMPLOYEE MEMBER

편집부
네 명의 에디터들로 구성되어 있으며 답사를 다니는 시간이 많아 출퇴근시간이 자유롭다.

Editorial Department
Composed of four editors and they have flexible commuting time due to a lot of field investigation.

디자인부
두 명으로 이루어져 있으며 기획부터 함께 참여하기 때문에 근무시간이 가장 길다.

Design Department
Composed of two designers. They participate from the planning stage so they have the longest working time compared to the others.

마케팅, 총무부
총 네명으로 구성되어 있으며 전체 총괄을 담당한다.

Marketing/General Affairs Department
Composed of four people and they are responsible for overall work and management.

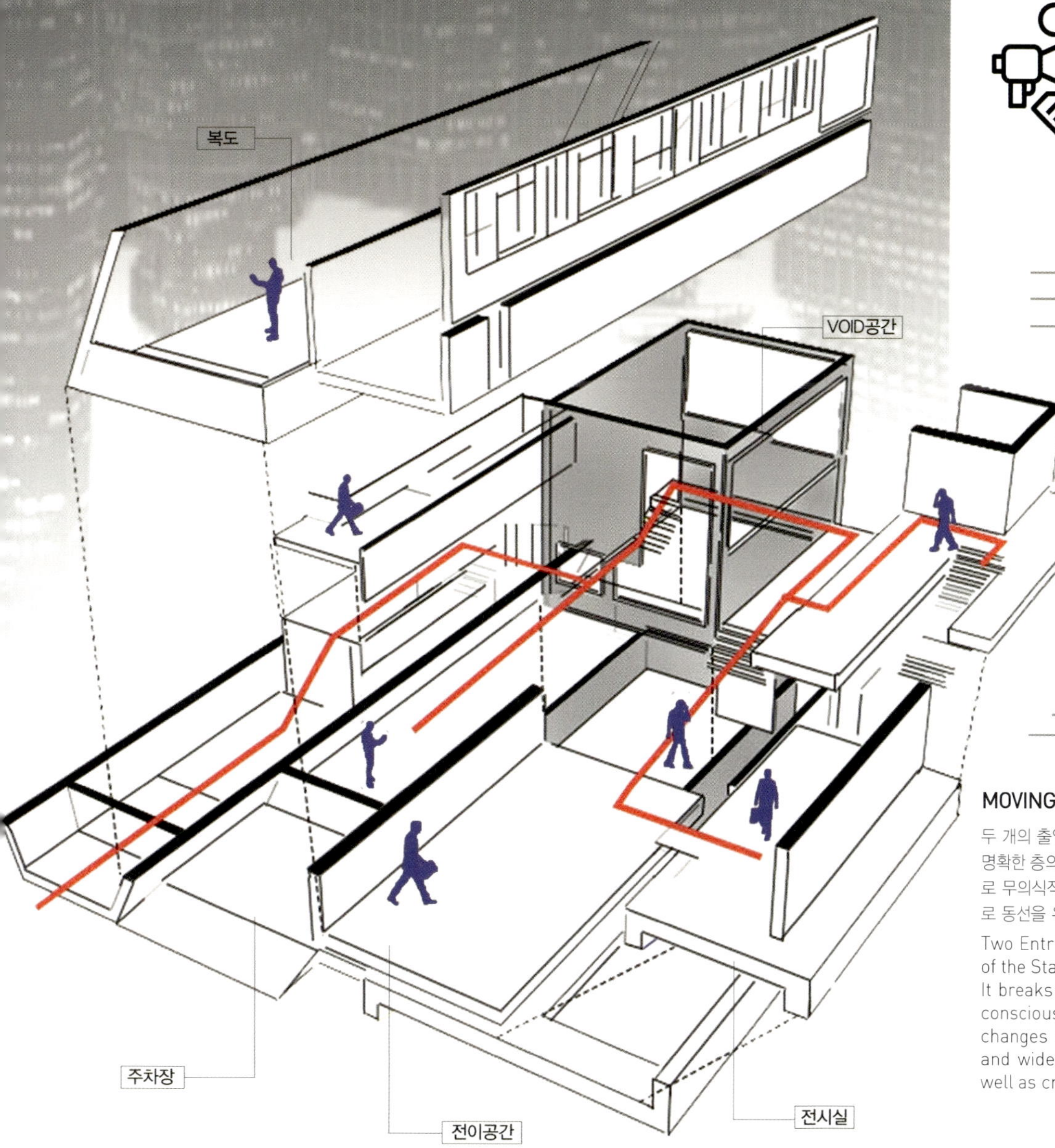

MOVING LINE

두 개의 출입구로 직원과 외부인들의 동선이 다양하게 구성됨 명확한 층의 개념을 허물어 공간 사용자에게 다양한 레벨의 변화로 무의식적 혼란을 야기한다. 또한 좁고 넓은 공간들의 반복으로 동선을 유도하는 동시에 다양한 긴장감을 연출한다.

Two Entries Help to Comprise Various Moving Lines of the Staffs and Visitors.
It breaks down the notion of the floor and gives unconscious confusion to the users through diverse changes of the level. Also, repetition of the narrow and wide spaces induces moving line of people as well as creates various tension.

DESIGN PROCESS

MASS PROCESS

도시파사드의 실루엣을 다양한 형태로 변형시켜 건물의 벽, 천장, 바닥이 구성된다.
Silhouette of City Facade is Transformed to VTarious Forms to Construct the Wall, Ceiling and the Floor of the Building.

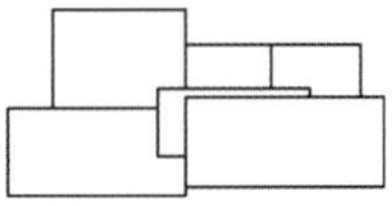

버스에서 바라보는도시 형태들을 기하학적으로 단순화하게 표현한다.

City shape viewed from the bus is geometrically simplified and represented into shapes.

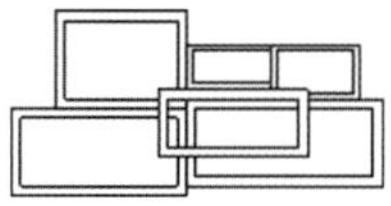

형태마다 두껍고 얇은 두께를 설정한다.

Apply thick and thin lines on each shape.

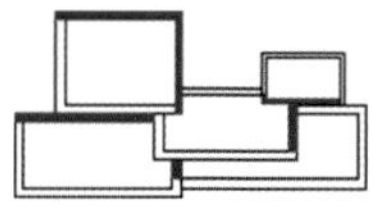

각 MASS들의 면을 무작위로 선택한다.

Randomly choose the face of each mass.

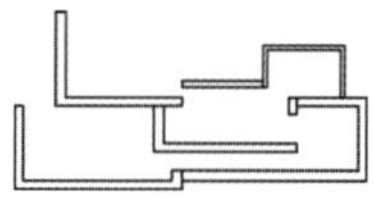

선택한 부분을 제거하여 각 MASS들이 연결되도록 한다.

Remove the selected faces to connect the masses.

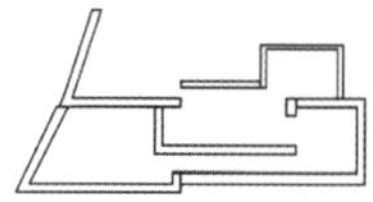

왼쪽 부분에 기울기를 주어 전체적으로 한 MASS로 통일감을 준다.

Make a slope on the left side to give unity as one mass.

SPACE COMPARTMENT

선과 면으로 공간들을 구획하였다.
The Space is Constructed with the Lines and Faces.

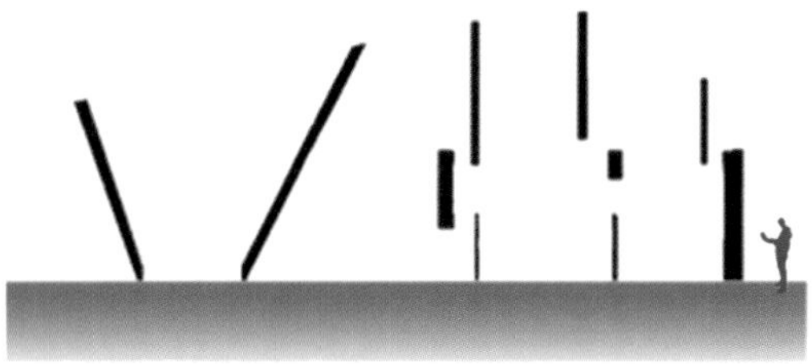

공간의 수직적 요소 (벽)
Vertical elements of the space (Wall)

공간의 수평적 요소 (천장과 바닥)
Horizontal elements of the space (Ceiling and floor)

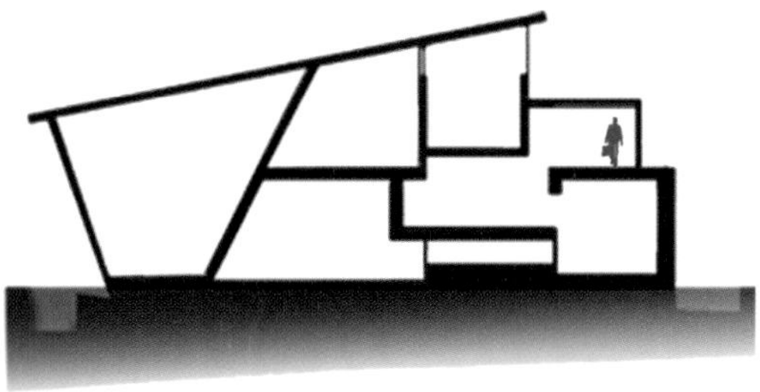

수평적 요소와 수직적 요소가 합쳐져 전체적인 MASS의 공간을 구획함

Horizontal elements and vertical elements are combined to construct overall space of the mass

선적 요소들로 인해 면, 부피 형성 (공간감)

Elements of the lines create surface and volume (Sense of space)

VOID SPACE

다양한 두께의 슬라브로 층의 경계를 모호하게 하였으며 각 공간들이 매스 덩어리로 이루어져있지 않아 동선의 연속성이 나타난다.

Slabs with various thickness make the boundary of the floor vague, and each space shows continuity of a moving line because they are not composed of a chunk of masses.

PROGRAM

PUBLIC SPACE

EMPLOYEE SPACE

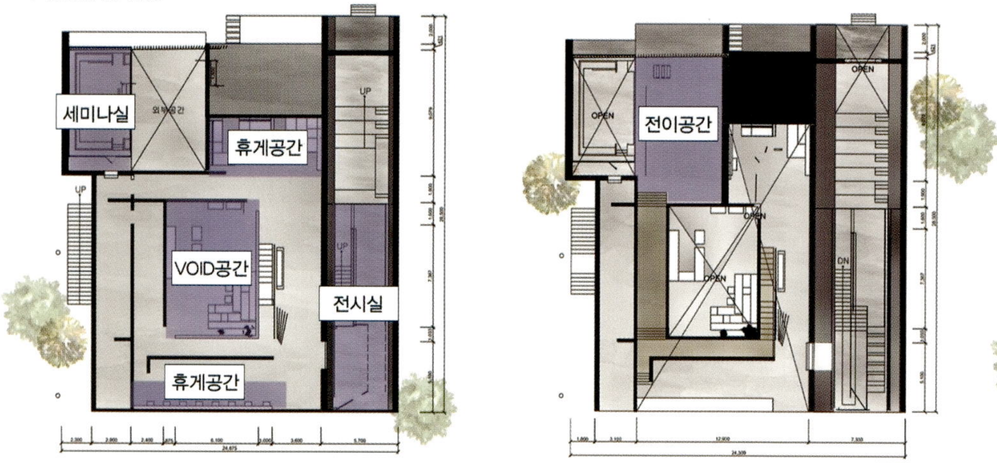

ELEVATION

파사드의 각도와 태양의 고도에 따라 외부로 나가는 빛과 내부로 들어오는 빛이 시시각각 변하는 파사드 구상 연결을 상징하는 파사드의 라인을 흰색 콘크리트로 마감하여 더욱 강조되게 나타난다. 각 매스, 공간마다 개방적, 변화적, 폐쇄적으로 나누어 입면을 구성한다.

Outgoing Light and Incoming Light Instantaneously Change Depending on the Angle of the Facade and the Altitude of the Sun. White concrete highlights the symbol of the facade, which is connection.
Each mass and space is divided into 'open', 'transitional' and 'closed' to construct the elevation.

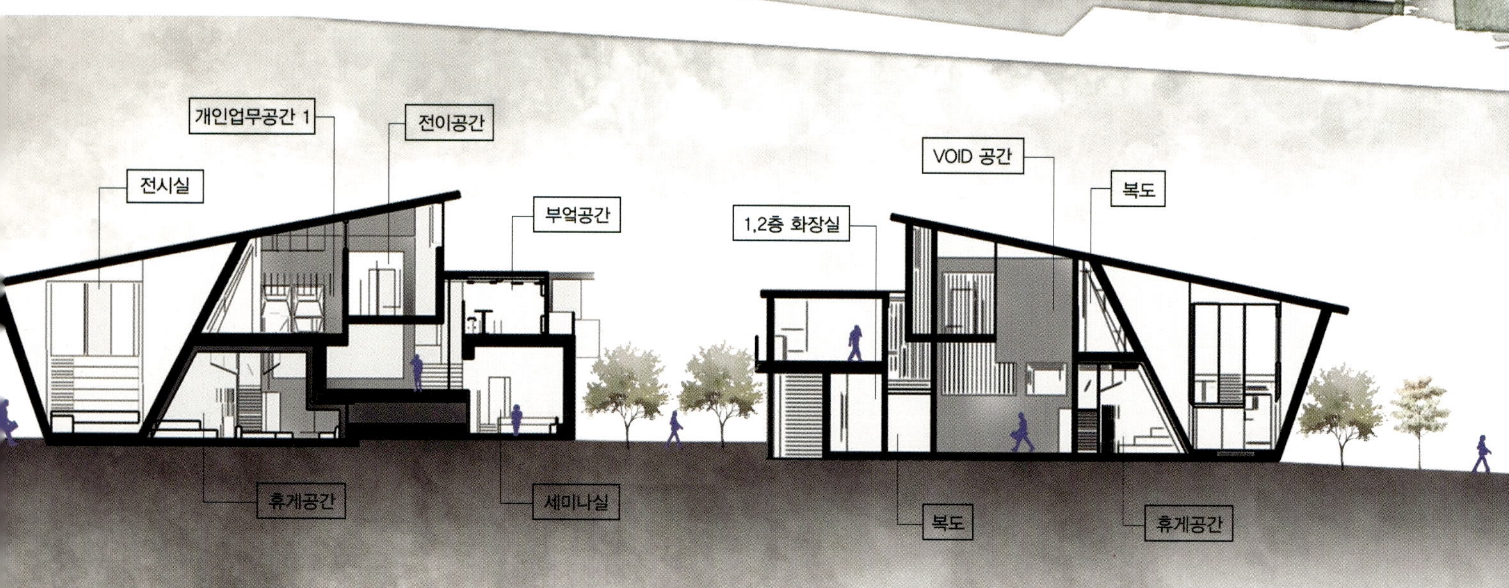

LIGHTING

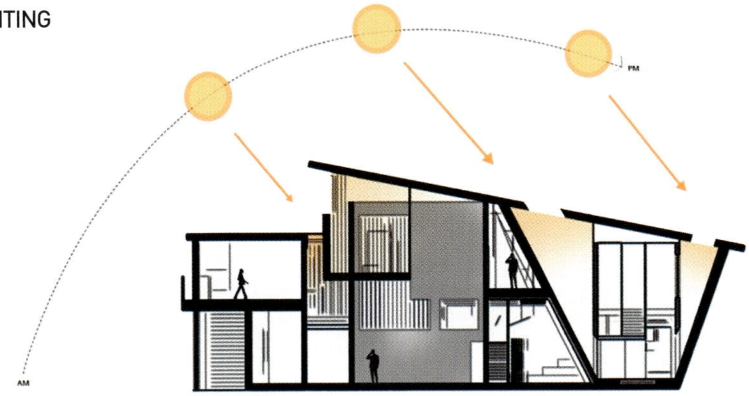

태양의 조도에 따라 공간, 복도에 떨어지는 빛의 양이 달라진다. 이에 따라 빛에 의해 시간대별로 다양하게 공간들이 구획된다.

The amount of light coming into the space varies according to the illumination intensity of the sun. Therefore, the spaces are constructed according to the light intensity, changing by time.

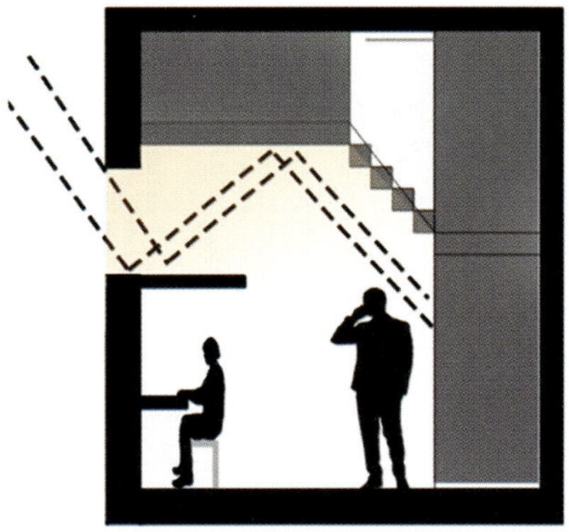

1층 휴게공간

앉아 있는 사용자와 서 있는 사용자가 받는 빛에 차이를 두어 무의식적으로 공간감을 느낄 수 있도록 한다.

Resting Area in the 1st Floor
Person sitting and standing will receive different light so that the user unconsciously feel the sense of space.

SKETCH

직원 출입구 쪽 복도

낮은 천장과 어두운 마감재를 사용하여 공간적 몰입감을 조성하였다. 왼쪽 틈새에서 들어오는 빛과 도시 파사드의 실루엣이 복도 동선을 유도한다.

Entrance Hall for the Staffs
The low ceilings and dark finishing materials are used to create spatial centralization. The light coming in from the gap on the left wall and the silhouette of the city facade lead the moving line.

1층 복도 입구

왼쪽 VOID 공간과 오른쪽에 휴게공간 사이에 어두운 마감재와 낮은 천장으로 주변과 대비되는 빛과 색으로 동선을 유도한다.

Entrance to the Hall in the 1st Floor
Dark finishing materials and low ceiling are used between VOID space at the left side and resting area at the right side. Colour contrasting with surrounding area is leading the moving line.

1층 출입구 쪽에서 바라본 휴게공간

책상에 앉았을 때와 서서 걸어갈 때 받는 빛의 양을 달리하여 무의식적으로 공간의 변화를 느낄 수 있도록 한다.

Resting Area Viewed From the Entrance of the 1st Floor
Amount of light received is different when the person is sitting on the chair and walking. It is planned to help the users unconsciously feel the change of the space.

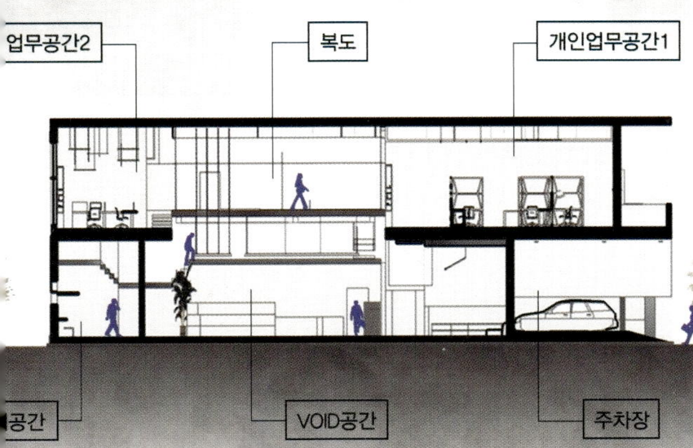

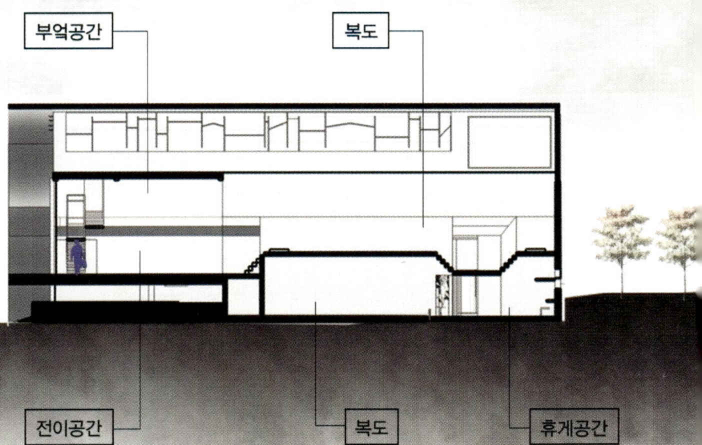

부엌공간 (KITCHEN SPACE)
다른 바닥 마감재로 동선을 유도하고 푸른색의 포인트로 '매거진 파노라마'와 통일감을 형성한다.

Different finishing material leads the moving line and the blue points form the unity with 'Magazine Panorama'.

전시실 (Exhibition Room)
빛으로 동선을 유도하고, 사용자의 움직임과 작품의 형태의 인식이 동시에 이루어지는 공간으로 무의식과 의식의 흐름을 표현한다.

Light leads the moving line. This area is where the movement of the users and recognision of the shape of the artwork happen simultaneously, and also expresses the flow between unconsciousness and consciousness.

BIRD EYE VIEW

보통 수직적인 층으로 전체 영역을 구분하는 개념에서 벗어나 수평적인 매스 형태의 조합으로 다양한 단차이가 나타나도록 한다.
대지에 맞춰 오른쪽으로 살짝 기울어 들어가게 되어 있어 건물 입면에서 바라보았을 때 호기심과 동선을 유도한다.

It is different from the normal concept of dividing the whole area by a vertical layer, but horizontal combination of mass shape creates difference of the level.
The building is tilted to the right side so leads movement and curiosity when viewed from the top of the building.

VOID 공간 (Void Space)
큰 스케일의 천장으로 엄청난 개방감과 동시에 압박감을 느끼게 한다.

Ceiling of the large scale creates a sense of openness and pressure.

Personal Work Space 1

창틀, 조명, 그림자가 모두 도시형태를 형상화하여 버스에서 창 밖을 바라보았을 때의 서울 도시의 느낌을 공간화하였다. 또한 이동식 책상과 의자로 1:1의 회의나 다른 부서와 함께 작업을 해야할 때 등 다양한 상황에 따라 다양하게 변화할 수 있는 특징이 나타난다. 한편, 자유롭지만 개인의 프라이빗을 지키기 위해 사방 중 한 면을 완전히 막았으며, 다른 한 면은 간이 벽으로 변화가 가능하도록 구성하였다.

The window frames, lightings and shadows altogether form a shape of city and make the space look like a Seoul city viewed from the bus window. Mobile desks and chairs allow the space to be changed depending on the situations, such as 1:1 meeting or working with other departments. On the other hands, one side of the space is completely blocked to ensure privacy and another one side is able to be changed into the wall.

Personal Work Space 2

버스의 단차이를 책상의 단차이로 업무공간에 공간화하여 레벨차이를 느낄 수 있도록 하였다. 수납장과 창틀은 수직과 수평의 교차로 도시 형태의 단면화를 더욱 극대화하였다. 또한 천장의 전구는 아래에서 보는 시선과 측면에서 보는 시선에 따라 입면 파사드가 자유롭게 변한다는 특징이 있다.

Different levels in the bus is expressed as different levels of the desks in the work space. The cabinet and the window frame express the city silhouette through the intersection the vertical and horizontal line. Also, facade of the lighting on the ceiling changes diversely according to the sight from below or from the side.

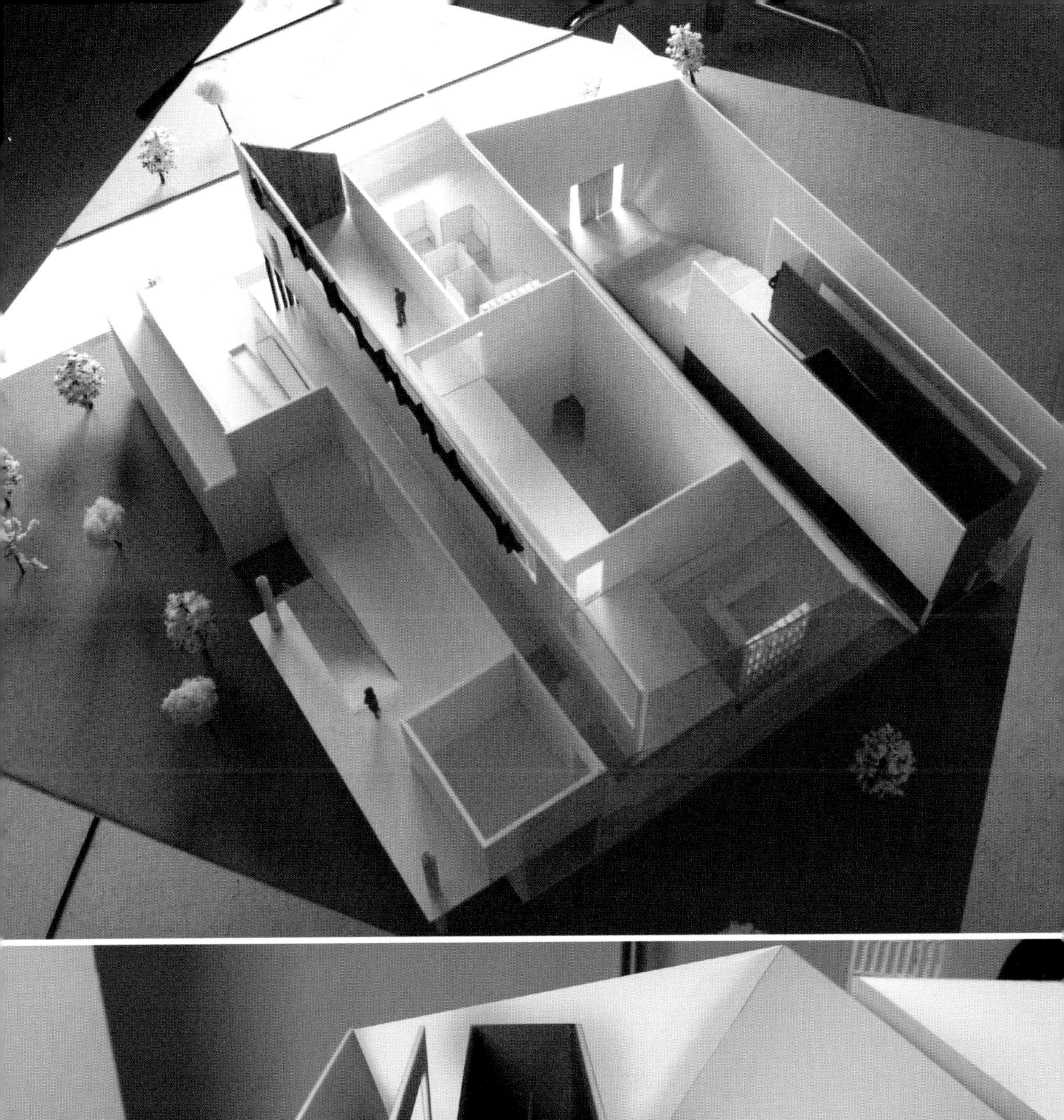

오정현 I Oh, Jyeong Hyun 박정현 I Park, Jeong Hyun

BACKGROUND

직장인이 하루에 소비하는 커피의 양은 과연 얼마나 될까요?
사회에서 직장인에게 커피란 도저히 떼어낼 수가 없는 존재입니다. 통계에 따르면 하루에 한 잔을 마시는 사람은 약 78%, 두 잔을 마시는사람은 약 35%, 세 잔 이상 마시는 사람은 25%나 됩니다. 그리하여 우리는 커피 한 잔의 여유, 달콤한 휴식, 친한 친구와의 담소 등 많은 긍정적인 면에서 벗어나 커피를 찾을 수 밖에 없는 이유에서 접근하였고 바로 커피의 카페인과 그에 따른 커피의 중독성에 대해서 생각하게 되었습니다. 커피의 뻔한 이미지에서 벗어나 선뜻 얘기하기 어려운 부정적인 이미지에서 과연 어떠한 이야기가 나오게 될까요?

How much coffee would an office worker consume a day? In modern society, coffee is inseparable for office workers. According to statistics, 78% of office workers drink a cup of coffee a day, 35% of them drink two cups of coffee a day, and 25% of them drink three cups of coffee a day. So, we approached the reason for drinking coffee aside from the various positive aspects such as relaxing time, sweet break time, and chat with close friends. We focused on caffeine in coffee and addictiveness of such caffeine. Since coffee is consumed by many people, it is hard to get away from obvious images of coffee and talk about its negative parts. What stories would come out?

The Other Balance

SITE

서울특별시 구로구 중앙로6길 16 동양미래대학교 실습실
16, JUNGANG-RO 6-GIL, GURO-GU, SEOUL, SOUTH KOREA

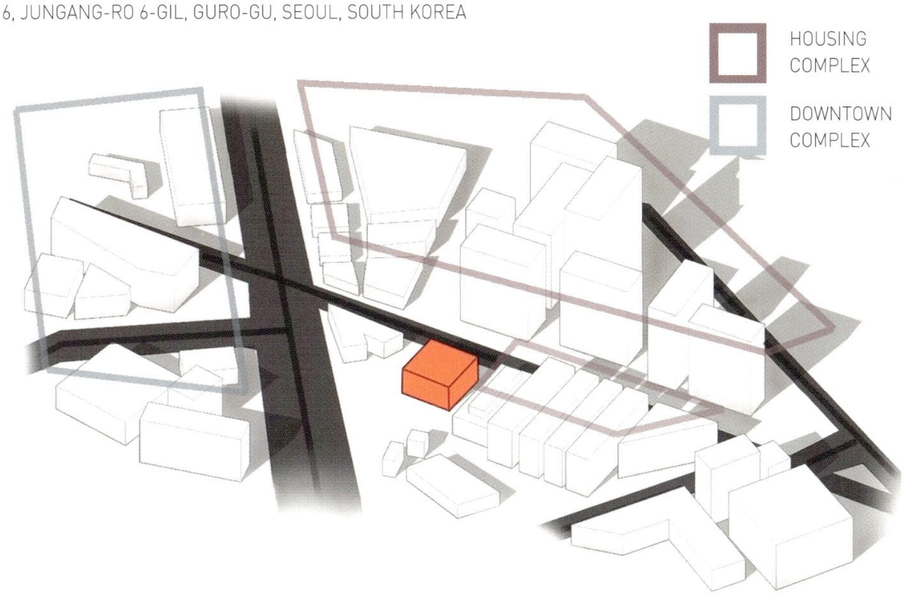

'감각과 지각'

– 우리가 평소에 느끼는 것은 감관에 의존하고 도출되어 나오는 것을 감각이라고 한다. 우리들이 흔히 얘기하는 시각, 촉각, 후각, 청각, 미각의 오감을 느끼면서 그에 따른 반응을 하게 되고 그러한 감각들을 건축적으로 풀어나가기 위해 벽, 바닥, 천장 등을 이용하여 어떠한 감정을 이끌어내는 가에 대해 이야기 하고자 한다. 여러가지 감각에서 불완전함이라는 완전하지 않은 상태에 초점을 맞춰 진행하였다. 불안정한 상태, 혼동, 어지러움, 쏠림, 불안함 등을 어떤 형식의 지각으로 나타낼 것인지 고민하고 스튜디오를 시작하였다.

'Sense and Perception'

We usually feel 'sense' with our sensory organs and we feel five senses of sight, touch, smell, hearing, and taste with our sensory organs. To talk express those senses architecturally, I used wall, floor, and ceiling and examined what kind of emotions would come out from them. For the work, I focused on the state of incompleteness of various senses. I thought about what type of perception would be suitable for expressing unstable state, confusion, dizziness, leaning, and anxiety. Based on such thoughts, I started the studio.

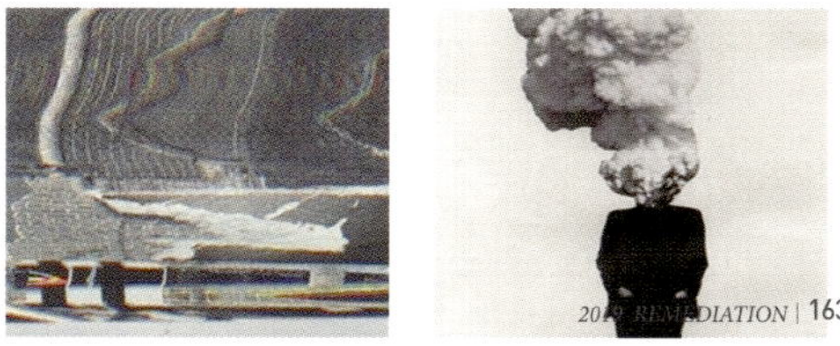

월간 [COFFEE]

커피인의 필독서이며 커피와 차를 중심으로 하는 F&B 전문 정보지입니다. 커피산업과 카페문화를 이해하고 사랑하는 이들과 커피마니아들에게도 매달 참신하고 흥미로운 기획으로 새로운 읽을거리를 제공해오고 있습니다.
단순히 커피뿐만이 아닌 커피와 디저트, 커피와 역사, 커피와 향기, 커피와 행사, 커피와 인테리어 등 커피와 관련해서 모든 분야를 다루기 때문에 커피마니아라면 필독서라고 할 수 있다.

A must-read F&B information magazine for coffee lovers. Every month, this magazine provides unique and interesting planning for providing fresh reading materials to people and coffeeholics who understand and love coffee industry and café culture.
This magazine is highly recommended for coffee lovers as it covers all fields related to coffee such as coffee and dessert, coffee and history, coffee and scent, coffee and event, and coffee and interior.

CONCEPT

커피의 카페인과 중독성

모두가 아는것처럼 커피에는 카페인이라는 성분이 존재한다. 그러한 이유로 커피를 과도하게 섭취하게 된다면 불안함, 불완전함, 두통, 두근거림, 불면증 등 부작용이 나타나게 되는데 이러한 것 들을 토대로 불완전한 상태라는 것에 초점을 맞춰 컨셉을 찾아나갔다.

Caffeine and Addictiveness of Coffee

As we all know, coffee has caffeine in it. If you drink excessive amount of coffee, you may experience side effects such as anxiety, incompleteness, headache, palpitation, and insomnia. Based on these side effects, I focused on the incompleteness for finding the concept.

불완전함이 들려주는 이야기는 무엇일까?

우리 모두는 완벽하려하면서도 불완전하다.
역설적인 문장의 의미를 되짚어본다면 우리는 대부분 스스로가 불완전하다는 것을 안다. 그러나 불완전한 모습 속에서도 가능한 그렇게 보이고 싶어하지않고 결국 불완전함 속에서도 완벽하고 싶어하며 흐트러지지 않으려 노력한다. 그렇기 때문에 우리는 불완전함이 보여주는 다양한 이야기가 있을 거란 생각이 들었고 컨셉의 방향을 불완전함 속의 이야기로 선정하였다.

What are the stories of incompleteness?

Although we all try to be perfect, we are incomplete.
In detail, most of us know that we are incomplete. However, we do not want to show our incompleteness and we end up trying to be perfect and stay in order even in the incomplete status. Therefore, we thought that there would be various stories from the incompleteness and we decided to talk about stories of incompleteness for our concept.

CEILING

2층 업무공간
– 원래 1개로 되어있는 플랫한 형태의 천장이었으나 총 5개로 나누었고 나뉜 천장을 불완전한 기울어진 형태의 천장으로 변형시켜 그로인해 생긴 틈사이로 들어오는 외부의 빛이 불규칙하게 들어오도록 하였다. 에 초점을 맞춰 컨셉을 찾아나갔다.

2F Work Space
- At first, the ceiling had a flat form and it was integrated as one. Then, I divided the ceiling into 5 parts and made the ceiling to have an incomplete slanted from to allow irregular lights to come in through the gaps.

MASS

As we focused on remodeling the existing building, we worked on the studio while maintaining the framework of the building

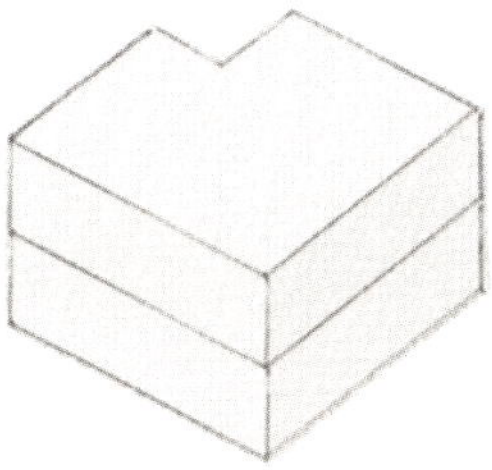

Basic Site Form

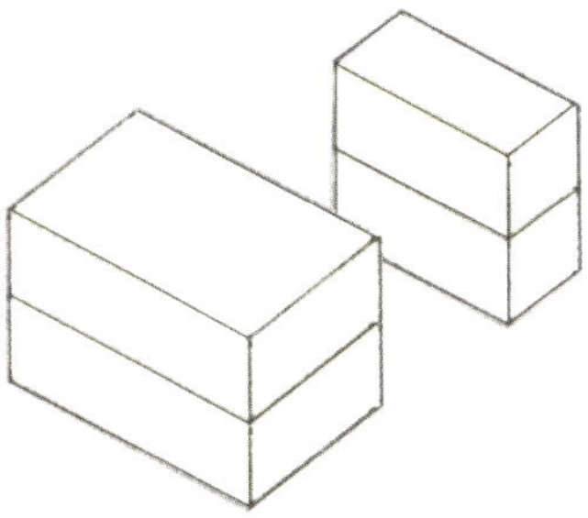

Structural Differentiation of Existing Site Building

Difference in Layer and Increase in Number of Stories based on Height Change

DESIGN PROCESS

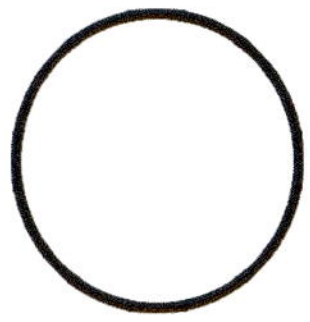

Perfection

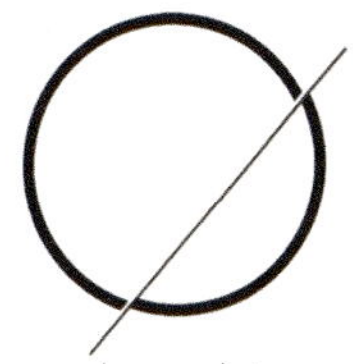

Incomplete

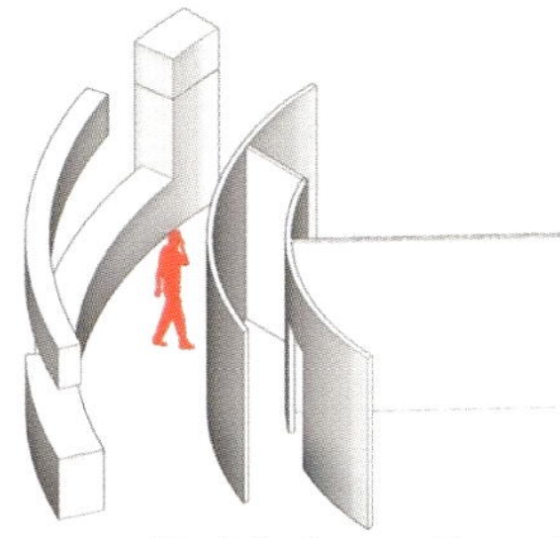

1F - Utilization as café space/ Space between gap

square

Twisting

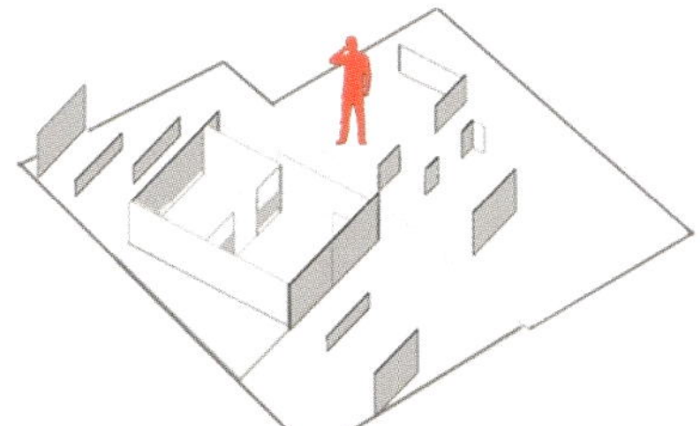

2F - Distortion /Utilization as work space / Distorted wall and furniture

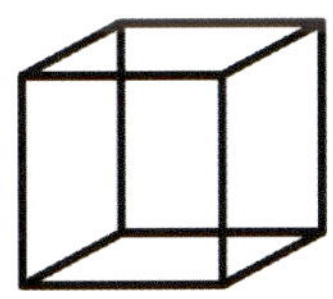

a cube

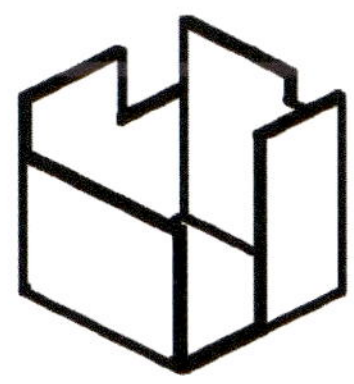

a height change

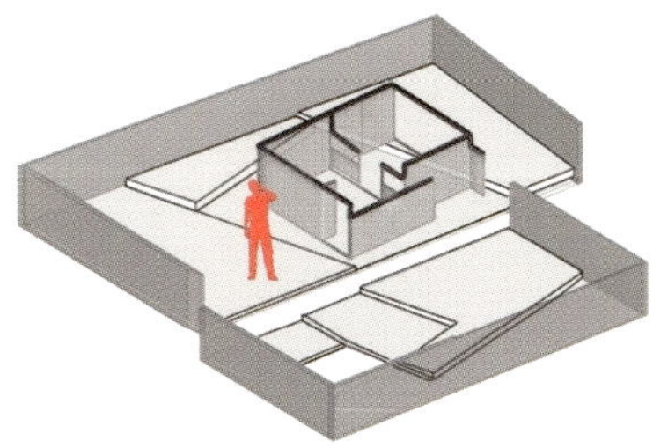

Hight of wall and layer – Utilization of all spaces in 2nd floor / Layer difference in slab and height of partition

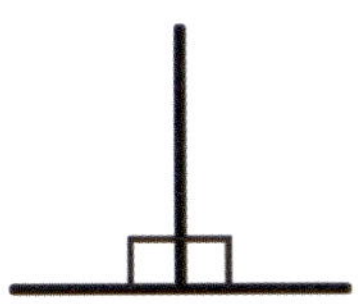

a vertical line

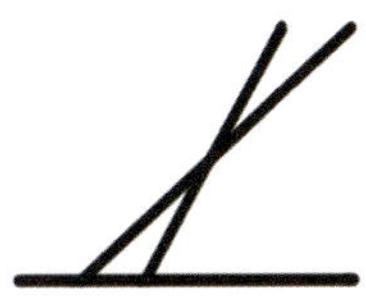

a tilted line

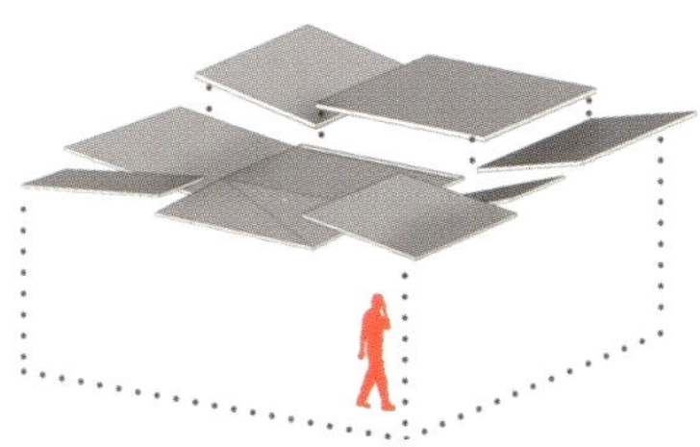

Slanted ceiling/ Ceiling part / Slanted gap – Form of light

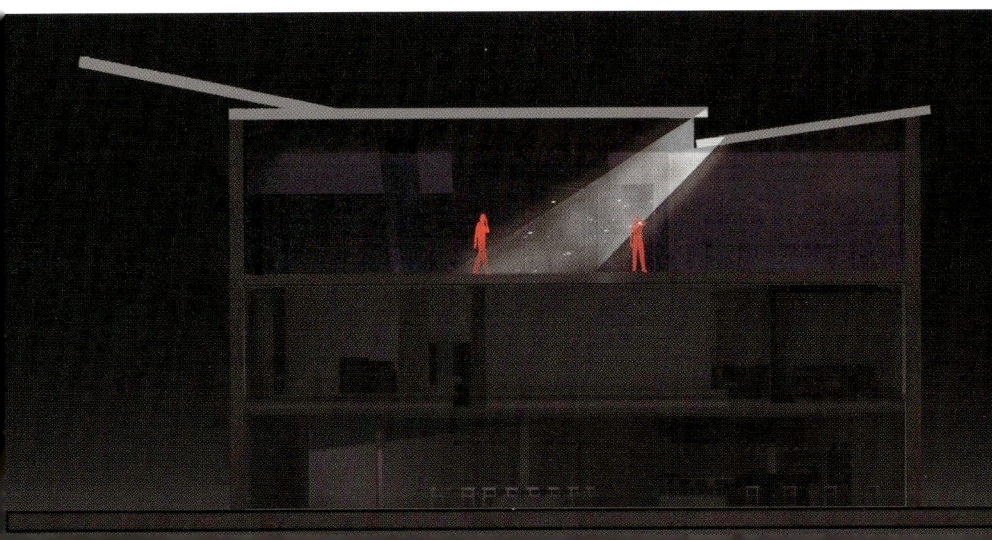

2-1/2층 전시공간
– 뒤 건물의 천장을 중간에 끊어냈고 비스듬히 기울여 천장을 안으로 파고들게해서 틈사이로 외부 빛이 들어오게 하였다.

2-1 / 2F Exhibition Space
- I disconnected the ceiling from the building in the back and made it slightly slanted to allow external lights to come in through the gaps.

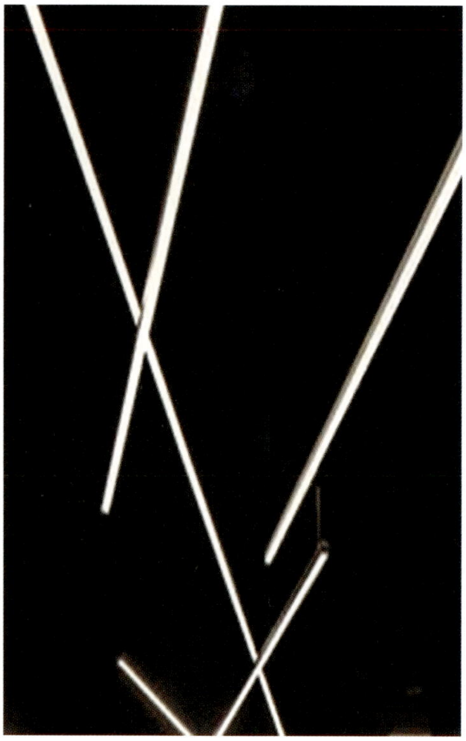

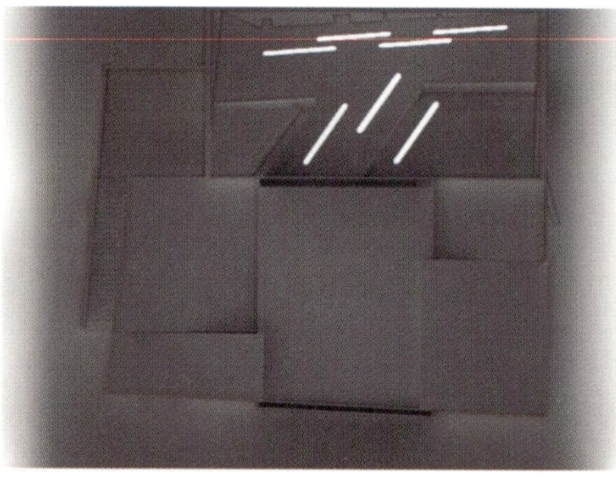

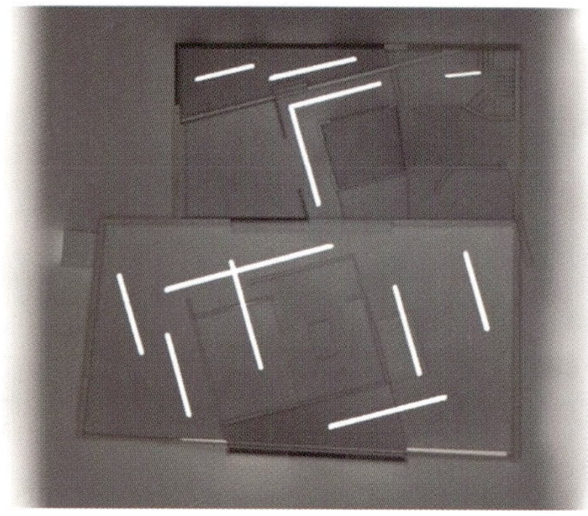

LIGHTING

불규칙한 형태의 조명을 모티브로 삼았고 2층 전체공간과 3층 전시공간에 적용하였다. 그러나 단순한 불규칙함이 아니라 내부 공간의 비틀어진 구성에 맞추어 조명을 배치하였다.

I was inspired by the lighting of the irregular forms and I applied the lightings in whole space in the 2nd floor and exhibition space in the 3rd floor. Also, I arranged the lighting based on the distorted composition in the inner space.

2F WORKSPACE 1

2F WORKSPACE 2

EDITOR IN CHIEF

EDITING TEAM

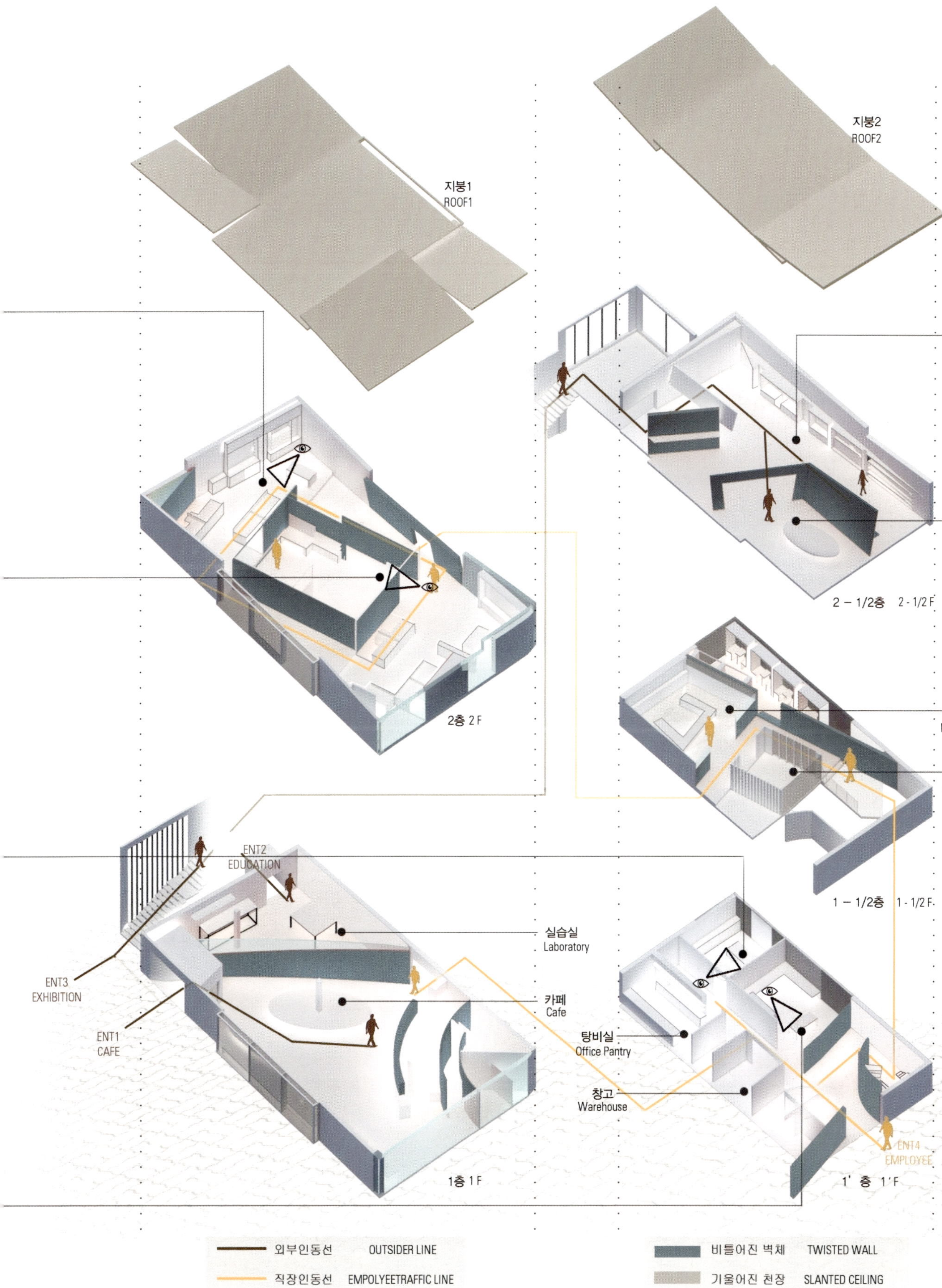

지붕1
ROOF1
지붕2
ROOF2
전시공간
Exhibition Space
커뮤니티실
Comunity Room
2 – 1/2층 2 - 1/2 F
2층 2 F
회의실
Meeting Room
탕비실
Office Pantry
1 – 1/2층 1 - 1/2 F
ENT2
EDUCATION
실습실
Laboratory
ENT3
EXHIBITION
카페
Cafe
ENT1
CAFE
탕비실
Office Pantry
창고
Warehouse
ENT4
EMPLOYEE
1층 1 F
1' 층 1' F
외부인동선 OUTSIDER LINE
직장인동선 EMPOLYEETRAFFIC LINE
비틀어진 벽체 TWISTED WALL
기울어진 천장 SLANTED CEILING

SECTION

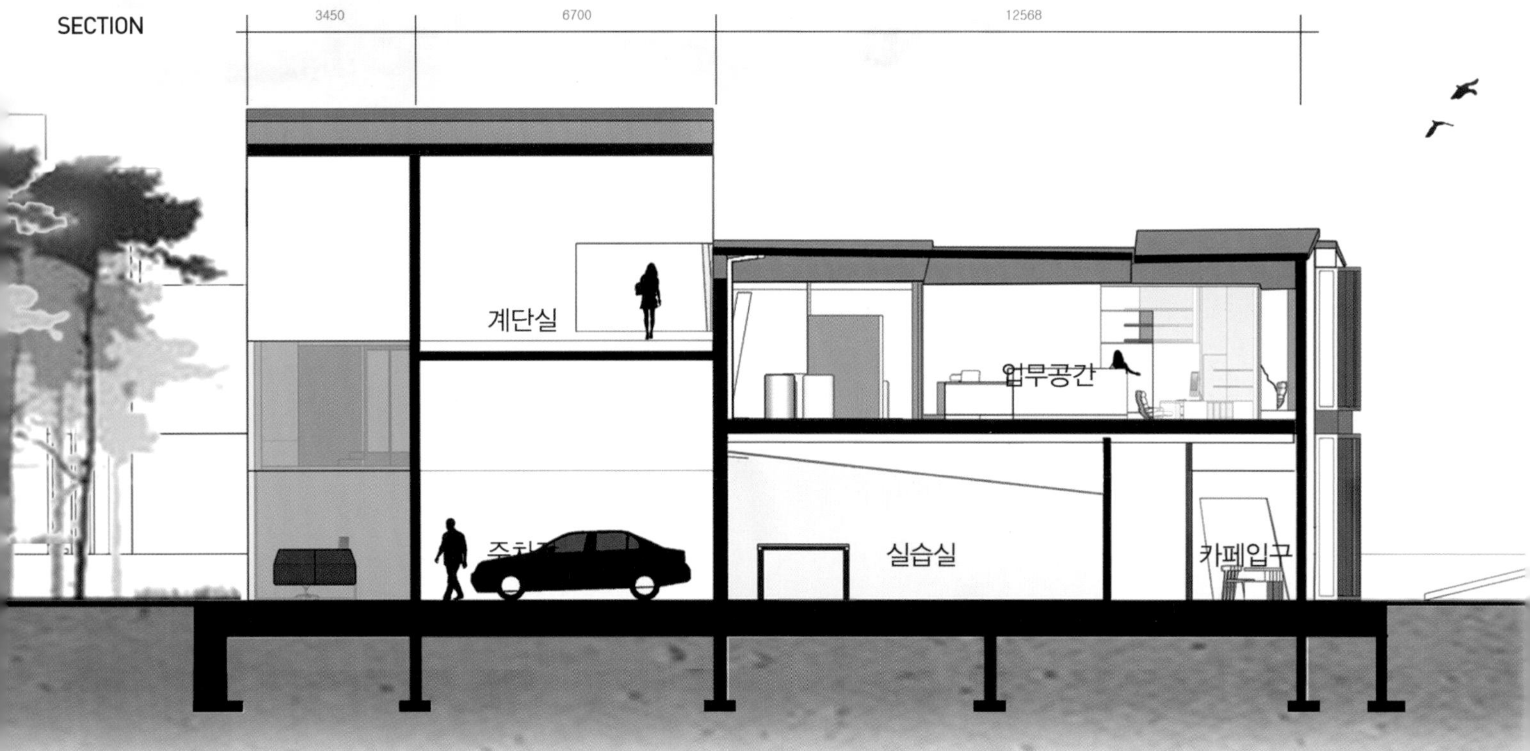

SECTION – A

PLAN

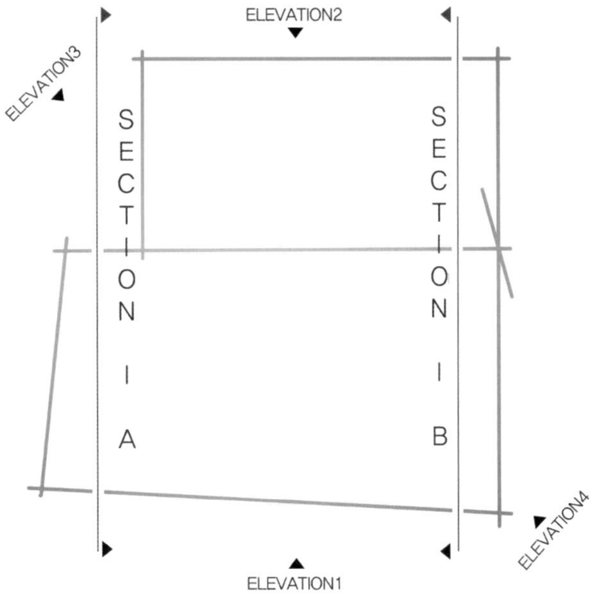

1FLOOR PLAN

ELEVATION

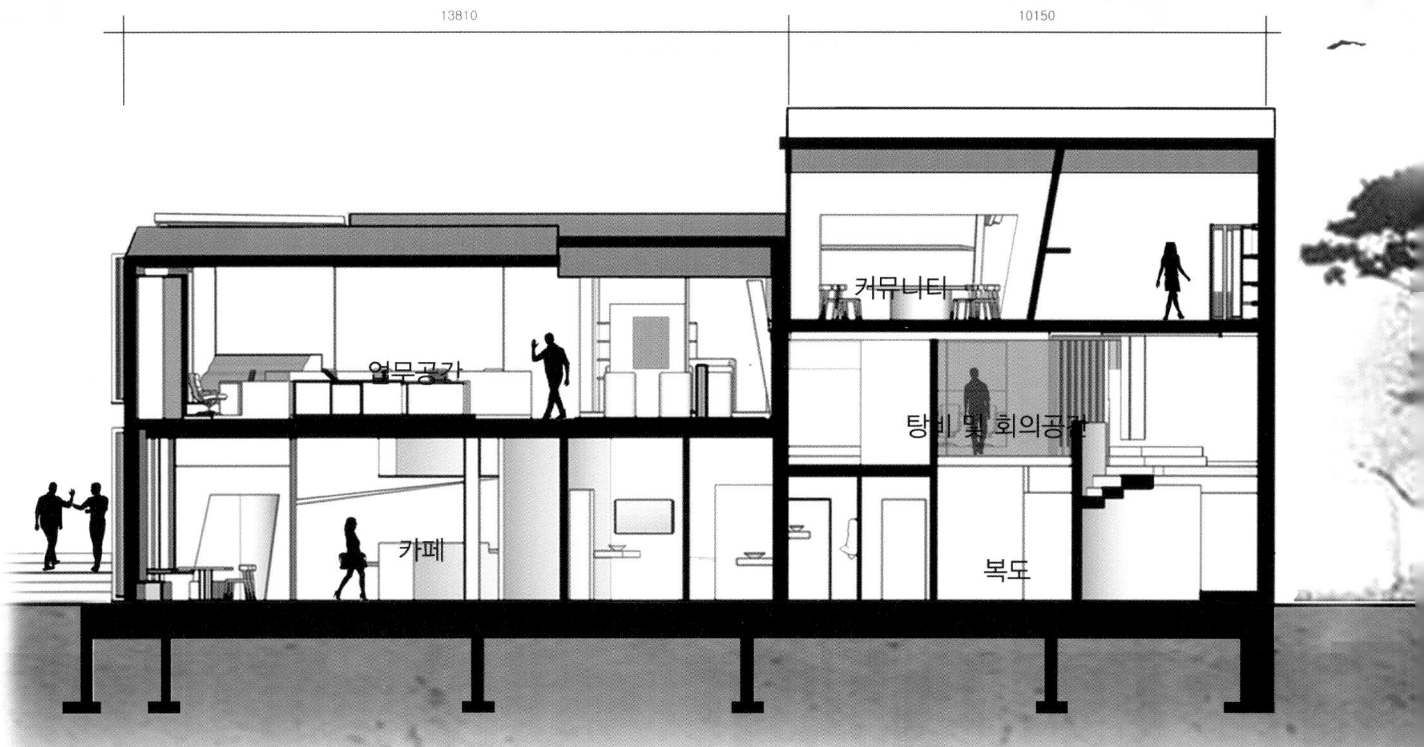

SECTION - B

2FLOOR PLAN

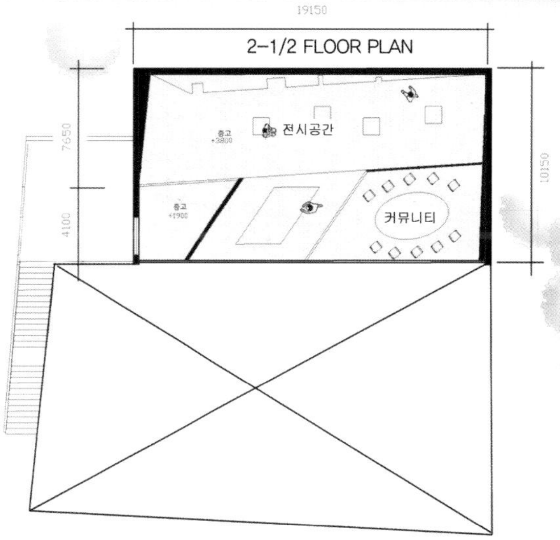

Exhibition Space

The entrance to a cafe

Exhibition Space Entrance

1F Cafe

불완전함의 끊어진 호를 겹쳐 활용한 공간과 그에 대비되는 완전함의 정사각형 형태의 창문 프레임의 조화.

1F Café - Harmony of overlapping incomplete disconnected arcs and complete square-type window frames.

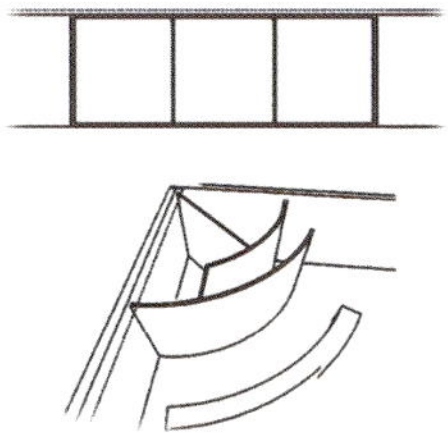

2F Retiring room

사각형의 벽을 비틀어 건물 내부로 들였고 그로 인해 생긴 공간구분으로 새로운 공간감을 주도 하였으며 휴게공간으로 사용하였다.

2F Rest Area - Distortion of square wall to create new space division and spatial impression and to provide rest area.

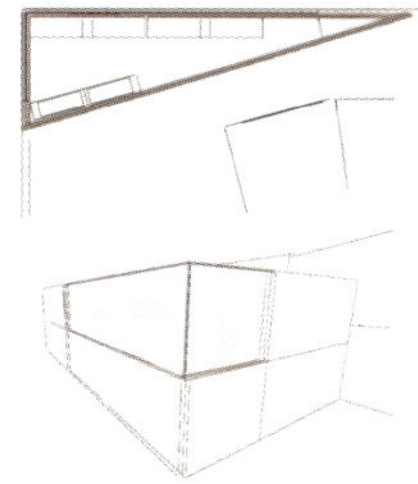

2F Pantry

사방을 완전히 막아두지 않았고 복층형태로 구성해서 심적으로 탁 트여있는 느낌을 주고자 하였다.

2F Pantry - The space is partially open and it has a double-layer to provide feeling of openness.

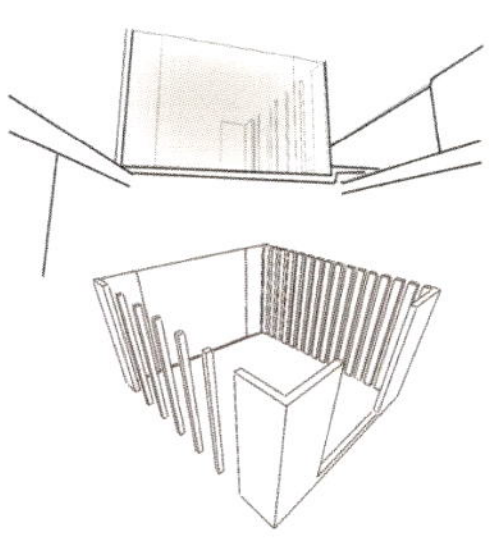

2F Hallway

건물의 단차이로 인해 업무공간으로 들어가는 입구가 순간적으로 1900mm의 낮은 층고를 형성하고 벽이 기울어져있어 몰입감을 주도록 의도.

2F Hallway - The entrance to work space instantly creates low height of 1900 mm due to difference in building height and the wall is slanted to provide feeling of immersion.

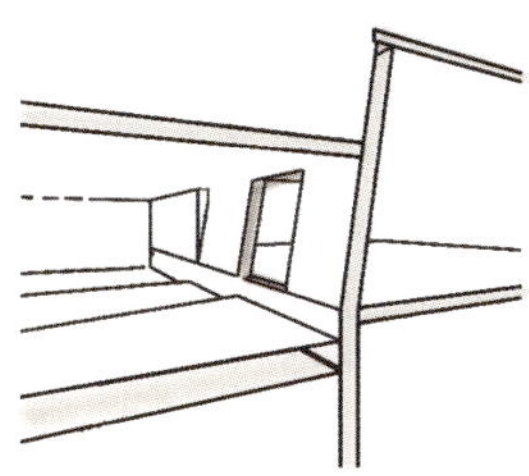

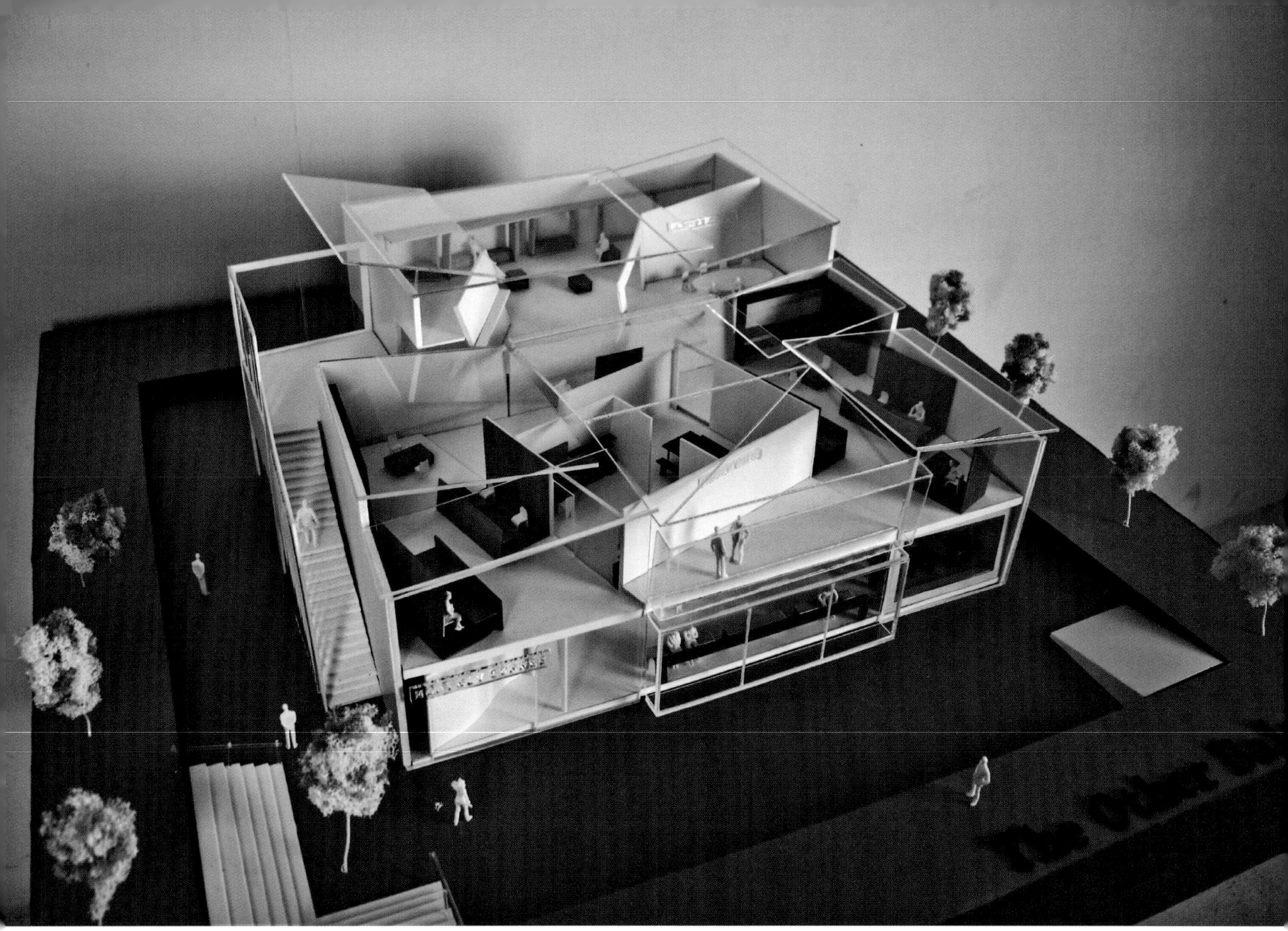

COFFEE

이채림 I Lee, Chae Rim · 마정연 I Ma,Jeong Yeon

JAZZ PEOPLE

CONCEPT

미국 뉴올리언스에서 시작된 음률의 시(詩) 재즈를 동양적 시각으로 재해석한 재즈의 공간. 이 공간 속에서 사용자는 속세의 고통 속에서 잠시 벗어나 마음껏 유희를 즐기는 신선(神仙)이 되어 비로소 재즈의 공간과 화합한다.

A space of jazz reinterpreting the New Orleans, U.S. - oriented jazz from an Oriental perspective. In this space, users get out of their pain of the world for a while and become the Shinsun enjoying what they wished to play as much as they like to harmonize with the space of jazz.

BACKGROUND

신선(信仙) Shinsun

동양의 도가 사상을 바탕으로, 인간이 수행을 쌓아 도달할 수 있는 신의 경지를 말한다. 신선은 몸이 가벼워져 자유롭게 하늘을 날아다니며 불로불사의 몸을 가진 자로 거듭난다. 우리는 이러한 신선의 경지를 재즈의 묘미를 깨닫는 즐거움과 연결시켜, 재즈를 처음 접한 이에게 강렬한 영감을 주는 공간을 구성하였다.

It means the state of God, that is achievable as human through constant practising of asceticism based on the Oriental Taoist thought. The Sinsun becomes lighter to fly freely in the sky and is reborn as the being with the body of immortality. We linked this state of the Shinsun with the joy of realizing the attraction of jazz to create a space presenting a strong inspiration to the people who encounter jazz for the first time.

JAZZ PEOPLE

2006년 창간된 재즈 전문 월간지 재즈피플은 대중적인 전문지를 표방합니다. 국내 최고의 재즈 평론가들이 참여하여 재즈계의 소식을 빠르고 정확하게 전달합니다. 마니아에겐 깊이있는 정보를, 초보자에겐 신선한 읽을거리를 제공합니다.

The 'Jazz People' as a magazine founded in 2006 features a popularized professional magazine. As the best jazz critics in the country joined in, it delivers accurate news from the jazz world quickly. It provides thoughtful information to manias, while fresh reading to beginners.

STORY

애환을 리듬으로, 한恨을 흥으로.

Sublimation of a spirit into a rhythm

고통 끝에 자유를 얻었지만 당장 먹고 살아야 할 것을 걱정해야 했던 노예 출신 흑인들의 애환은 유럽 악기의 음악적 틀과 결합해 즉흥적 리듬의 재즈로 표현되었다.
이러한 재즈의 정서는, 언뜻 한국과는 그 결이 다른 듯하지만 우리는 그들의 애환의 정서를 한국 고유의 정서 '한' 과 연결시켜, '리듬' 으로 풀어낸 애환의 정서를 또 다른 한국적인 요소인 '흥' 으로 해석하였다.

The woe of black slaves, who had to worry about their means of living while they could gain the freedom after suffering, was expressed as improvisatory rhythm of jazz in combination with the musical frame of European instruments.
Such jazzy emotions may seem to have some different texture from the one of Korea by all appearances, but we have linked their painstaking emotions to a Korean indigenous emotion, 'han', and interpreted the painstaking emotions expressed in 'rhythm' as another Korean element, 'heung.'

어떻게 우리는 멀고 먼 타지의 음악 재즈와 한국 고유의 흥을 연결할 수 있었는가? 바로 '즉흥성' 이라는 키워드였다.

How were we able to link jazz as a music from a faraway foreign country with the Korean indigenous 'heung'? That was a keyword of 'improvisation'.

즉흥적 변주와 강렬한 리듬, 서로 어울리며 연주자와 청취자가 하나로 화합되는 미국의 재즈와 한국의 국악은 공통적으로 민족 고유의 혼을 담고 있다.
이러한 점을 바탕으로, 재즈를 한국에 전하고 널리 알리기 위해 재즈 전문지를 발간하는 '재즈피플' 의 오피스를 한국적 정서가 담긴 공간으로 표현하기로 하였다.

American Jazz and Korean traditional music, in which improvisatory variations and powerful rhythms are going well together to make players and audiences into one, contain their own ethnic souls in common.
Based on this point, the office of the 'Jazz People' to publish the professional jazz magazine will be described as a space filled with Korean emotions in order to introduce and publicize jazz to Korea.

ARCHITECTURAL ELEMENTS

누하진입

Nu-ha-Jin-Ip : Enters under a gazebo

누각 아래로 진입해, 사용자에게 공간의 클라이막스를 보여준다.

As entering under a gazebo, It shows the climax of the space to users.

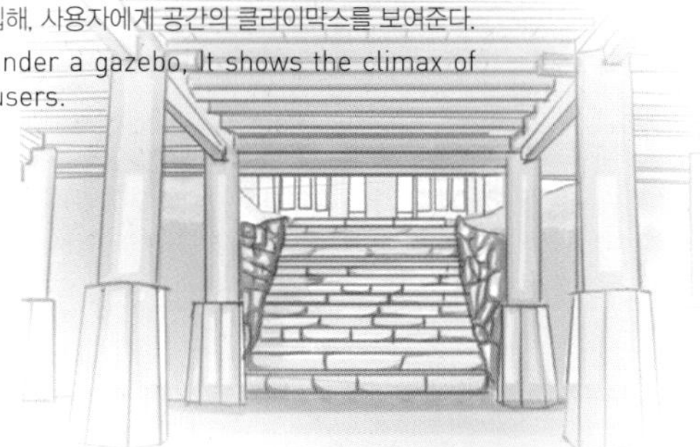

전이공간

Transition space

높은 천장고의 스케일에서 초월감을 느끼게 되는 영감의 공간이다.

It is a space of inspiration to make the transcendental feeling due to the scale of high ceiling height.

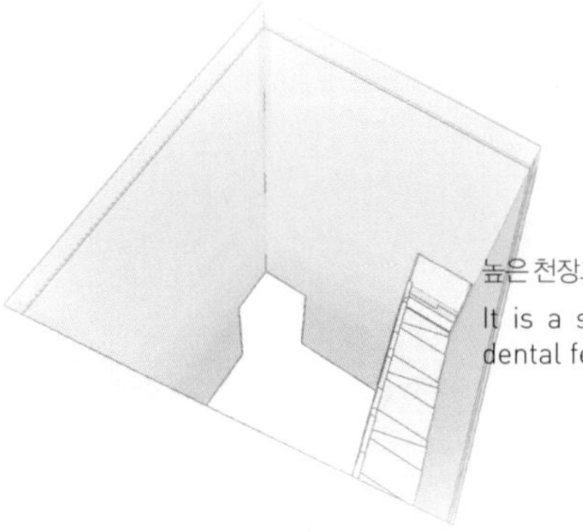

좌식형 생활공간

Sitting life space

온돌 구조를 바탕으로 한 한국 고유의 생활 공간 특성인 좌식형의 공간을 오피스에 접목, 레벨 차로 공간을 구분하고 정의하였다.

As the Korea-specific life space based on 'ondol' (Korean floor heating system) structure, the sitting life space was grafted onto the office to divide and define spaces by the differences in steps.

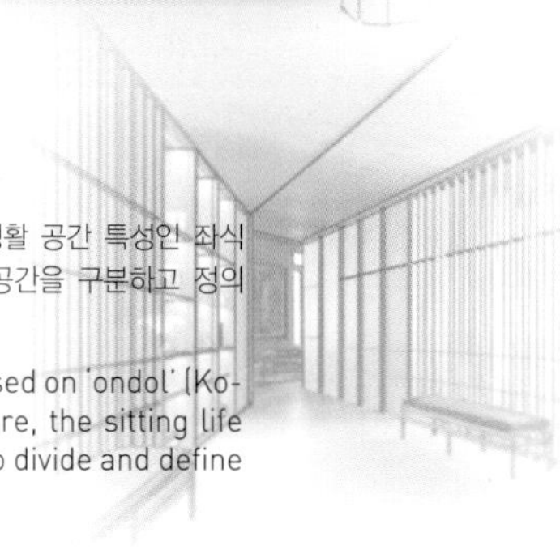

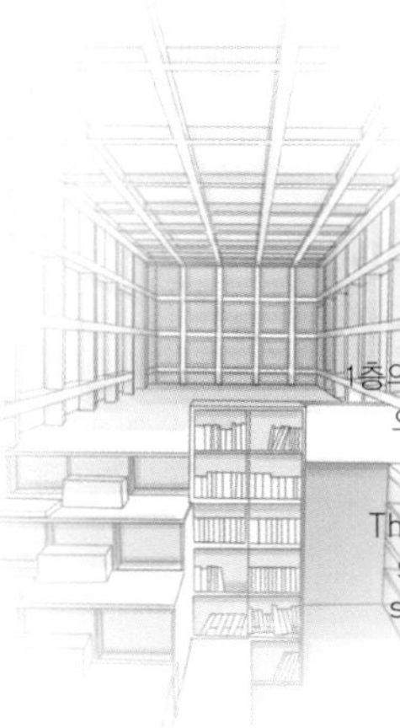

커뮤니티 공간

Community space

1층의 서재와 연결되며, 각각 폴딩도어와 전통 형식의 미닫이문으로 외부 마당 공간과 조우하는 개방적인 커뮤니티 공간은 쉬운 접근성과 공간 시퀀스의 묘미를 보여준다.

The community space, where is connected with the study on the first floor and meets in with the yard space outside each by folding door and traditional sliding door, shows the subtlety of easy accessibility and space sequences.

MASS PROCESS

일과 휴식은 다르다
: 공간 분리

Work and rest are different: Space Separation

우리는 내부의 벽체를 이용해 공간을 구분하는 것을 넘어서, 외부의 매스 분할로 공간의 역할을 부여하고자 하였다. 가장 먼저 고려한 것은 바로 사무공간인 오피스와 지역주민들의 참여공간인 커뮤니티실과의 실질적인 분리였다. 분할의 방식으로 매스를 나눈 뒤 그것을 다시 재결합해 새로운 공간 구성을 창조하였다.

We wanted to give the role of spaces by the mass segmentations of the outside going beyond the space division using the wall of the inside.
The first consideration was the actual separation between the office as the working space and the community room as the participatory space for local residents. We divided the mass by the way of segmentations and then reassembled them to create a new spatial structure.

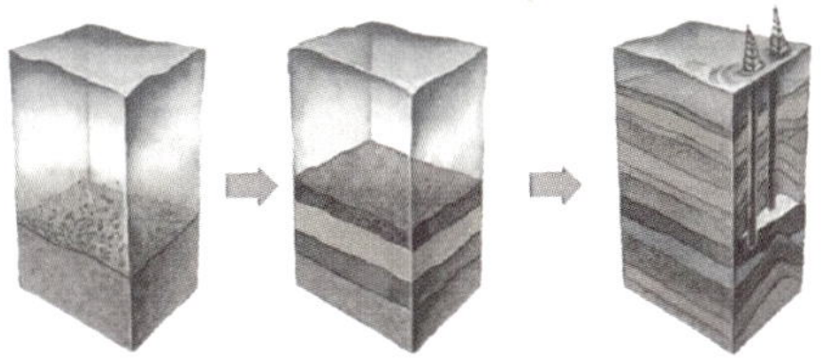

외부의 영향으로 자연스럽게 변화하는 지층의 형성을 매스 프로세스에 이용

Utilize mass processed for the foramtion of layers thatnaturally change due to external influences.

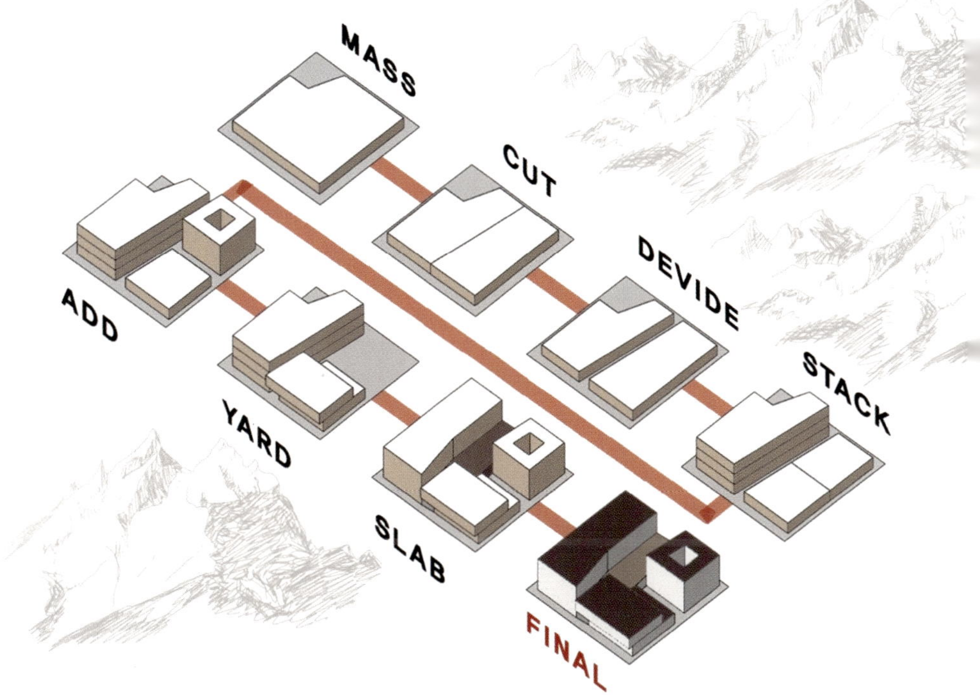

SITE

서울시 구로구 중앙로 6길 16 두드림센터

16, Jungang-ro 6-gil, Guro-gu, Seoul, Republic of Korea

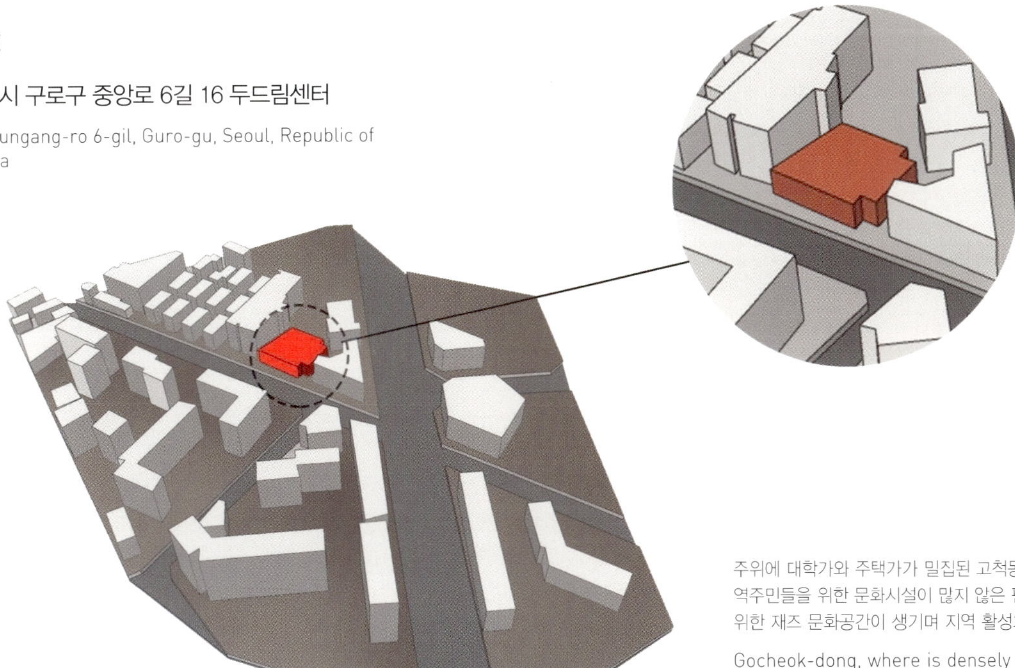

주위에 대학가와 주택가가 밀집된 고척동은 고척돔을 중심으로 상권이 밀집되어 있지만, 지역주민들을 위한 문화시설이 많지 않은 편이다. 재즈피플 사옥 인테리어를 통해 지역 주민을 위한 재즈 문화공간이 생기며 지역 활성화에 의미있는 효과를 가져올 것을 기대할 수 있다.

Gocheok-dong, where is densely populated with a university town and residential areas, has the commercial supremacy concentrated with the Gocheok Dome as the center, but there are relatively not many cultural facilities for local residents. Through the office interior of the 'Jazz People', it makes a cultural space of jazz for local residents and also can be expected to bring some meaningful effects on the revitalization of the region.

SPACE COMPOSITION

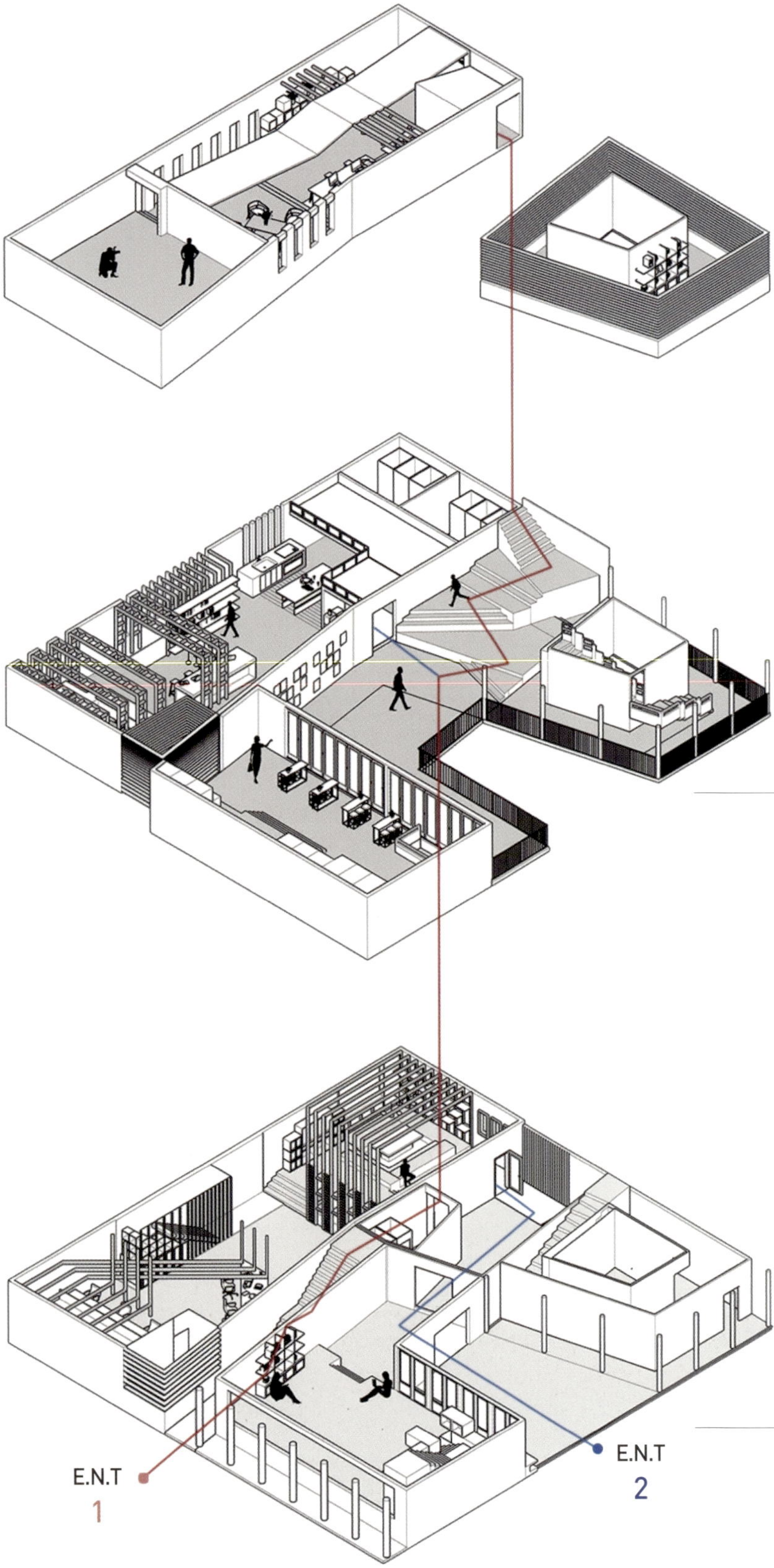

MOTIVE

시선의 교환
Exchange of sight

도담 삼봉
Three masses

처마 밑 공간
The space under the eaves

1F

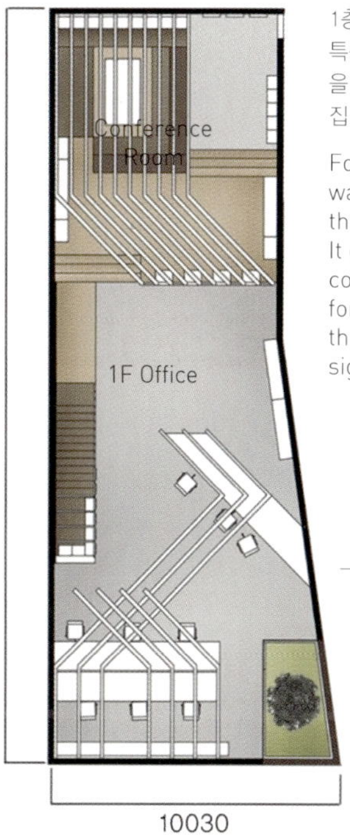

1층 오피스는 회의실을 단을 높인 좌식형 마루로 공간의 역할과 특성을 부여하였다. 업무가 집중되고 과열될 수 있는 회의 공간을 오히려 자유롭고 편안한 공간으로 디자인한 것이다. 또한 편집 · 디자인팀의 창의적 업무를 위한 자연 공간을 내부로 들였다.

For the office on the first floor, the conference room was made with the sitting floor by elevated step to give the role and characteristics of the space.
It designed the meeting space, where works could be concentrated and overheated, to be a free and comfortable space. It also brought the natural space into the inside for creative works of the editorial and design team.

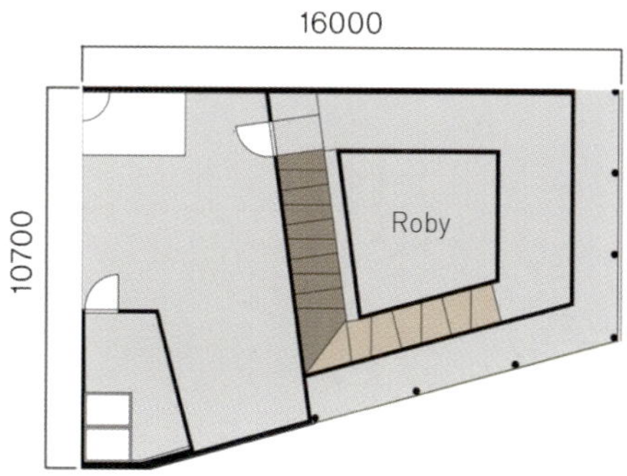

전이공간의 1층 공간은 외부와 흐름이 연결되는 공간으로 오피스와 커뮤니티 공간으로 가는 로비가 위치해있다.

The first floor of the transition space is a space where the flows are connected with the outside and the lobby going towards the office and the community space is located.

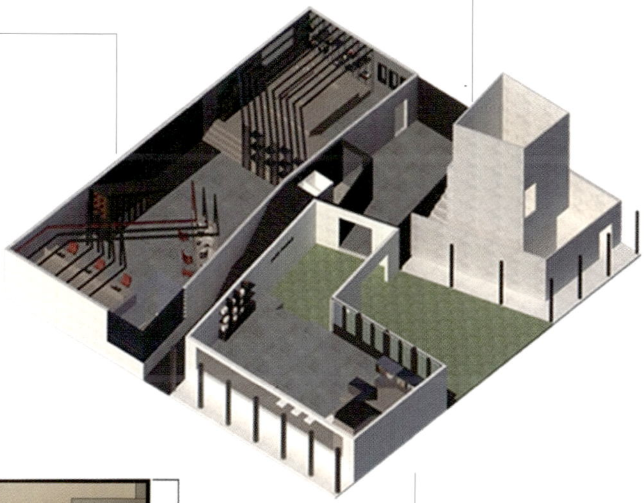

1층 서재 역시 1층 오피스처럼 전통적 느낌의 마루 디자인을 도입해 사용자가 편안한 분위기에서 서적을 읽을 수 있도록 하였고, 매시브한 책장들 속 계단을 배치해 2층 커뮤니티 공간과 자연스러운 동선의 흐름을 이어가도록 하였다.

For the study on the first floor, floor design of traditional feeling was introduced like the office on the first floor so that users can read books in a relaxed atmosphere and stairs were arranged within those massive bookcases to continue the natural flows of moving line with the community space on the second floor.

2F

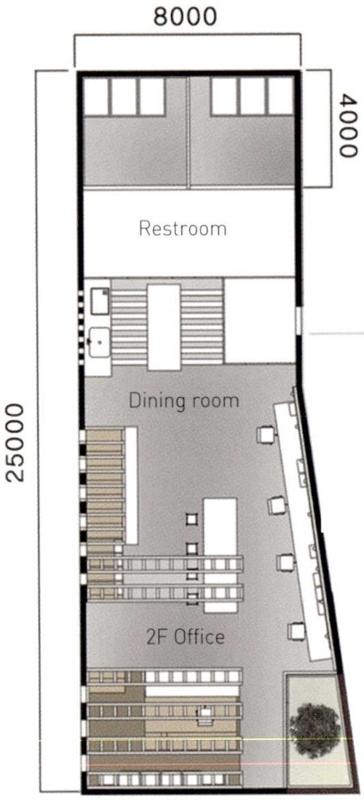

전이공간의 2층은 2층 슬라브와 1층 마당이 훤히 보이는 전경을 가진 공간으로 동선의 중앙에 위치한다.

The second floor of the transition space is located in the center of the moving line, with a wide-open panoramic view of the second floor slab and the first floor yard.

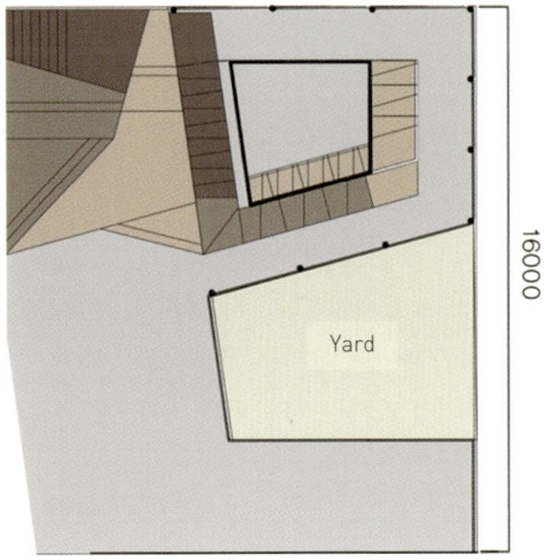

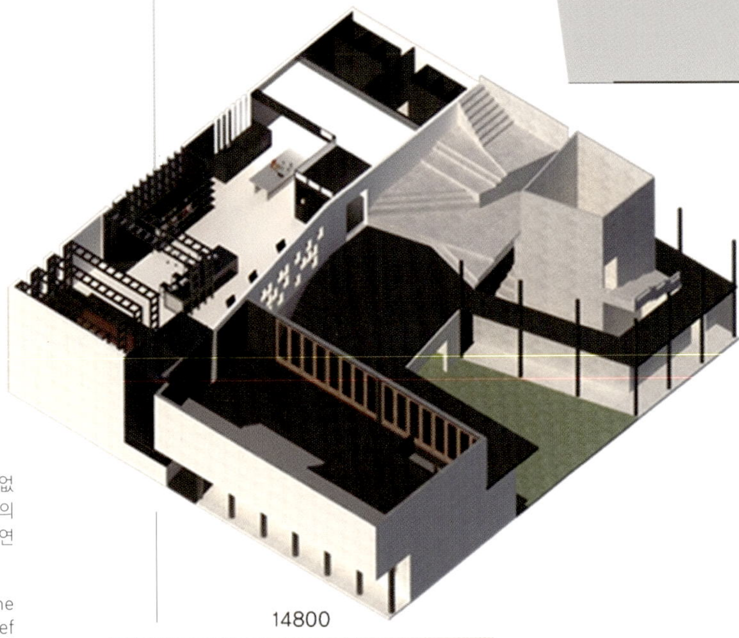

2층 오피스는 편집장의 공간을 제외하고 바닥의 레벨 차는 없지만, 탕비실과 입구쪽 천장을 낮춰 조명박스를 배치해 공간의 분할을 시도하였다. 자연공간을 통해 1층과 공간의 흐름이 연결된다.

The office on the second floor has no difference in the steps on the floor except of the space for the chief editor but it attempted segmentations of the space as arranged a lighting box onto the lowered ceiling at the office pantry and the entrance. The flow of the first floor and space is connected through the natural space.

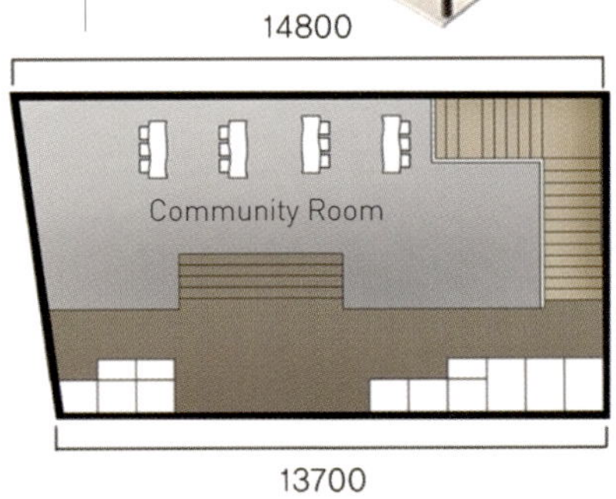

서재 계단을 통해 올라온 2층 커뮤니티 공간은 또 한 번 마당과 연결되며 전통 건축에서 발견되는 내부 안의 외부적 요소를 사용자에게 인지시킨다.

The community space on the second floor coming up through the library stairway is once again connected to the yard and let the users recognize the external elements in the inside foundin traditional architecture.

3F

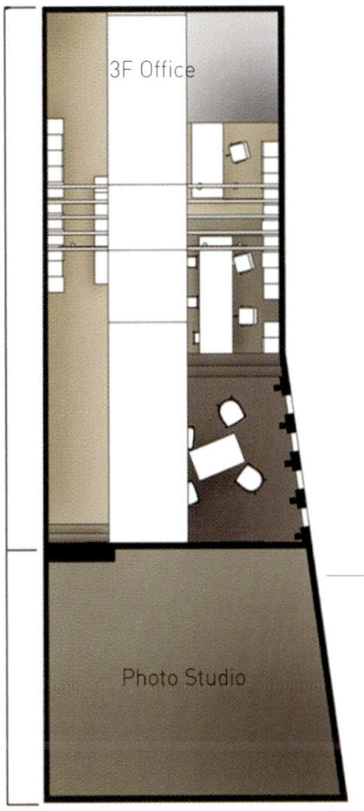

전이공간의 3층부에 도달하면 마침내 닿는 곳은 재즈 청음공간이다. 에디터가 추천하는 음악을 들으며 외부 풍경을 감상할 수도 있다.

Arriving at the third floor of the transition space, finally it reaches to the listening space of jazz. Listening to musics recommended by the editor, it is able to admire the outside landscape.

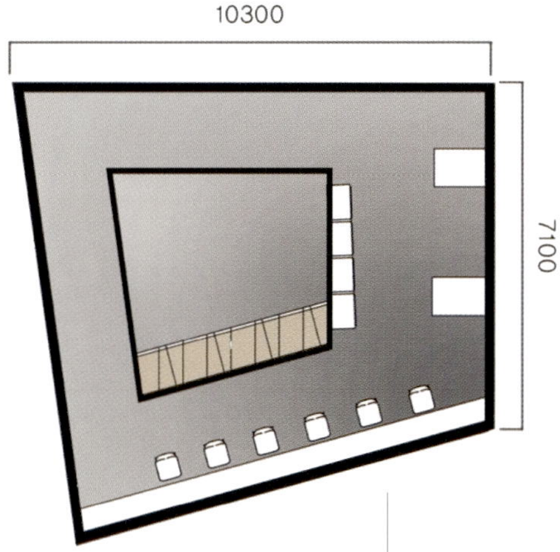

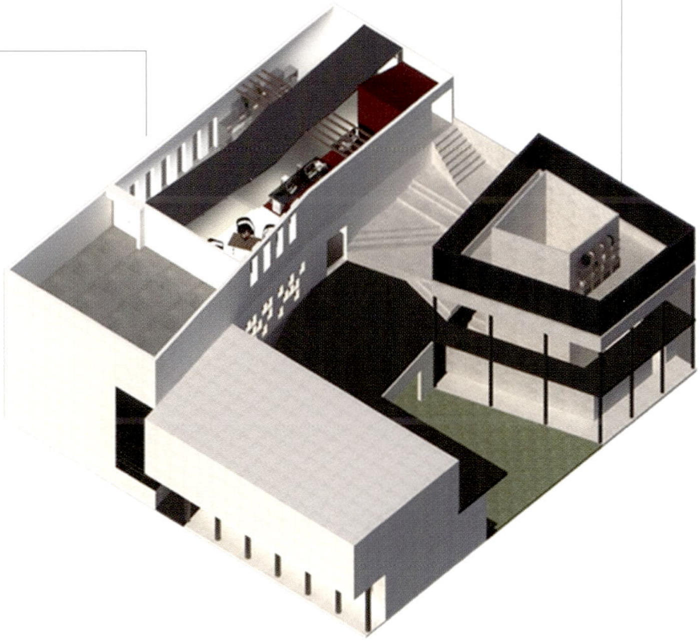

3층 오피스는 1 · 2층 오피스 디자인의 연장선상으로, 솔리드한 매스로 천장과 바닥의 레벨차를 주어 공간을 다채롭게 구성하였다.

The office on the third floor is an extension of the office design on the first and second floors, and the space is made up diversly by providing a level difference between the ceiling and the floor with a solid mass.

내밀 内密

Private

서재 계단을 통해 올라온 2층 커뮤니티 공간은 또 한 번 마당과 연결되며 전통 건축에서 발견되는 내부 안의 외부적 요소를 사용자에게 인지시킨다.

The secret facade unable to know of the inside. As the entrance is located between the column and the rouver, it arouses a curiosity about the space inside.

진입 進入

Entry

낮은 천장과 두터운 벽체, 좁은 계단을 오르는 수축의 공간, 누하를 경험하고 마침내 진입한다.

After experienced the low ceiling and thick wall, the space of contraction going up narrow stairs, under the gazebo, finally it enters.

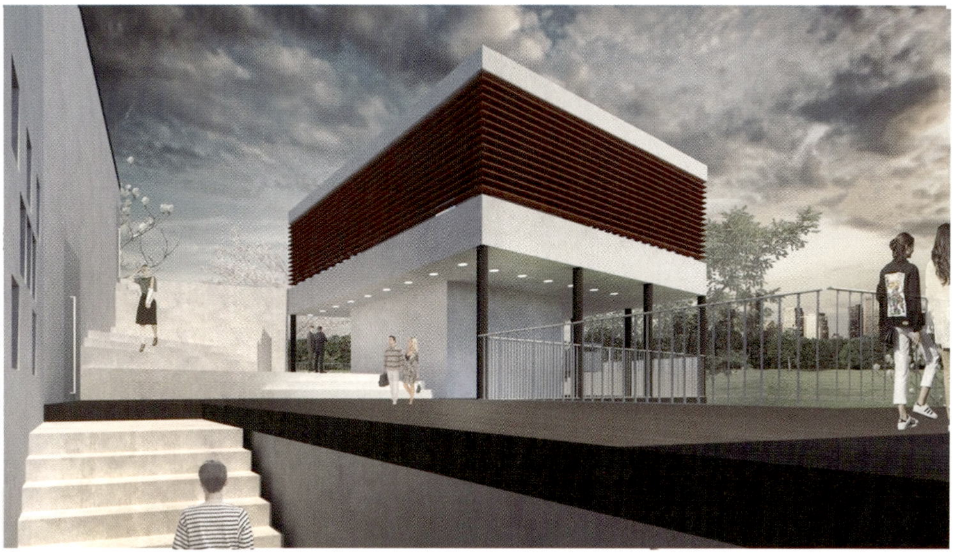

확장 擴張

Expansion

좁은 계단을 지나 다다르는 곳은 넓고 탁 트인 2층의 슬라브와 마당 공간이다.

The places arriving after passed through the narrow stairs are the spacious slab on the second floor with a fair view and the yard space..

즉흥 卽興

Inprovisation

넓은 내부 안 외부의 마당 공간에서 사용자는 자신이 원하는 대로 즉흥적인 공간 사용을 할 수 있게 된다.

In the large inside space and the outside yard space, the users are able to use the spaces improvisatori-ally as they please.

SENSORIUM

BRIEF

Sensorium 이란 두뇌, 정신, 마음 등을 포괄하는 의식의 세계이다. 여러 감각기관으로부터 입력되는 모든 정보를 재정리하여 처해 있는 환경에 대한 체계적이고 의미 있는 이해를 가능하게 한다. 이 개념을 디자인에 접목해 1층 바 공간은 유리 레이어와 계단으로 이어진 가구들, 2층 업무공간은 기능적 역할별로 단 차이를 두고 선으로 공간을 하나로 잡아주어 전체적으로 정리하고 안정적으로 디자인하였다.

Sensorium is the conscious sphere including brain mind and spirit as a whole. It reorganizes all the information input from the various sensory organs and can do a systematic and meaningful understanding of the environment in which we are in. By applying this concept to design, the bar on the ground floor was designed with glass layers and stairs, and the office on the first floor has different floor levels depending on the functional roles, and used thin pipes to hold this space together.

잡지 디캔터는 영국에 본거지를 두며 90여 개 나라에서 매월 발행되는 잡지이다. 산업 관련 뉴스, 와인 추천, 빈티지 와인 가이드 등을 게재한다. 고대에서 현대에 이르기까지 인생을 함께해 온 와인의 역사를 소개하며, 세계적으로 유명한 와인을 만나볼 수 있다. 2004년부터는 영국의 와인 평가대회인 디캔터 월드 와인 어워드(Decanter World Wine Awards)를 개최해오고 있다.

Decanter is a wine and wine-lifestyle magazine, published in about 90 countries on a monthly basis. The magazine was founded in London in 1975. It is the oldest consumer wine publication in the United Kingdom. It includes news histories, vintage guides and wine and spirits recommendations from the industry. Decanter also organizes the annual Decanter World Wine Awards.

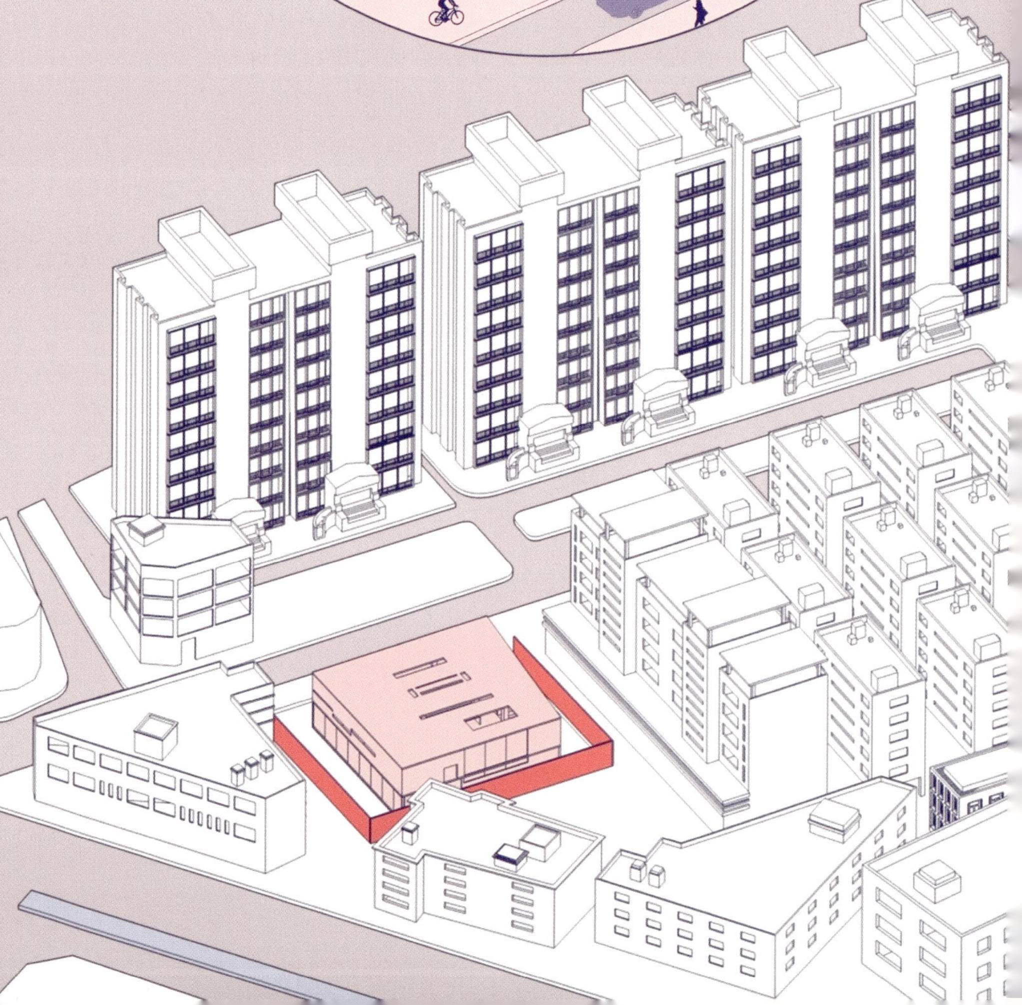

서울특별시 구로구 중앙로6길 12 (고척동 78-2)
12, Jungang-ro 6-gil, Guro-gu, Seoul, Republic of Korea

주변에 주거 단지가 조성되어 있어 대지의 대부분이 아파트로 둘러싸여 있다. 인근에 지하철역이 있으며 학교를 비롯해 병원, 야구장 등이 있고 동쪽에는 안양천이 흐르고 있다. 하지만 주거시설의 규모에 비해 커뮤니티 시설이 부족하다. 이에 대한 해결책이 필요하다고 느껴진다.

Most of the site is surrounded by apartments because residential complexes are built around it. There is a subway station nearby, and there are schools, hospitals, a baseball stadium, and Anyangcheon Stream flows to the east. However, it lacks community facilities compared to other residential facilities. We feel that a solution to this is needed.

PROGRAM AIMS AND PROCESS

정동을 디자인하는 것은 의사소통을 설계하는 것과도 같다. 그리고 이는 신체적 관계를 통해 성취할 수 있다고 본다. 따라서 몸으로 구조를 느낄 많은 기회를 만들어 내려고 노력했다. 예를 들면, 만지거나, 앉거나, 기대는 것처럼 말이다. 사람들이 계속 보고 싶게 만들고 가까이 있고 싶게 만드는 것이다. 결국 정동을 넘어서는 것은 재료, 사람 그리고 공간으로부터 나오는 힘을 끌어내는 것이다.

Designing affection is the same as designing communication. And this can be achieved by valuing physical relationships. So we tried to create many opportunities for people to feel the architecture with their bodies (touching, sitting on, leaning against the architecture). It makes people want to keep looking at the architecture or want to stay in. Eventually, 'Beyond affection' is about drawing on the power that comes from the design of materials, people and spaces.

간단하고 자연스러우며 자유롭지만, 공간의 원형과 핵심만을 반영하는 공간을 디자인하려 했다. 균형 있고 건강한 환경을 조성하고 사람들 간의 많은 상호 작용과 의사소통을 할 수 있는 공간. 공간이란 단순한 명사, 고정된 실체가 아니라 계속된 제작 활동이라고 생각한다. 그 공간을 어떻게 꾸며나갈지는 사용자가 결정하는 것이다. 분위기, 정동은 주어진 것이 아닌, 우리가 디자인하고 만들어나가는 것이다.

We tried to design a space in which everything is simple and natural, but reflects the original and real status. The space that encourages a balanced, healthy environment and enables as much interaction and communication between people. The word 'space' is not a noun a static entity, but an ongoing activity of making. Atmosphere, affection is not something that is given, it's something we design and create.

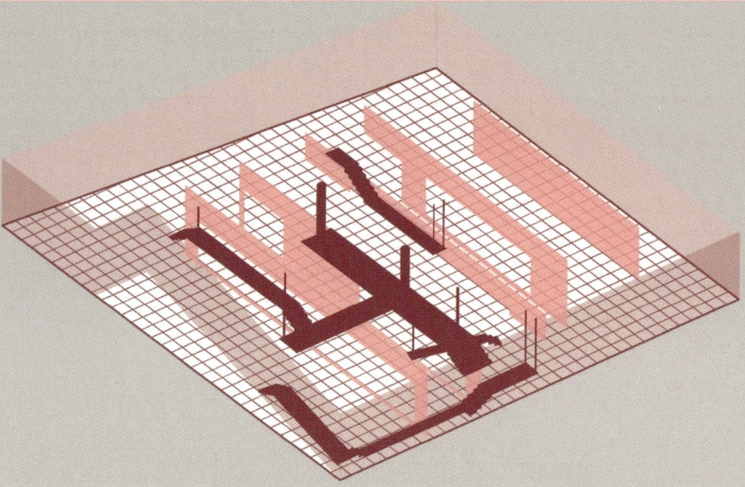

1층 바는 크게 유리 레이어와 하나로 이어진 가구로 구성된다. 벽이 아닌 유리 레이어로 공간을 구획했는데, 이는 공간이 나눠진다는 느낌을 없애고 유리 재료의 특징에서 오는 투명성으로 유리 뒤로 펼쳐진 공간에 대한 호기심을 자극한다. 나눠진 5개의 공간은 서로 조명의 밝기와 색이 다른데, 이는 각 공간마다 차별성을 주고 바 테이블을 강조하는 역할을 한다. 바 외에도 와인을 배울 수 있는 와인 클래스와 유명한 와인들과 서적을 만나볼 수 있는 라이브러리가 있다.

The ground floor bar is consisted of glass layers and furniture connected by stairs. The glass layers, not walls, divide the bar giving blurry boundaries. And the transparency from the glass material arouse curiosity about the space behind glass layers. The five divided space has different color and brightness of light. By this, each space has its own characteristics and it put emphasis on the bar tables. Besides the bar, there is a room where people can learn about wine and a library where you can meet famous wines and books.

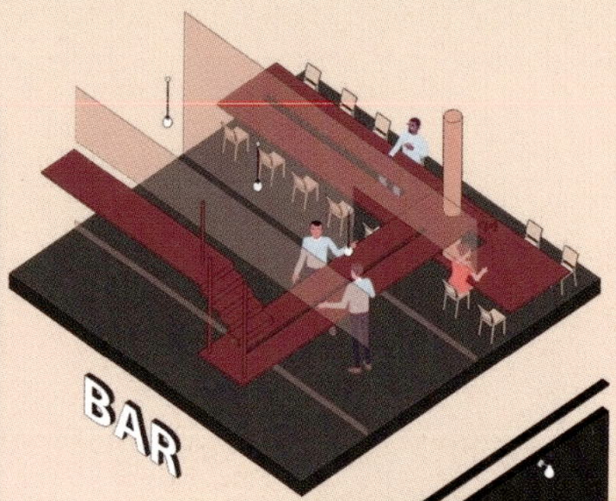

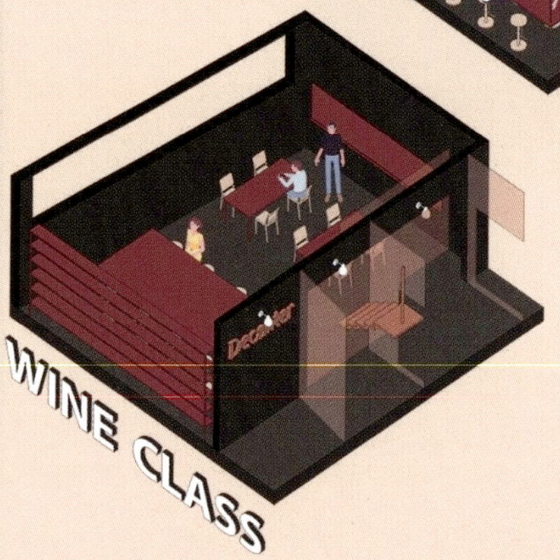

Ground Floor _ Bar

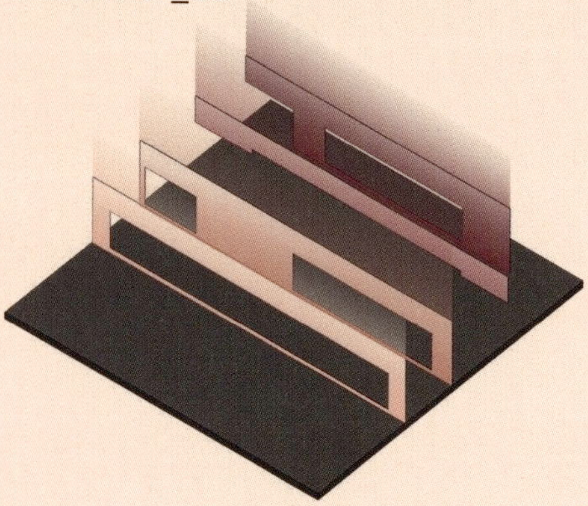

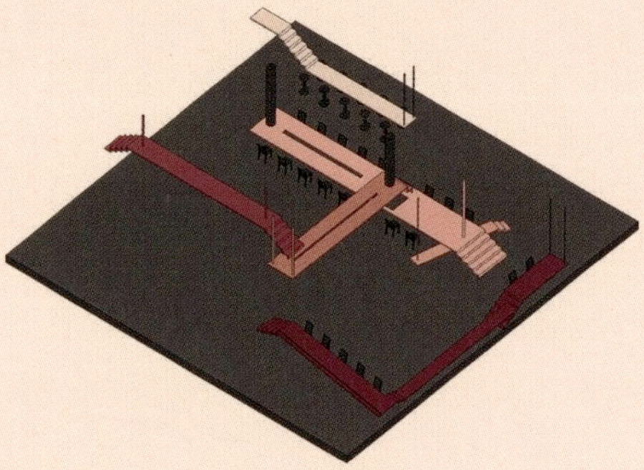

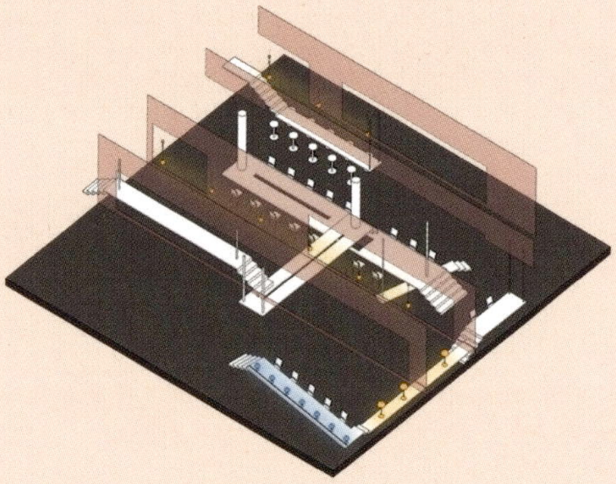

바닥 단 차이 | Floor Level

4개의 유리 레이어는 서로 다른 농도를 갖고 있는데, 공간의 내부로 갈수록 점점 짙어진다. 이는 공간의 깊이를 표현하고 내부로 갈수록 집중된 느낌을 주어 마치 와인의 풀 바디감을 연상시키며, 메인 바와 그 뒤 라이브러리를 더욱 부각해준다.

The four layers of glass have different concentrations, and get darker as you go inside. It expresses the depth of space and gives a more focused feeling inside, like a full bodied wine, and highlights the main bar and the library behind it.

얇은 파이프 | Thin Pipes

바 테이블은 가구처럼 인식되는 것이 아니라 하나의 공간처럼 다가온다. 본래 계단은 다른 층으로 가기 위해 사용되지만, 이 공간에서는 서로 다른 것들을 연결해주고 응집력 있는 전체를 형성한다. 공간 전체를 통해 흐르며 각 활동과 기능을 분할 없이 연결해준다. 결국 이는 공간의 정체성이 되어 건축적 개념에서의 인식을 넓히는 역할을 한다.

The bar tables are perceived as a space not as furniture. Traditionally used to connect disparate levels, the stairs in this case forms a cohesive whole. Flowing all the space, the stairs allow each activity and function to flow with no division. The stairs become the identity of the space, allowing visitors to broaden their own perception of an architectural conception.

창문 | Window

공간 전체에 같은 조명을 사용하지 않아 유리 레이어로 구획된 공간은 각각의 특색을 갖는다. 공간의 중심이 되는 바 테이블이 강조되도록 하였으며 바에 어울리는 주백색 조명을 주로 사용했다. 전체 조명보다는 부분 조명을 활용하여 대체로 어두운 분위기를 내며 메인 바 공간이 가장 밝다.

Because we used different lighting, the 5 spaces have their own characteristics. Amber lights are mainly used to match with the space and to make the bar table stand out. Partial lighting creates dark ambience that makes the bar look luxury.

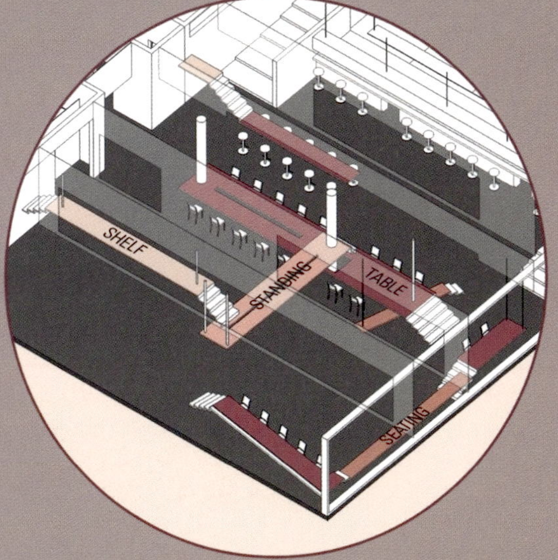

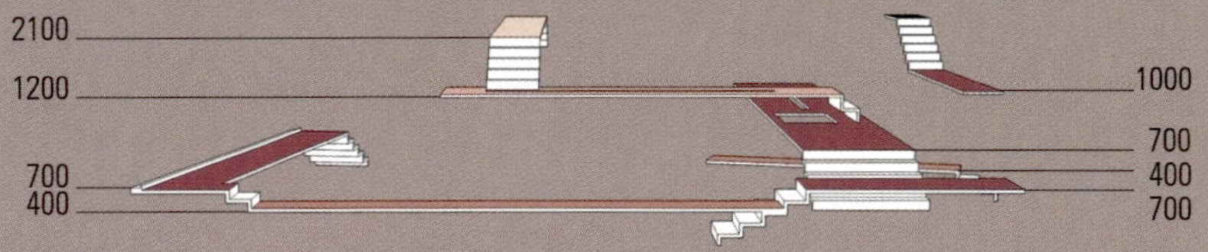

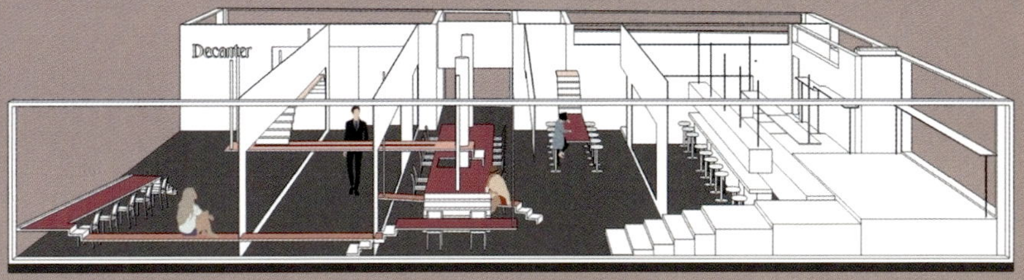

첫 번째 공간은 700mm 높이의 테이블과 2100mm의 선반 그리고 400mm의 의자로 시작한다. 두 번째 공간에서는 1200mm의 스탠딩 테이블이 있고 세 번째 공간에는 700mm의 긴 테이블과 각도가 틀어진 의자가 있다. 마지막 네 번째 공간에는 1000mm의 테이블이 있다. 2100mm 높이부터 700mm로 낮아지면서 공간에 적응하며 메인 바로 도달하게 된다.

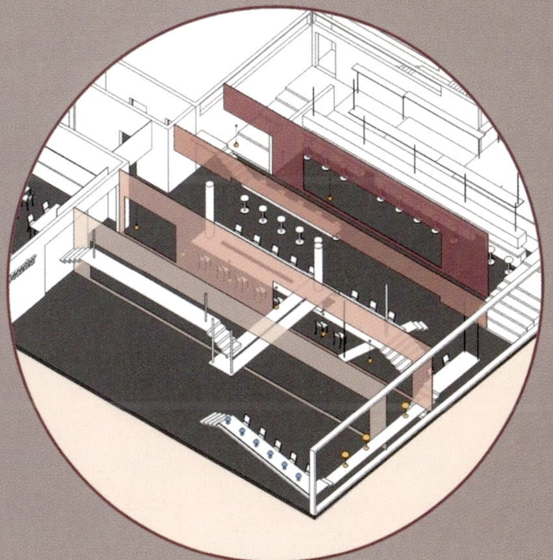

첫 번째 공간

푸른빛의 테이블 조명만이 공간을 밝히며 나머지 주황색 조명과 대조되어 신비감을 조성함

두 번째 공간 _ 유리 농도 10%

주황빛의 전체 조명이 스탠딩 테이블을 밝힘

세 번째 공간 _ 유리 농도 20%

천장의 핀 조명이 은은하게 공간을 밝히며 테이블 밑에 위치한 조명이 사람들의 시선을 사로잡음

네 번째 공간 _ 유리 농도 30%

테이블 위에만 3개의 조명이 은은하게 공간을 밝힘

메인 바

전체 조명 없이 테이블 바로 위에 위치한 가구에서 조명이 나와 고급스러운 분위기를 한층 더 높임

페인트 콘크리트 대리석

고급스러운 분위기를 내기 위해 어두운 재료를 사용하였다. 바 테이블을 대리석으로 하여 공간과 잘 어우러지게 하였다.

유리 구리 금

유리 레이어는 색은 없으며 내부로 갈수록 뿌옇게 하였으며 메인 바는 구리와 금을 사용하여 가장 두드러지게 하였다.

유리 레이어의 중첩이 일어나 공간의 깊이를 형성하고 다양한 유리 형태가 호기심을 자극

Glass layers creates a depth of space and various shapes of glass arouse the curiosity

넓은 테이블과 의자에서 와인을 마실 수 있는 공간으로 안정적임

A stable space where there is a wide table and a tilted chair where you can drink wine

입구를 지나 공간에 적응하기 시작하며 유리 레이어 농도가 가장 옅음

People start to adjust to the space and the density of the glass layers is the lightest of four

가장 사적인 공간으로 유리 레이어의 농도가 가장 짙으며 메인 바 공간에서 전체 공간의 분위기가 정리됨도 있음

It's the most private space and the density of the glass layer is the thickest of four. The atmosphere of the whole space is arranged at the main bar

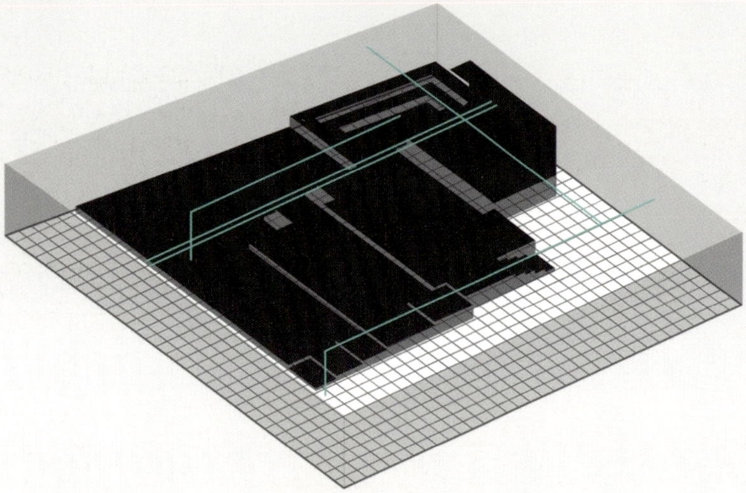

2층 업무 공간은 크게 단 차이로 나눠진 공간과 얇은 파이프로 구성된다. 벽이 아닌 바닥 단 차이로 공간을 구획했고 이는 동선의 유도, 공간감의 변화 등을 도모한다. 개인 지정 자리를 없애고 열린 공간에서 일하는 방식의 공간으로 디자인했다. 업무공간과 휴게공간을 정확히 구분해놓지 않은 것이 이 사무실의 특징이다. 공중을 가로지르는 아주 얇은 파이프는 가상의 공간감을 형성하며 동시에 단 차이로 나눠진 공간들을 하나로 잡아주는 역할을 한다.

The first floor office is consisted of spaces divided by different floor levels and thin pipes. The space is divided not by walls but by the difference of the floor levels, and this leads to the guidance of a flow and the change of space. It was designed to be an open space where people are not assigned a specific space. This office is characterized by its blurry distinction betwen work space and rest space. A very think pipes crossing the ceiling forms a virtual space and at the same time, holds divided space into one.

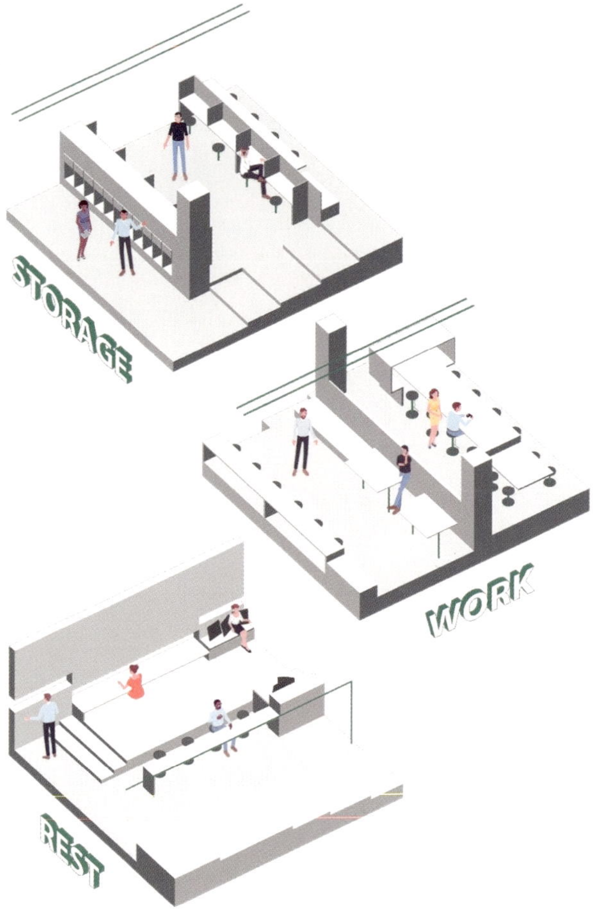

BRIEF

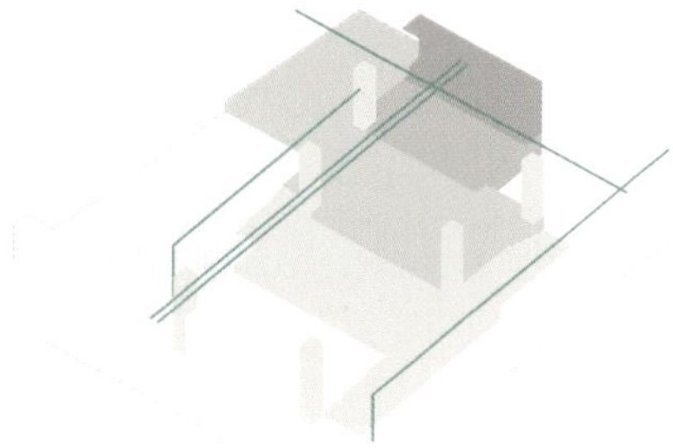

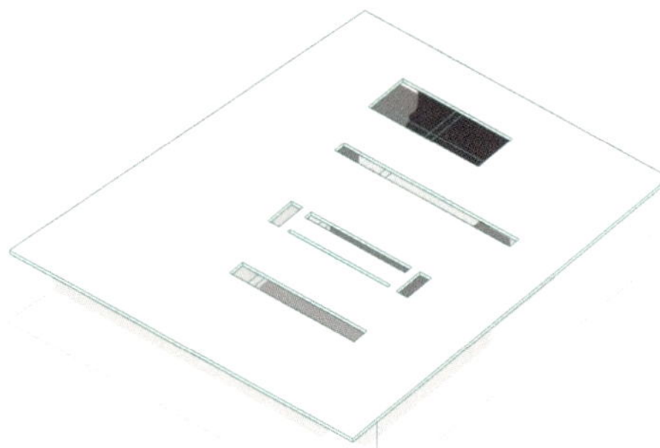

바닥 단 차이 | Floor Level

회사 내의 소통을 원활하게 하고 정해진 루틴을 벗어나고자 개인 지정 자리를 없앴다. 사물함에 개인 물품을 놓고 원하는 자리에 가서 업무를 보는 것이다. 개방형 사무실의 단점을 보완하고자 바닥 높이를 다르게 하였다. 300mm씩 높아지는 바닥은 제일 높은 공간의 바닥에서부터 천장까지 이어지는 구조물에서 절정을 이루게 된다.

To facilitate communication and to get away from everyday routine, we have removed personal space. People can put their belongings in the cabinet and are presented with options for working at a table, lounging or standing. To compensate for the disadvantages of open-type office, floor heights were made different. It increases by 300mm, and reaches its peak at the last space where there's a structure that connects with the ceiling.

얇은 파이프 | Thin Pipes

바닥의 단과는 달리 수평적인 이 선들은 바닥과 대조되면서도 같이 어울려 하나의 공간을 이루게 된다. 바닥에만 변화가 있어 자칫 단조로워 보일 수 있는 공간이지만, 이 파이프로 인해 공간의 단조로움을 피하고 나뉘어 있는 공간들을 하나로 통합시켜 보이게 하려 했다.

The combination of thin vertical pipes and horizontal floor levels make the space look complete. Only by the change of floor height, it could have been monotonous, but this pipes prevented this and united the dispersed space into one.

창문 | Window

천장에서 떨어지는 변화무쌍한 빛은 아침에는 청백색에서 낮에는 노란색으로, 저녁에는 빨간색으로 바뀐다. 빛의 색과 각도의 변화로 다양한 표현들을 만들어 내며 공간의 분위기를 시시각각 변화시킨다.

Ever-changing light that drops from above continues to change from blue-white in the morning to yellow at noon and to red in the evening, and it creates a variety of expressions with the shifts of the color and angle of the light.

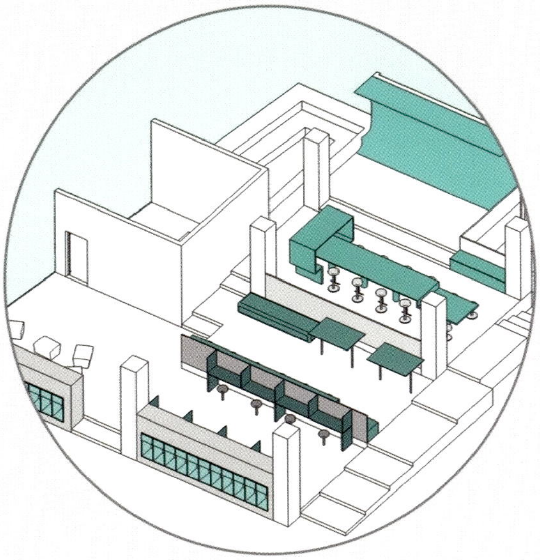

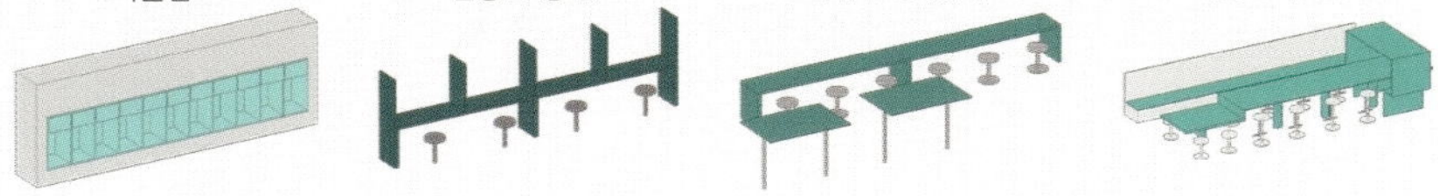

첫 번째 공간은 300mm 높아진 공간으로 의자가 바닥에 고정되어 있어 정적인 느낌을 준다. 두 번째 공간은 600mm 높아진 공간으로 앉아서 일할 수도 있고 서서 할 수도 있다. 세 번째 공간은 900mm 높아진 공간으로 카페처럼 열린 공간이다. 네 번째 공간은 1200mm 높아진 공간으로 정면 창문을 바라보고 있어 2층 공간 중 가장 개방감을 느낄 수 있는 공간이다.

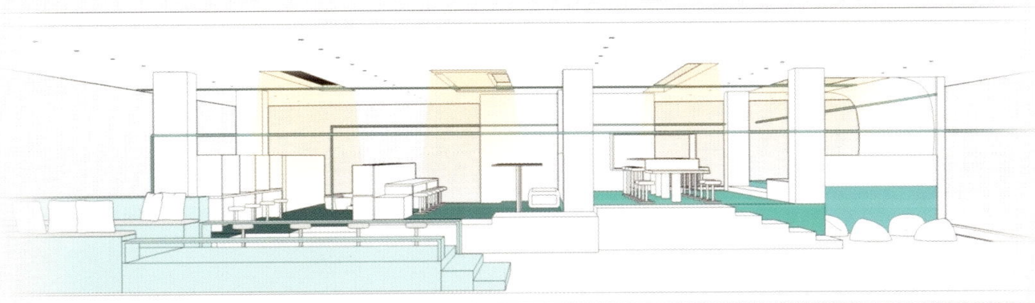

첫 번째 공간
직사각형으로 하여 루버를 설치하여 많은 빛이 들어오는 것을 피함

두 번째 공간
여러 창문을 만들어 자유로운 느낌을 줌

세 번째 공간
양쪽 계단까지 길게 창문을 냄

네 번째 공간
전체를 뚫어 하늘과 이어지는 느낌을 들게 하여 개방감 극대화

주 재료

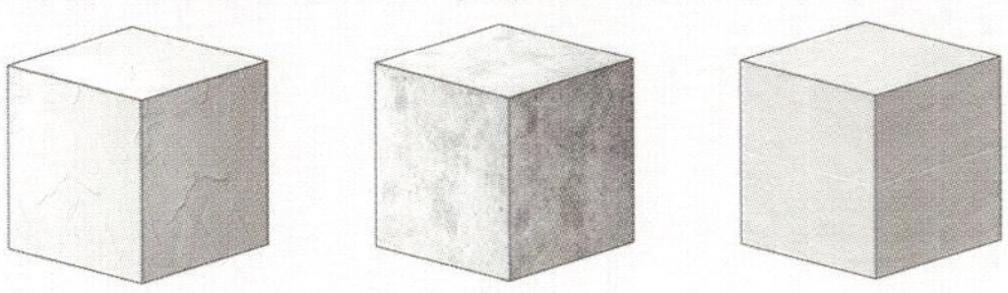

스투코 콘크리트 흰 타일

벽, 바닥, 천장을 흰색으로 통일하고 재질만 다르게 하였다. 흰색으로 칠해진 공간은 끊임없이 계속되는 공간을 통과하는듯한 경험을 제공하며 희미하게 인식되는 색의 차이는 미묘하게 사람의 감정에 영향을 미친다.

가구 재료

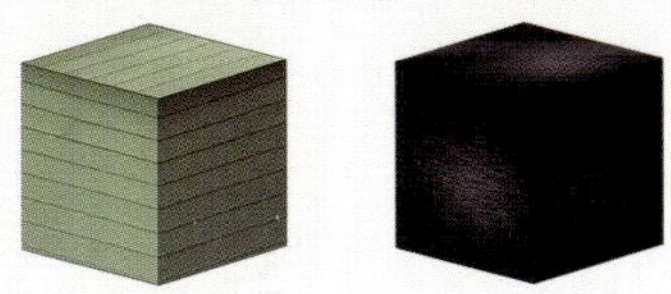

카키색 나무 메탈

단 차이에 의해 시선의 눈높이가 달라지며 얇은 파이프가 가상의 공간감을 형성함

By the difference of the floor height, the eye level changes and the thin pipes forms a virtual space

회의 공간으로 우물 모양으로 바닥이 파여있어 사람들의 시선 교환이 자유로움

Conference room shaped as a well where people can easily exchange looks

자유롭게 업무를 볼 수 있고 천장으로 이어지는 공간에 앉아 휴식을 취할 수 있음

People can work and take a break freely at wherever they want

시선의 높이에 맞게 창문을 배치했으며 쉴 수도 있고 편하게 앉아서 업무를 볼 수도 있음

Resting area where the windows are arranged to the eye level

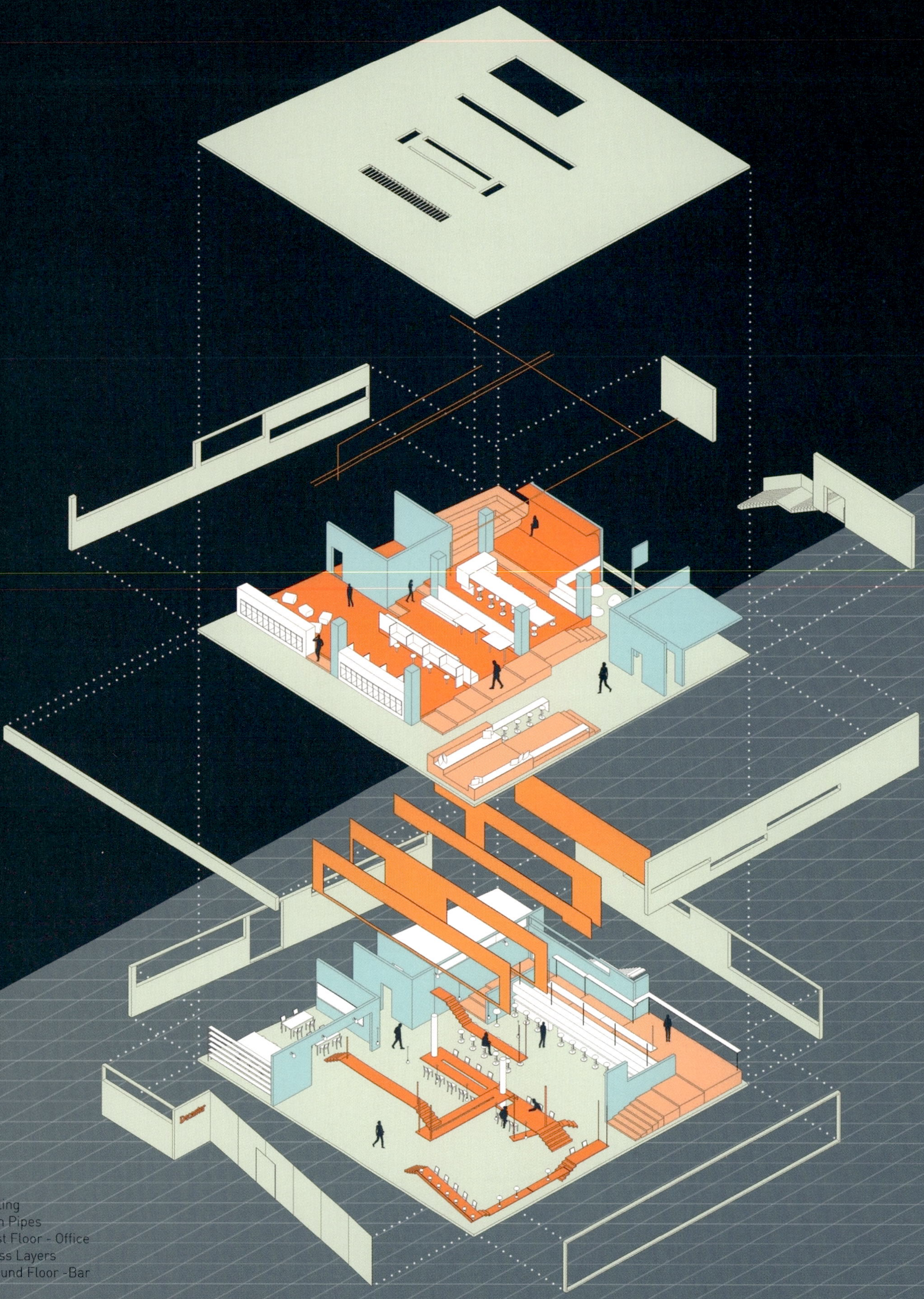

Ceiling
Thin Pipes
First Floor - Office
Glass Layers
Ground Floor -Bar

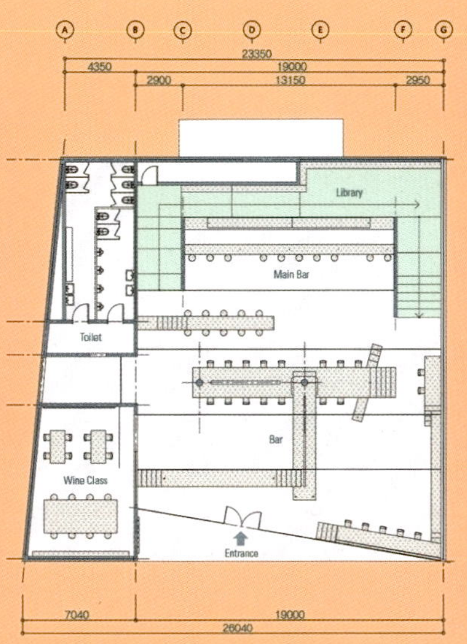

GROUND FLOOR PLAN

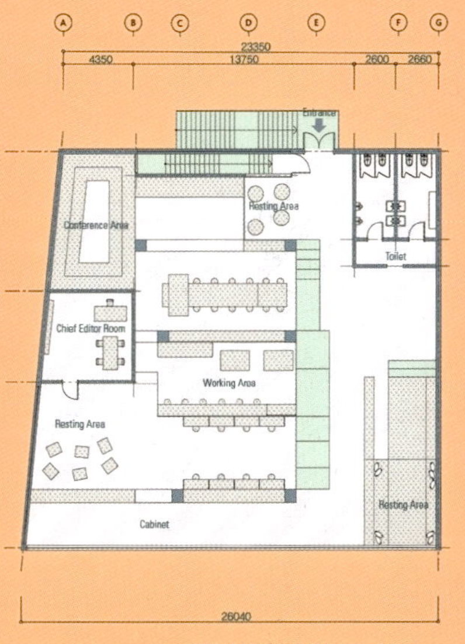

FIRST FLOOR PLAN

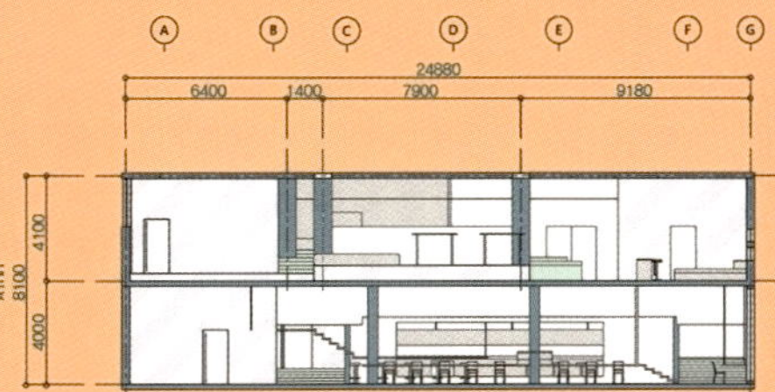

SECTION A - A'

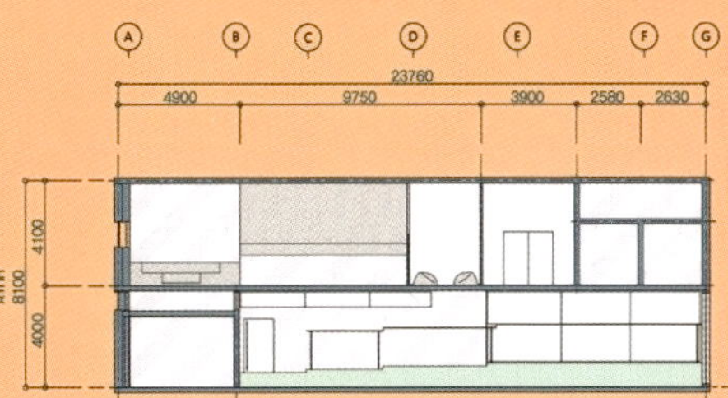

SECTION B - B'

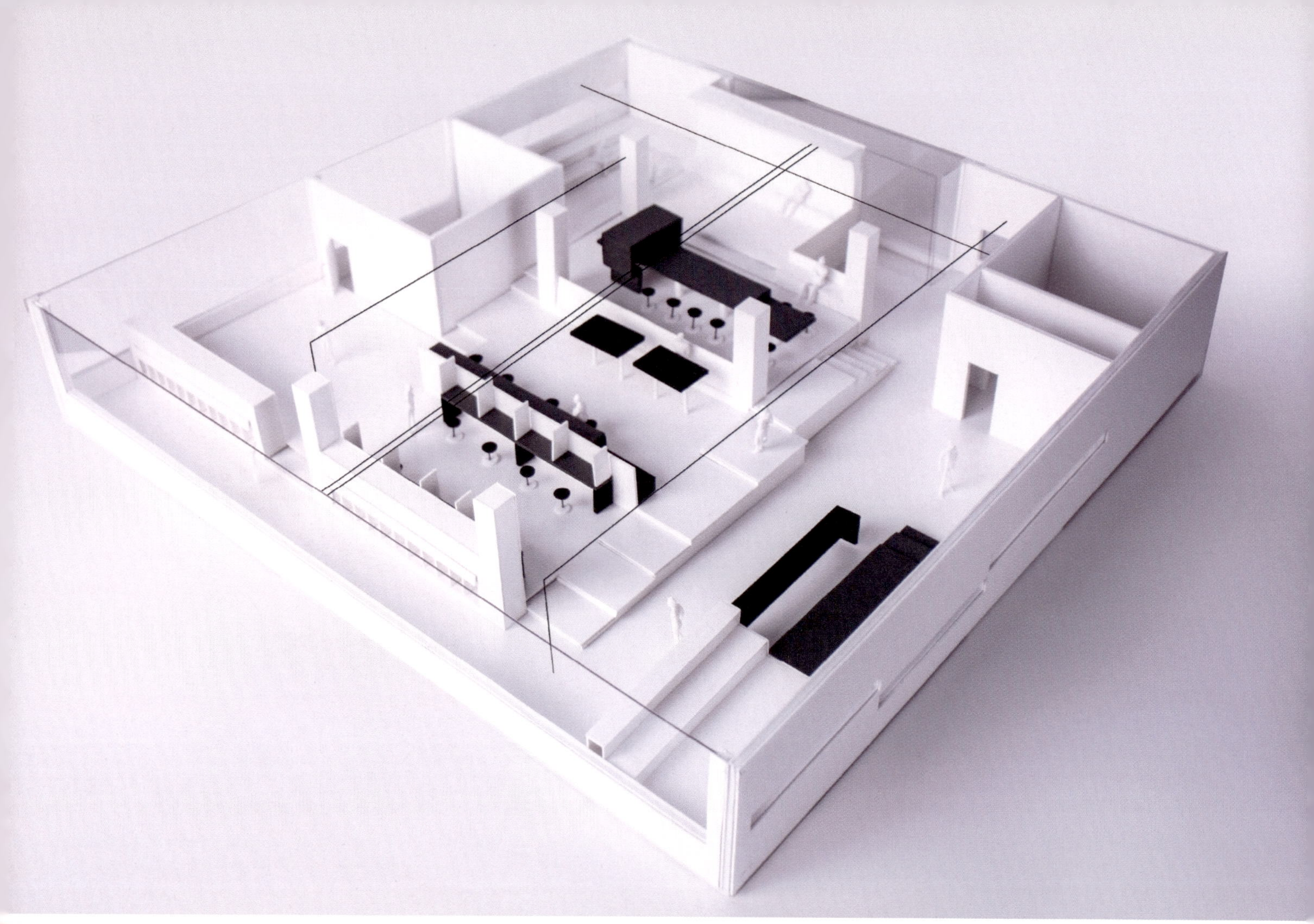

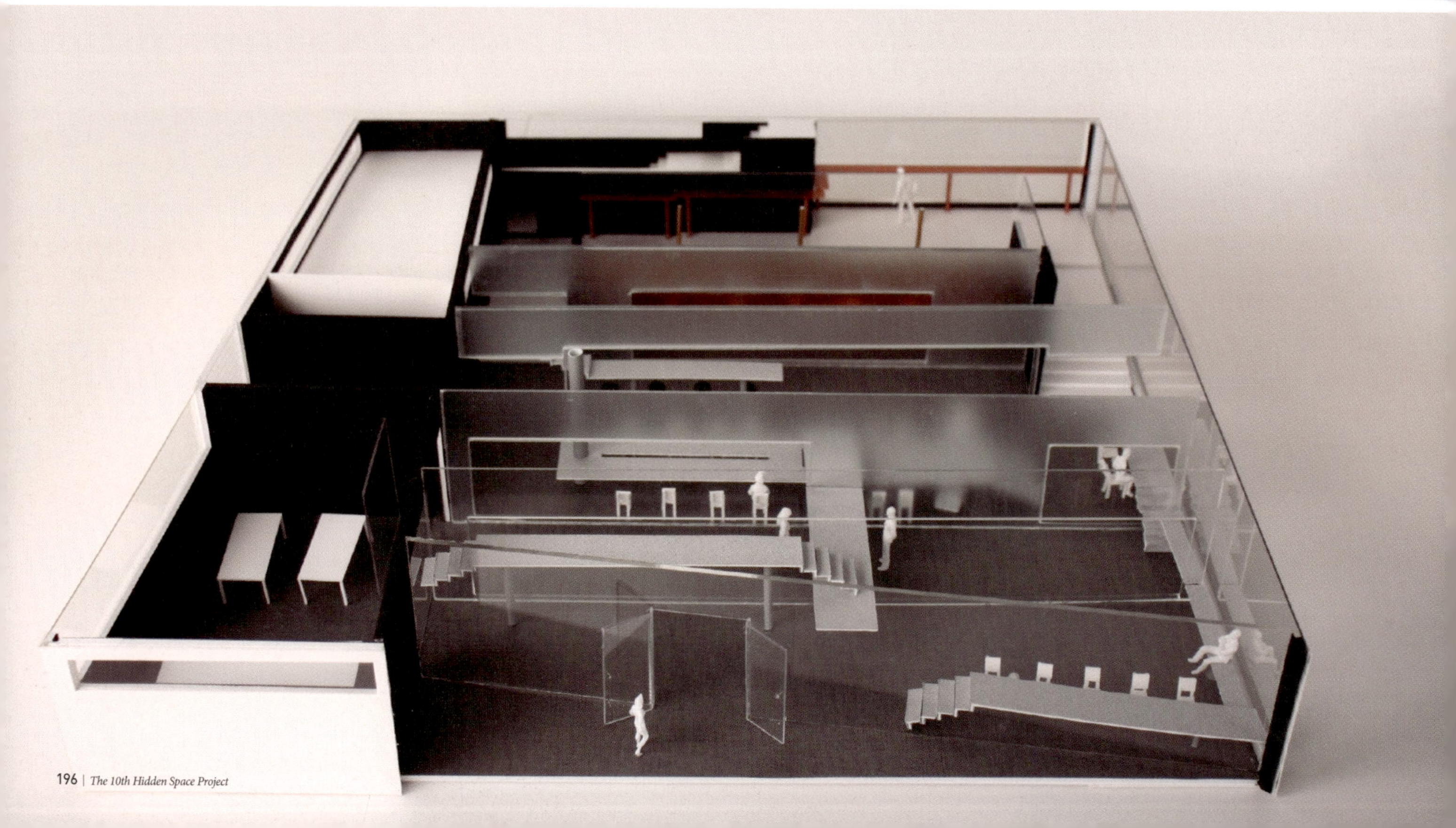

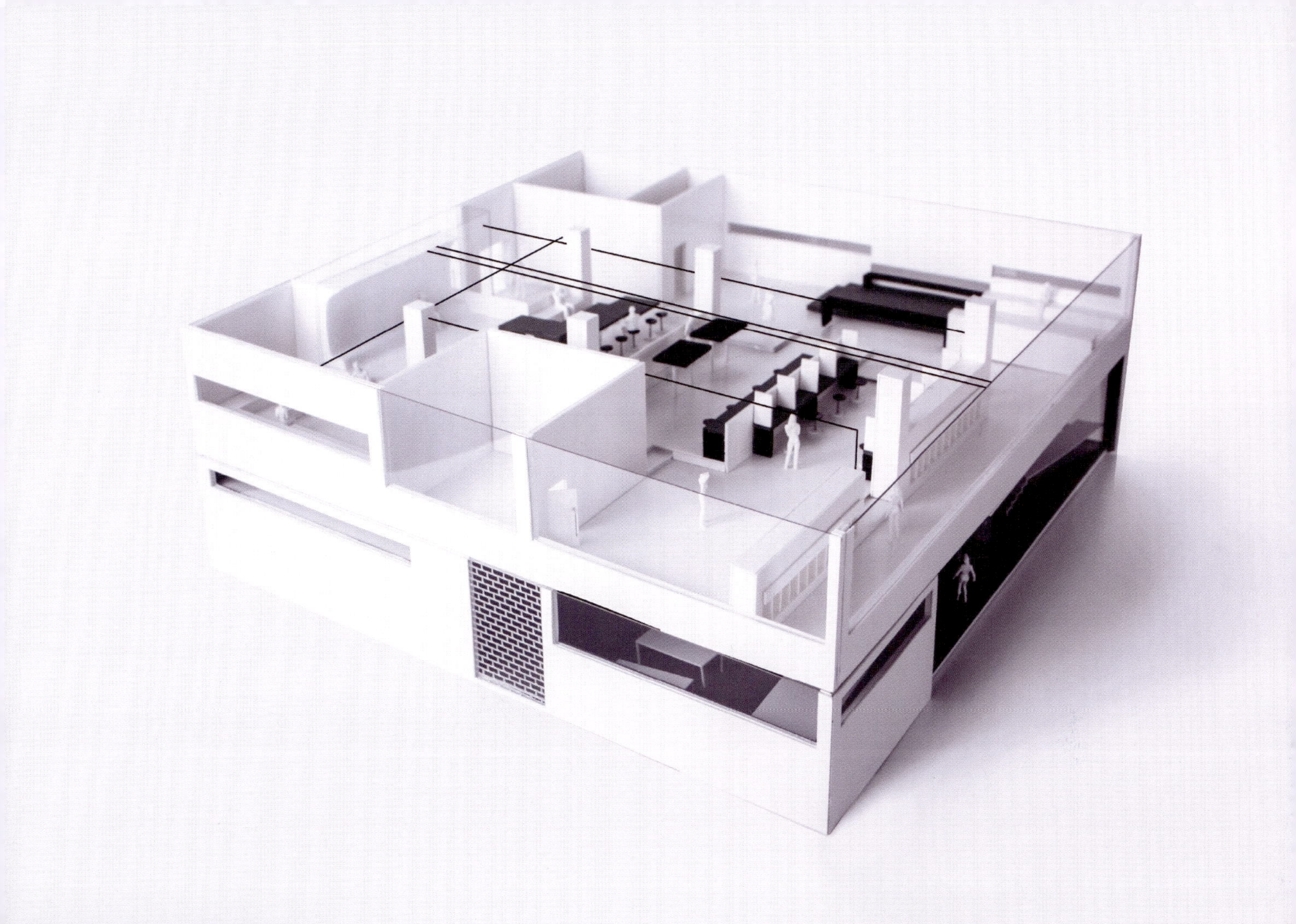

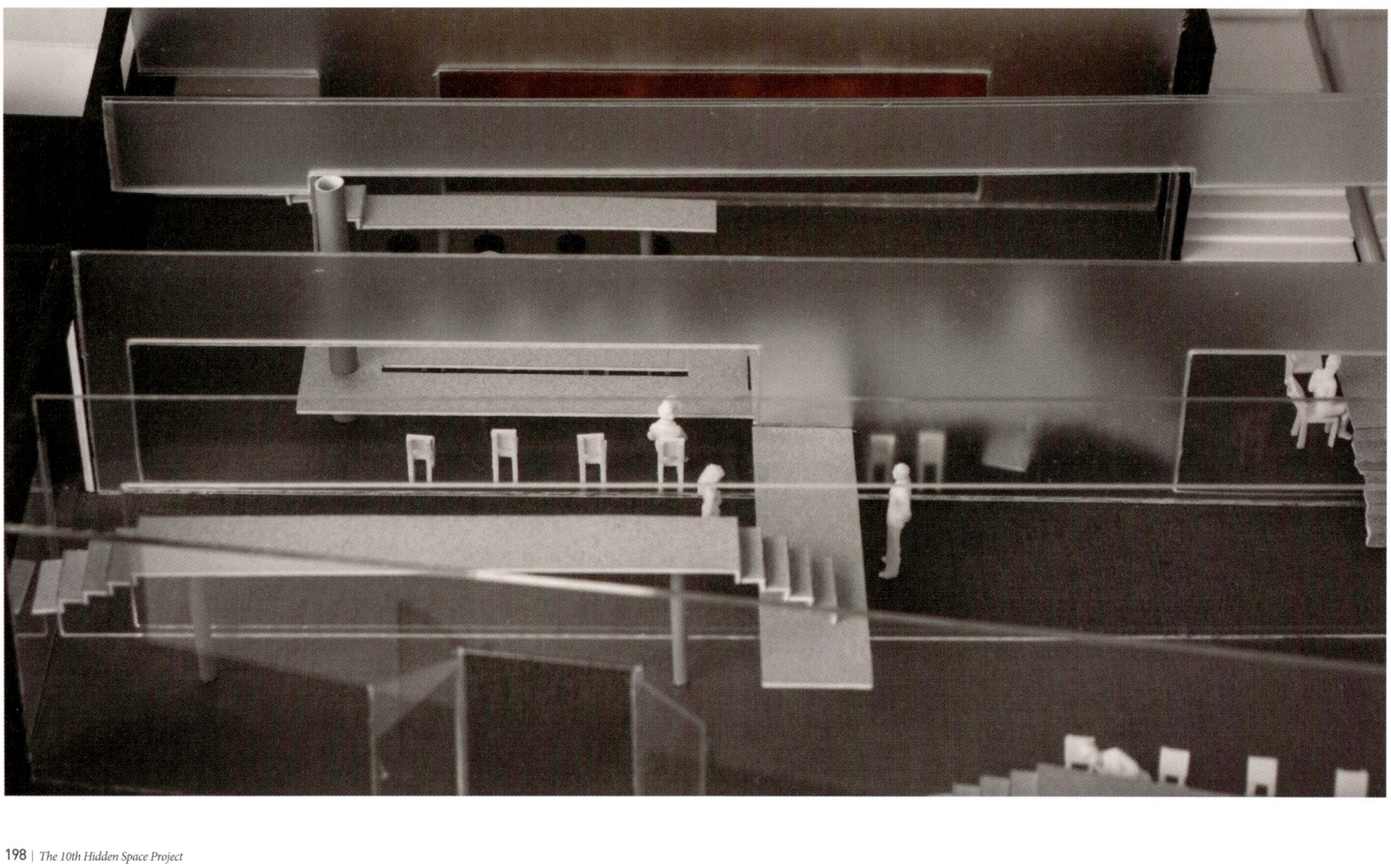

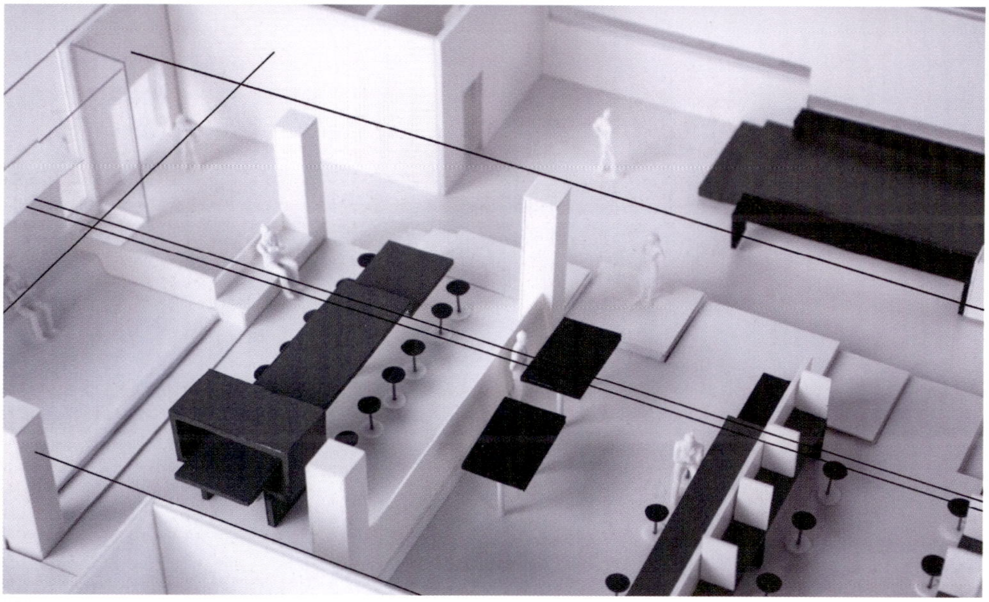

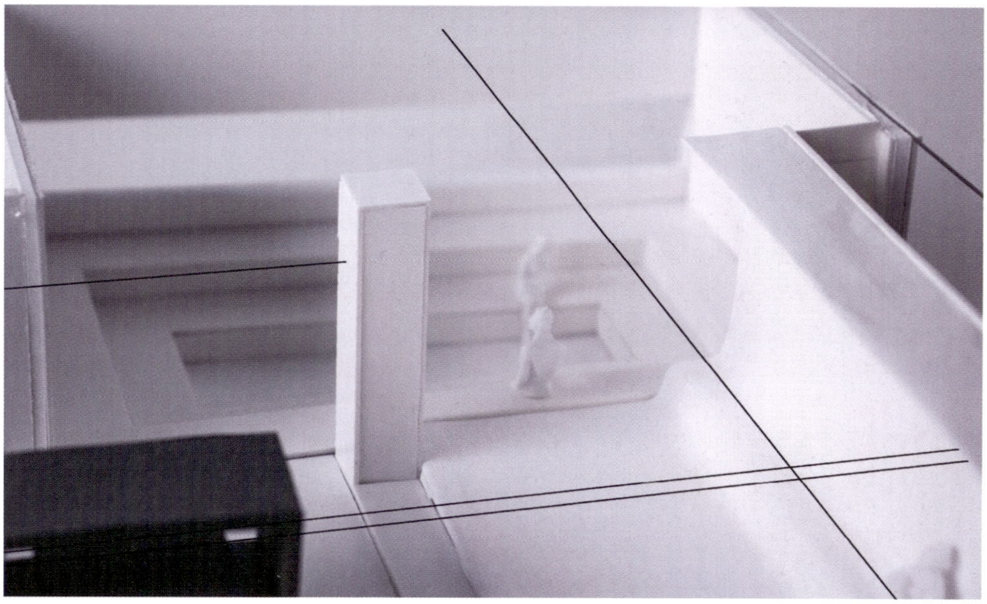

이소민 I Lee, So Min 음효원 I Eum, Hyo Won

BREATH

BACKGROUND

사람들은 각자 일하는 속도가 다르다. 또한 일을 할 때에 집중되는 공간도 다르다. 하지만 오늘날의 회사 업무 공간은 모두 같은 공간으로 이루어져 있어 지루하고 반복되는 회사 생활을 자아내며 각자의 업무 속도 차이에 의해 부딪히는 일이 생긴다. 이러한 문제점에서 조금이나마 탈피할 수 있는 회사를 설계하였다.

Each person has their own work pace, and also own place where they are able to maximize their work-force. But classic offices in companies and business-es of today do not consider about these differences, while the differences bring out all major issues that could be happened in their office. So, I have designed a business avoiding these classical issues.

BACKGROUND

서울 특별시 구로구 중앙로 6길 16

구일역에 근접하게 위치해 있어 교통이 편리하다. 주변은 주로 아파트와 대학교 기숙사로 이루어져 있으며 동네 주민들이 접근하기에 용이하여 주민들과 어우러질 수 있는 회사를 설계하고자 하였다.

Seoul, Korea
The office is nearby Gu-il subway station, which will comfort customers. Also, it is surrounded by several apartment complexes and dormitories for college students. I could, therefore, design a business which can be in harmony with those people.

RUNCITY

런시티는 러닝 & 라이프 스타일 미디어 플랫폼이다. 즐겁게 달리고 좋은 정보를 나누고 세련된 감각을 나누고 싶은 러너들을 위해 태어났으며, 러닝의 행복과 가치, 진정성을 전하는 잡지이다.

Run City is a media employing Running and Life style flat-form. It has been designed for runners desiring joyful running, sharing information, and sharing fine senses. It is a magazine communicating about a value and happiness of running.

CONCEPT

달리기를 할 때에는 많은 길을 만나게 된다. 평탄한 길, 오르막길, 산 길 등등의 장소와 사람에 따라 속도는 달라지기 마련이다. 이러한 개념과 달리기 할 때의 일정한 호흡, 역동적인 러닝의 이미지를 업무공간에 적용 할 계획이다.

When we are running, we meet a whole different kind of roads. It could be flat, inclind, mixed, and so on. This makes the difference in speed of each person. So, we are planning to apply the concept of running facing diffrent roads and the image of running facing different roads and the image of running not only being dynamic but also being steady.

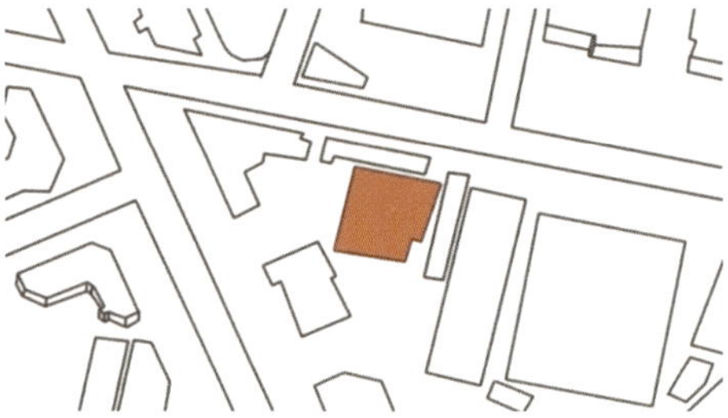

PROCESS1 경사로 A SLOPE WAY

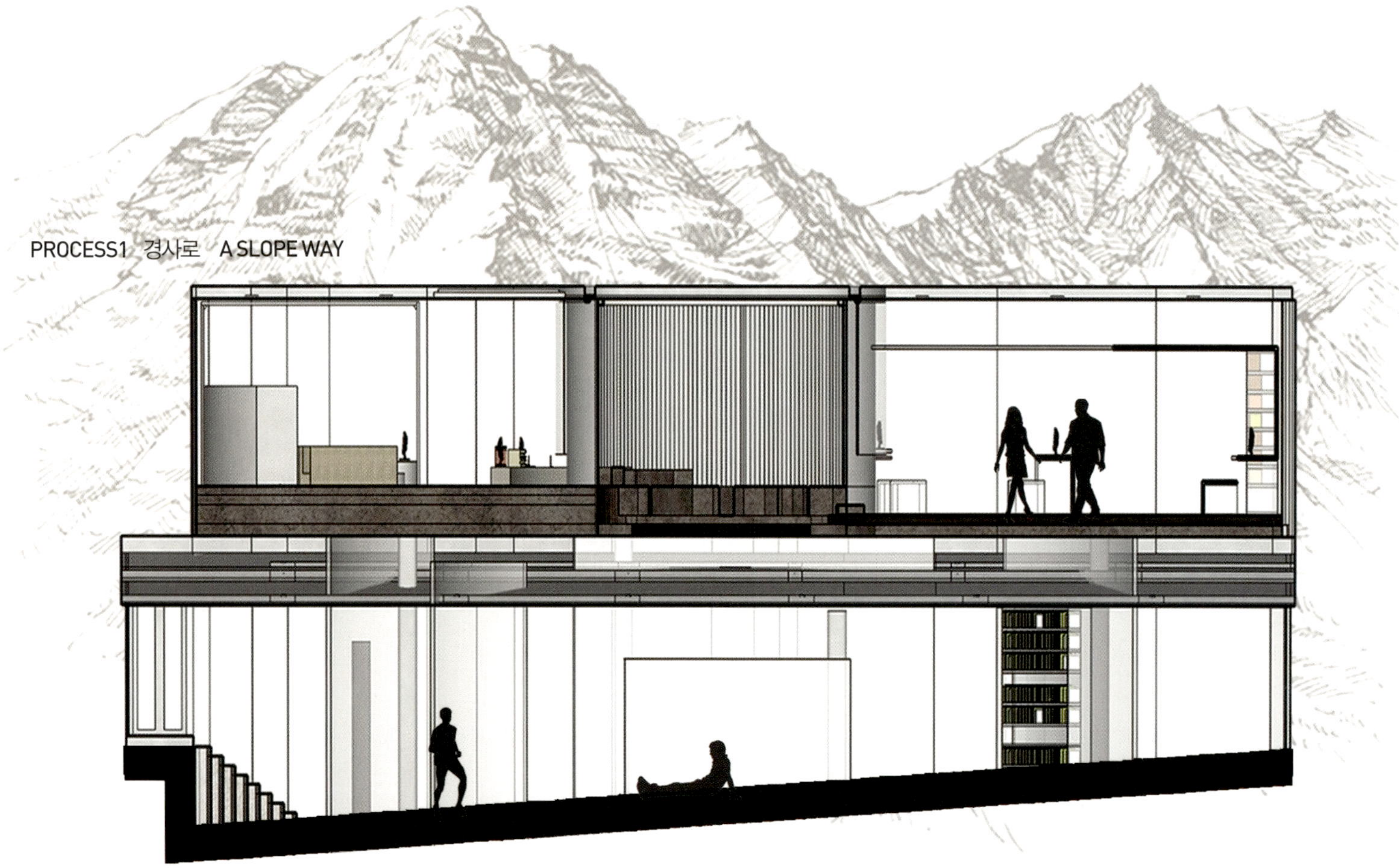

PROCESS2 공간의 높낮이 HEIGHT OF SPACE

UPHILL ROAD
자유로운 업무공간 Free work space

DOWN HILL ROAD
집중이 필요한 공간 A space needed concentration

FLAT ROAD
일반 업무 공간 General work space

PROCESS3 트랙 TRACK

공간 구획을 위하여 하나의 트렉을 반으로 분리하였다. 그 각각의 반원들은 다른 공간으로 사용된다.

For division of space, one track is divided into two half-circle. Each half is used as different space.

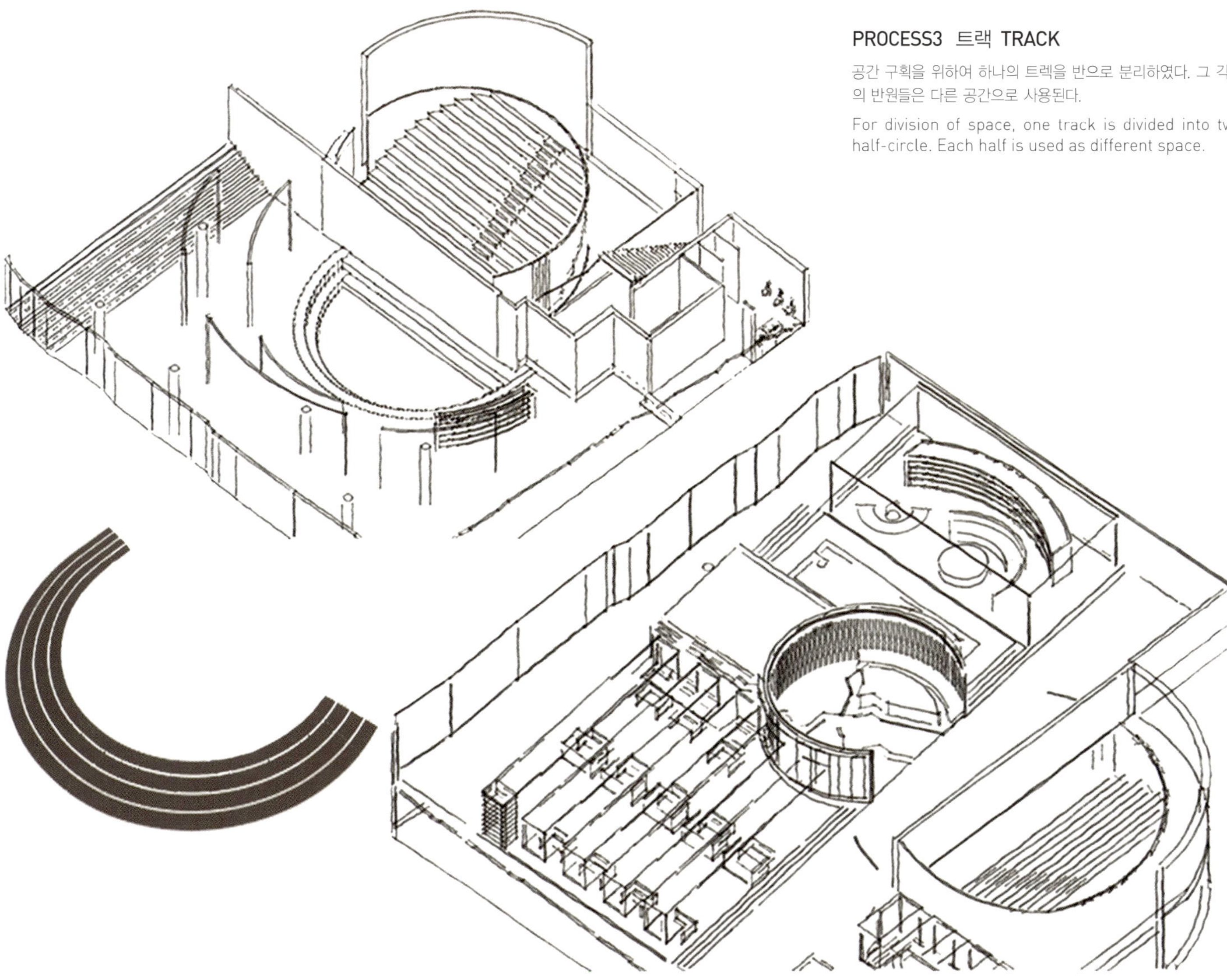

PROGRAM - FURNITURE

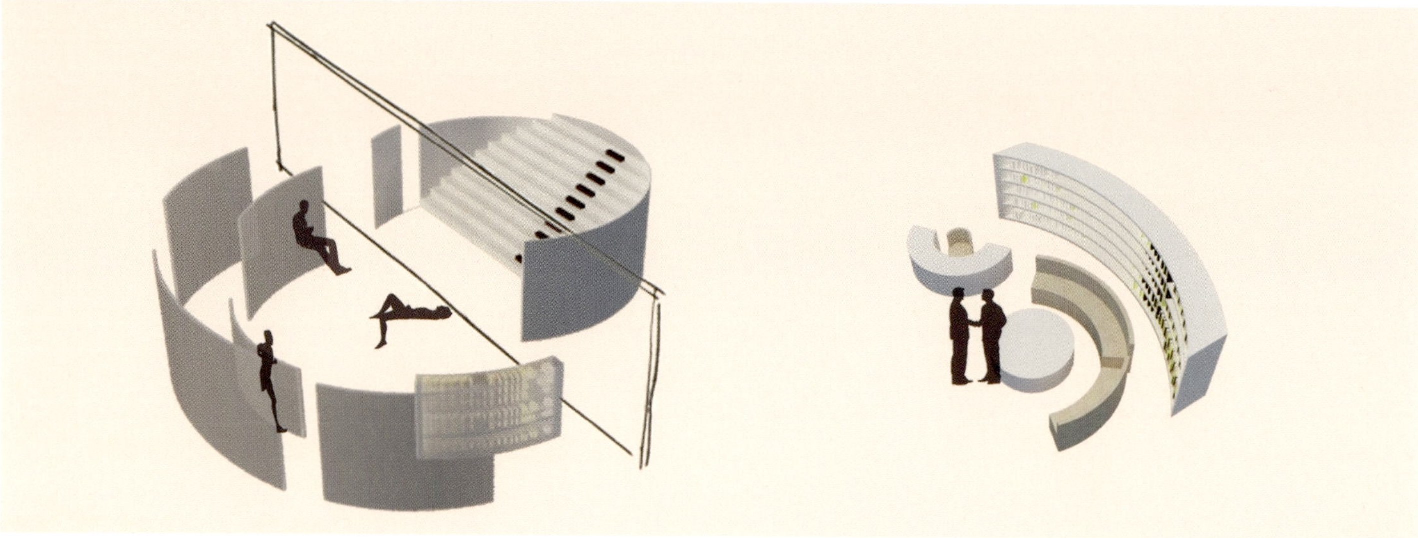

PLAN-1F

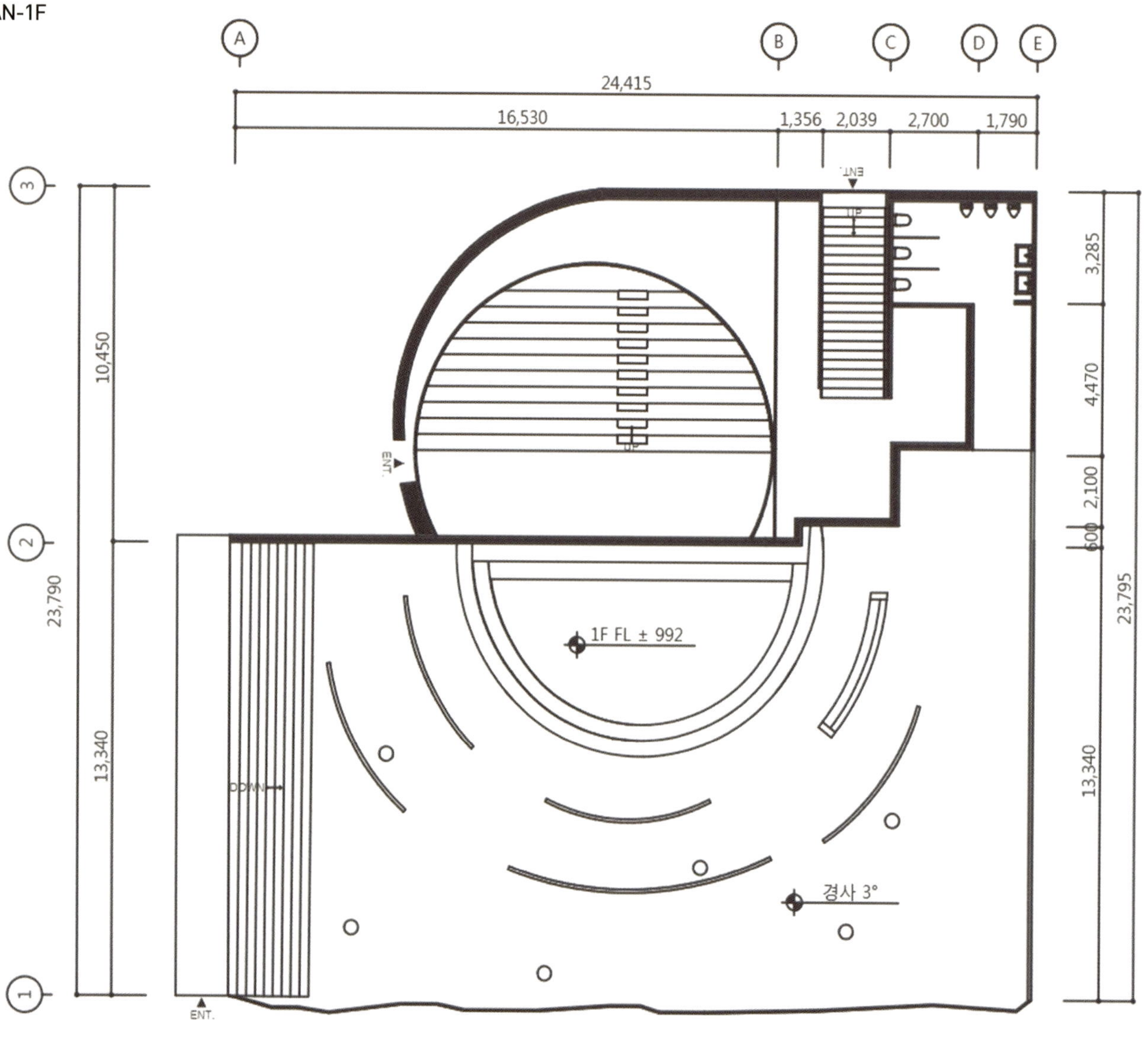

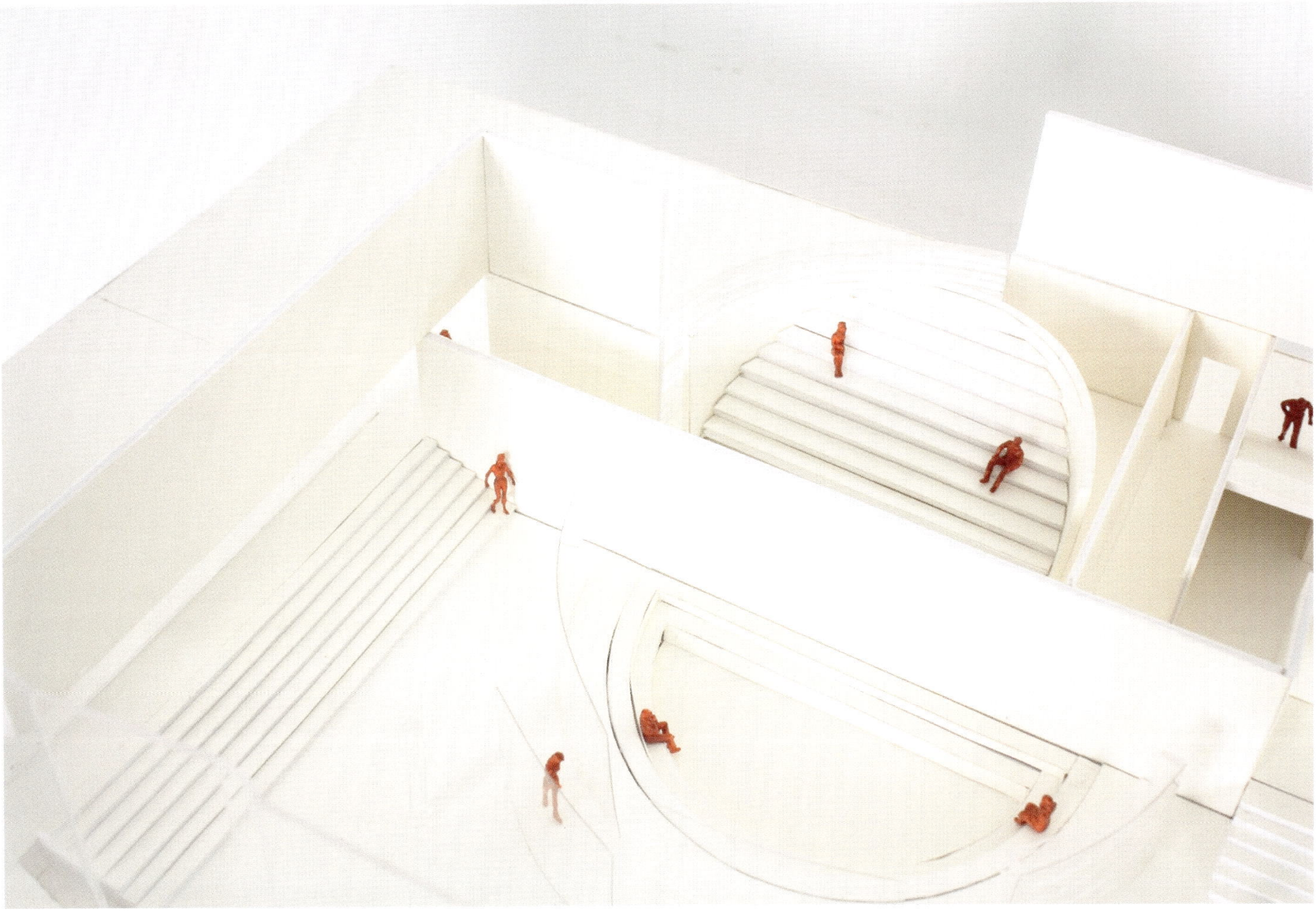

PROGRAM - 1F

달리기 트렉을 적용시켜 소통의 공간을 디자인하였다. 가운데 벽을 기준으로 원을 반으로 잘라 살짝 비틀어 배치를 하였다.

By employing running track, we could have designed a space of communication. We arranged by cutting a circle in a half and twisting it slightly based on the center wall.

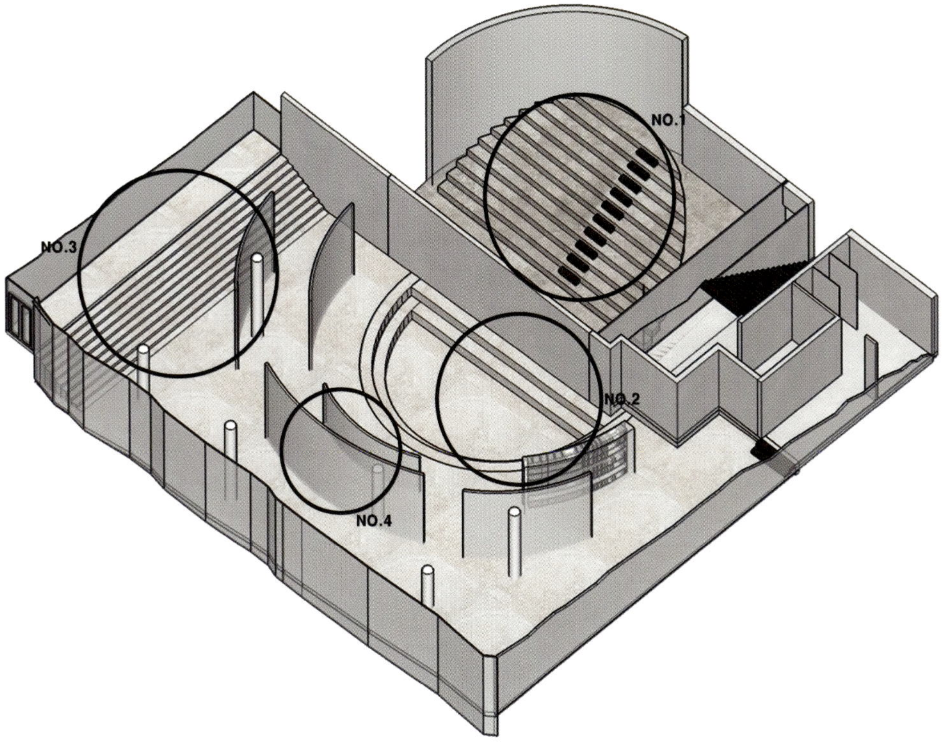

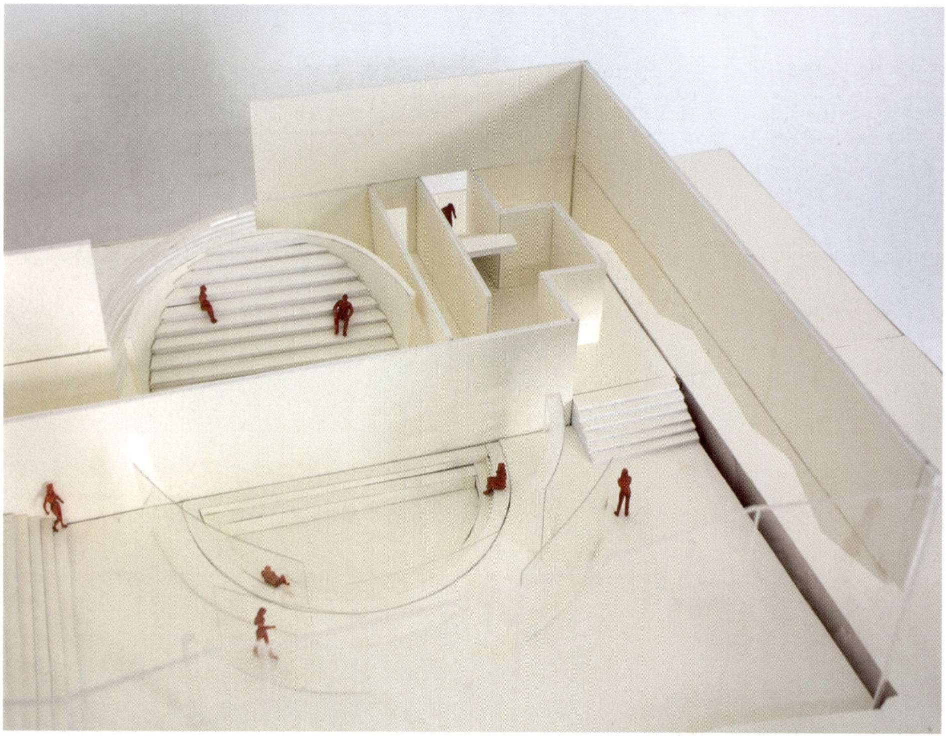

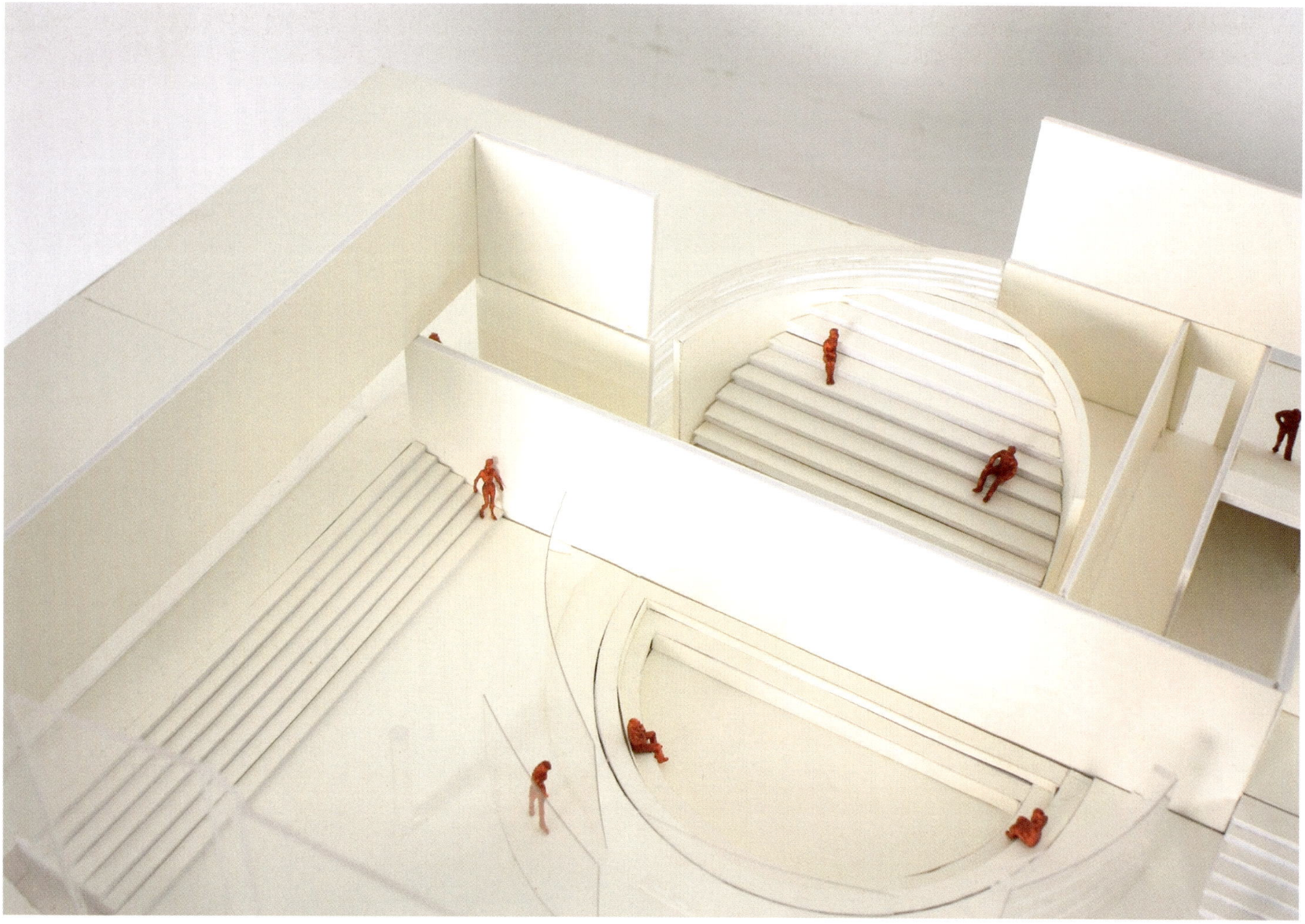

Communication space

Communication space

Conference space

Communication space

Facade

PROGRAM - FURNITURE

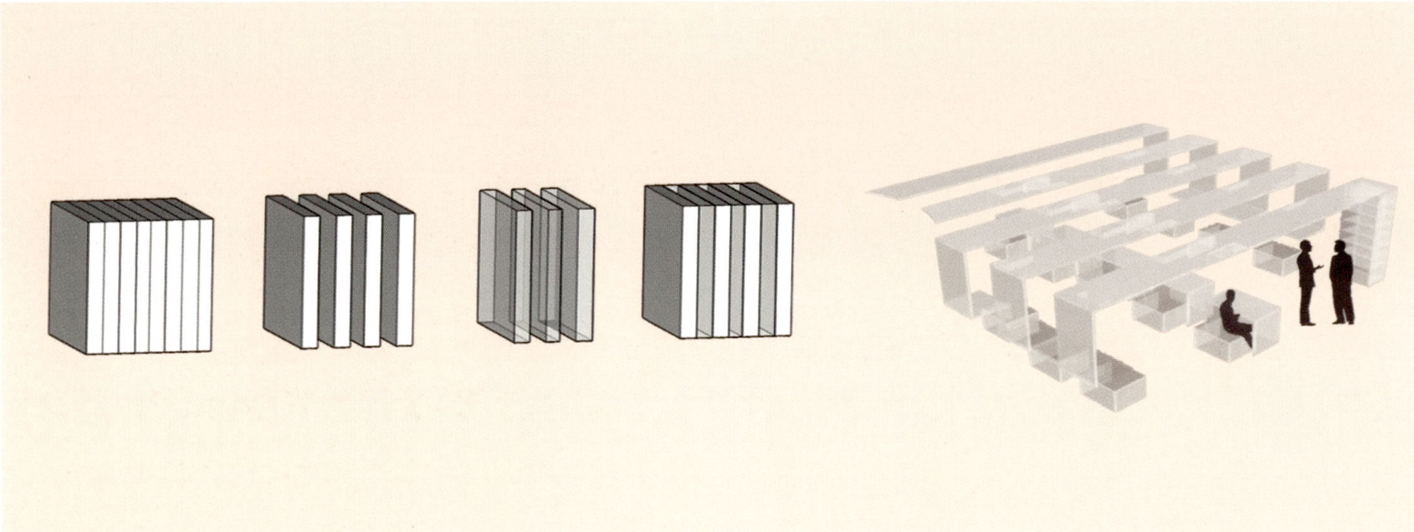

PLAN-2F

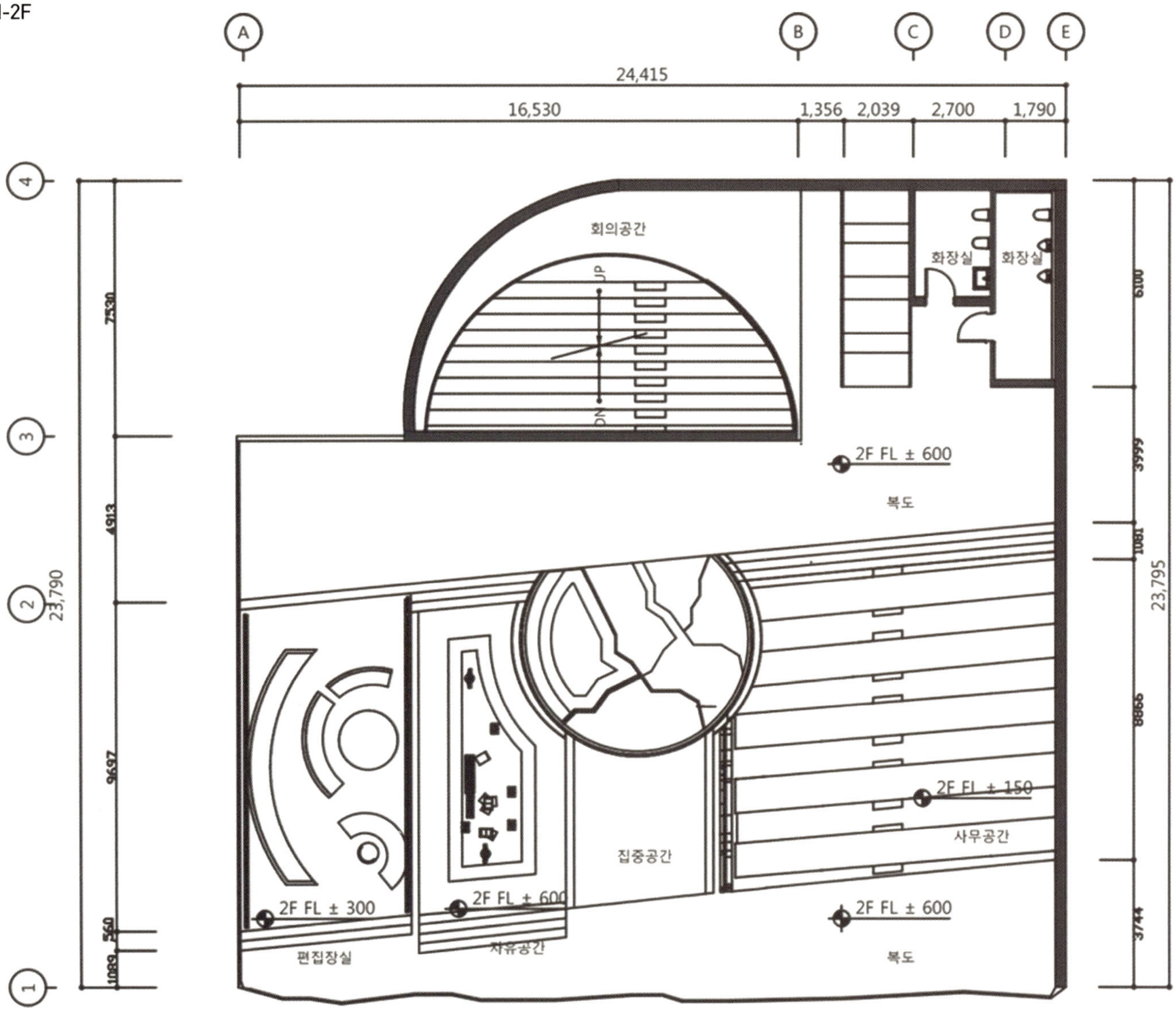

Hallway

Data room

Data room

Editorial office

Office space

Entrance

PROGRAM - 2F

달리기 할 때의 호흡의 일정한 간격과 틈에 집중된 공간이다. 업무공간의 집중도에 따라 단 차이와 재질의 무게감을 다르게 주었다.

It's a place focused on steady breath interval and its gap while running. We gave differences based on each level of focuses.

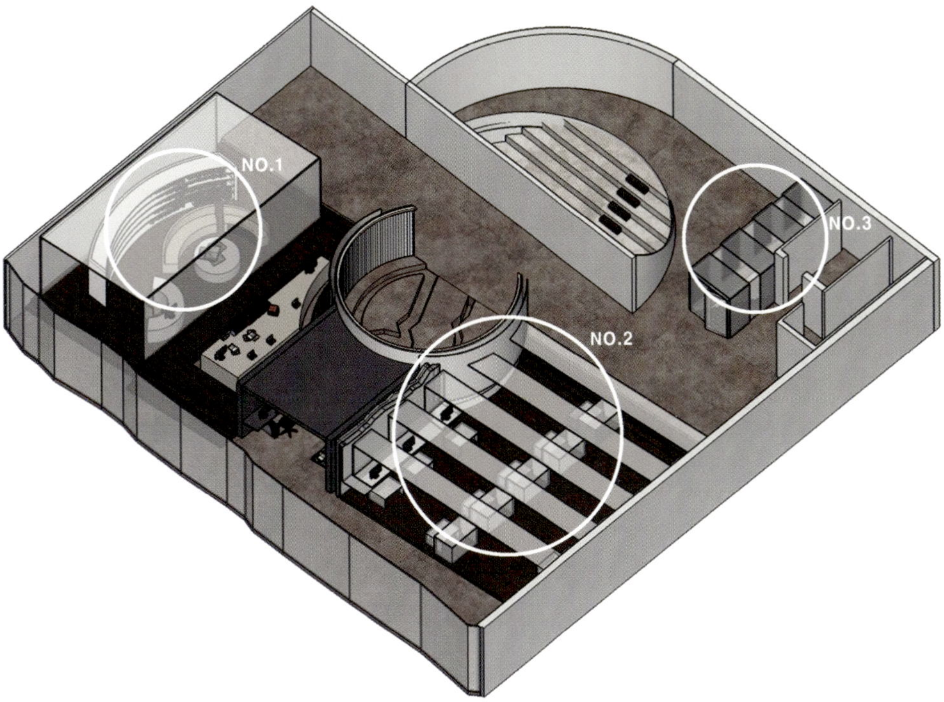

맘앤앙팡은 디자인하우스 M+에 속해 있는 육아잡지이다. 디자인하우스 M+의 사람을 중심에 두어 일상생활을 더 풍요롭게 한다는 회사모토를 배경으로 아이의 일상과 건강정보를 전달하는 잡지사이다. 아이들의 인생을 좌우하는 가장 중요한 시기는 태어나서 세살이 될 때까지라고한다. 우리는 이러한 아이들의 골든타임을 지각과 감각으로 일깨울 수 있게 맘앤앙팡 잡지사로 선택하여 공간을 구성하였다.

Mom&Enfant is one of the baby magazine in Design House M+. Based on the company motto of flourishing human life with man in the center, this delivers information on daily life and health of children. The most important period of deciding life of children is until their age of three. We chose Mom&Enfant magazine to perceive such golden time and toMom&Enfant is one of the baby magazine in Design House M+. Based on the company motto of flourishing human life with man in the center, this delivers information on daily life and health of children. The most important period of deciding life of children is until their age of three. We chose Mon&Enfant magazine to perceive such golden time and to provoke sensation to form this space.

SITE 서울특별시 구로구 중앙로 6길 16 (고척동 78-2)
면적 : 670m2 (202평)

16, Jungang-ro 6-gil, Guro-gu, Seoul, Republic of Korea
AREA : 670m2 (202PY)

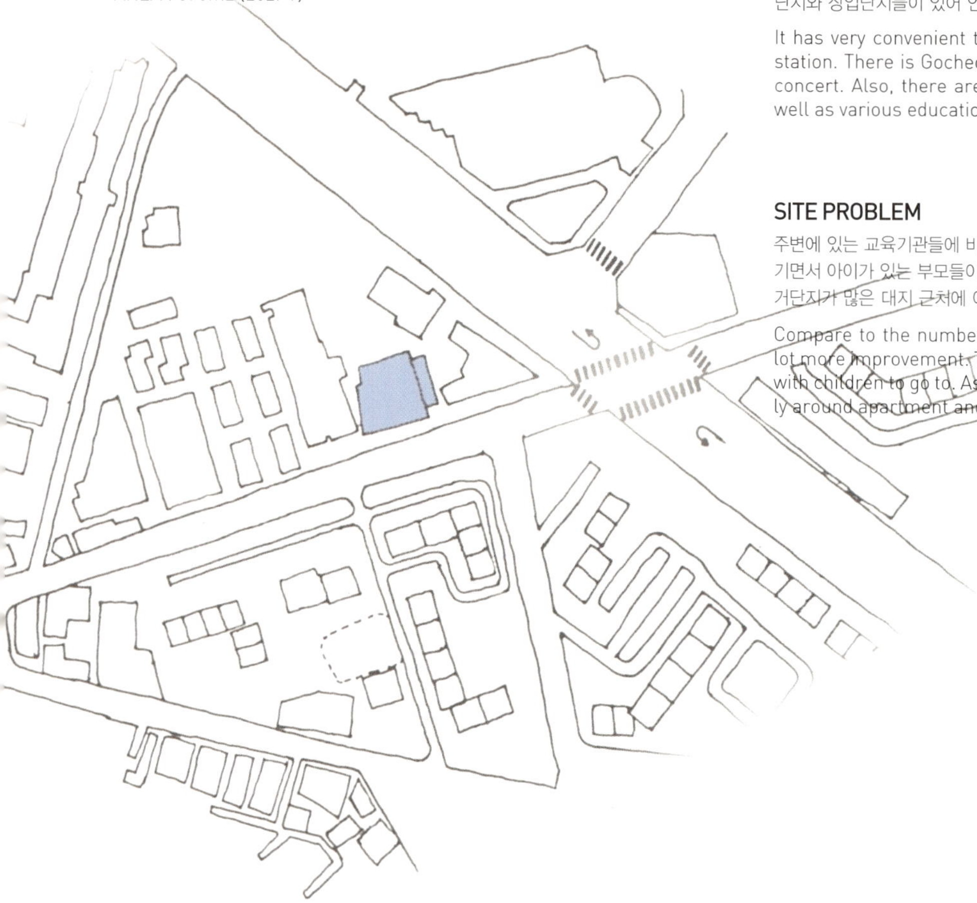

SITE ANALSIS

구일역 1번 출구 인근에 위치하고있으며 버스정류장이 가까워 교통이 편리하다. 주변에는 대표적으로 야구경기나 K-POP 콘서트 등을 개최하는 고척스카이돔이 있다. 또한, 대지근처에 많은 주거단지와 상업단지들이 있어 안전적이고, 초등학교부터 대학교까지 많은 교육기관이 밀집되어있다.

It has very convenient transportation, located near exit 1 of Guil station and bus station. There is Gocheok Sky Dome nearby which holds baseball game or K-Pop concert. Also, there are many resident and industrial areas, making it stable as well as various educational organizations like elementary school to universities.

SITE PROBLEM

주변에 있는 교육기관들에 비해 육아관련 시설은 많이 부족했다. 최근에 "NO KID ZONE" 까지 생기면서 아이가 있는 부모들이 갈 곳은 점점 줄어들고 있다. 이러한 이유로 아파트와 주택, 빌라등 주거단지가 많은 대지 근처에 아이와 함께 할 수 있는 공용공간이 필요할 것이다.

Compare to the number of educational organizations, facilities for children need lot more improvement. There are even "no kid zone" limiting the places for parents with children to go to. As this is the case, more spaces for children to play especially around apartment and main resident areas are required.

CORE KEYWORD

지각(perception), 감각(sensation)

지각과 감각은 혼합되어 하나의 연속적인 과정을 이룬다. 환경에 있는 물리적인 에너지를 받아들여 신경정보로 전환하는 과정을 감각이라 하고 이러한 감각을 선택하고 조직화하여 해석하는 과정을 지각이라한다.

Perception and sensation are combined into one to form a continuous process. The process of turning received physical energy in environment into nerve information is called sensation and the process of selecting such sensation, organizing, and interpreting is called perception.

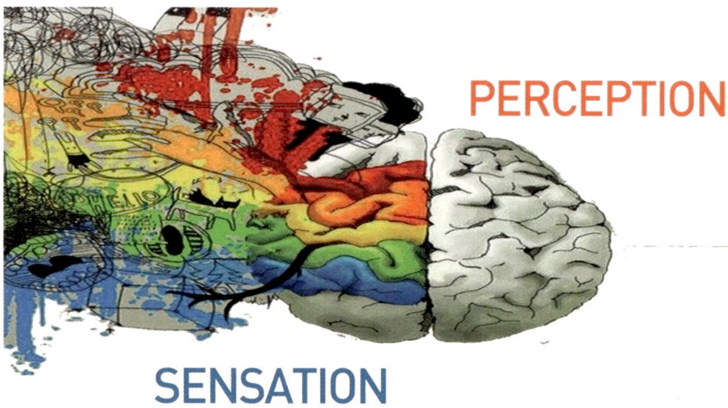

DESIGN KEYWORD

어른의 지각 - 현실
정제된 수평선 같은 일상속에서 보이는 그대로 해석을 하고 받아들인다.

아이의 지각 - 비현실
같은것을 보더라도 보이는 그대로의 해석이 아닌 다른것을 상상한다.

어른과 아이는 같은 것을 바라 보지만 다른 것을 상상한다.
그릇된 감각이 아닌 아이와 어른의 시점에 중심을 두었다.

Perception of adults-realistic
They interpret and accept what they see in daily life like refined.

Perception of child-unrealitic
Imagine something different instead of interpreting as they see.

Adults and children see the same but make different imagination.
It is focused on the view of child and adults rather than the on the wrong sensation.

CONCEPT - optical illusions

지각이란 감각을 통한 인지.
오감 중 시각과 촉각, 공감각에의해 공간을 감지하고 형태와 부피, 원근감을 인지한다. 공간에 대한 개념을 외적, 내적으로 보고 느끼는 감각. 이러한 감각을 어른과 아이의 시점에서 바라보았다. 어른의 지각인 보이는 그대로 받아들이는 것과 아이들의 지각인 같은 것을보고 다른 것을 상상하는것, 그리고 형태와 기능의 용도를 다르게 사용하는 행동을 통하여 어른과 아이의 시점으로 착시 (시각에 관해서 생기는 착각)의 개념을 나타냈다.

Perception means recognition through sense.
Out of five senses we have, we perceive space by sense of vision, touch, and space as well as its shape, volume, and distance. This internal and external sense on the space is viewed from the perspective of adult and child. The concept of optical illusion (illusion occurring from the view point) is developed from adults accepting what they see and child making different imagination from the same objective as well as different behaviors of using its shape and function.

DESIGN PROCESS

착시라는 것은 인위적이기도하고 자연스럽기도하며 호기심을 불러일으킨다. 또한 기하학적이기도 하면서 유기적이기도한 매우 매력적인 단어이다.
이러한 착시의 의미를 평면 계획에 적용시켜 일반적인 사각형의 평면을 분활하여 L과 사각형의 모습으로 평면을 계획하였다.

Illusion may sound very artificial or natural but whichever way, it provokes curiosity. Also, it is geometric and organic, which makes it very attractive. Such meaning of illusion is applied to flat design, regular square flat is divided into L shape and rectangular shape.

L과 사각형의 평면방식은 여러의미로 해석할 수 있다. 평면상에서 자연스러운 동선을 유도 할 수 있으며, 공간속에선 여러 매력적인 형태로 접근할 수 있다.

This L and rectangular shape can be translated into different meaning.
It may lead to natural pathway on the flat design and inside the space, this can be approached in various attractive ways.

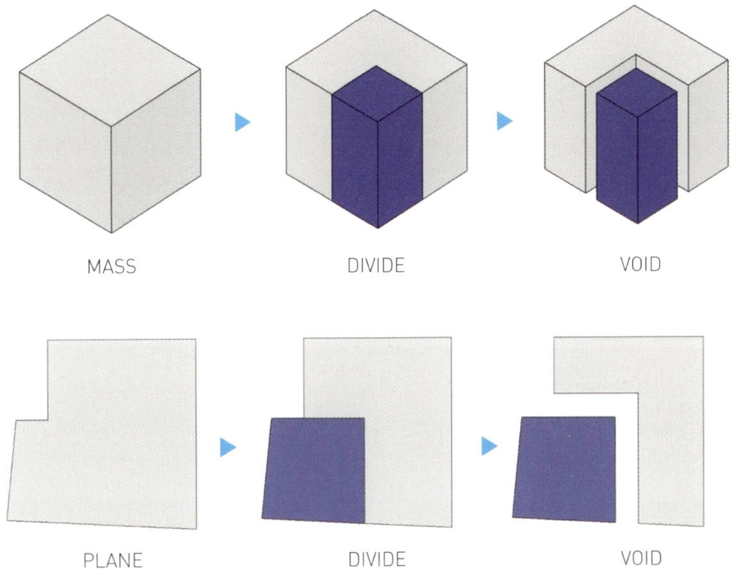

DESIGN MOTIVE

착시의 종류중 4가지 기법을 사용하여 공간을 구성하였다.

The space is organized by using 4 illusion techniques.

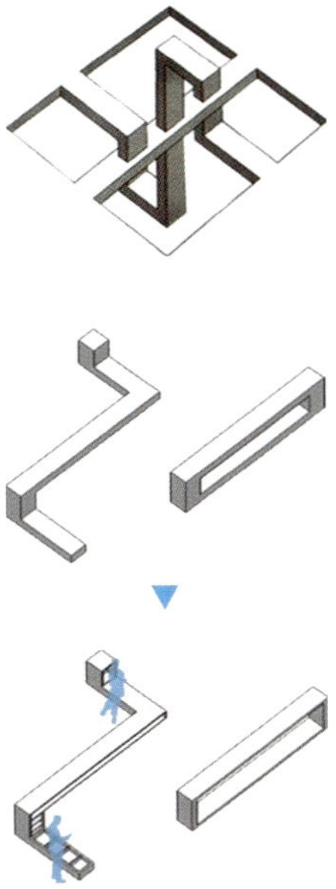
기하학T적 착시
(Geometric illusion)

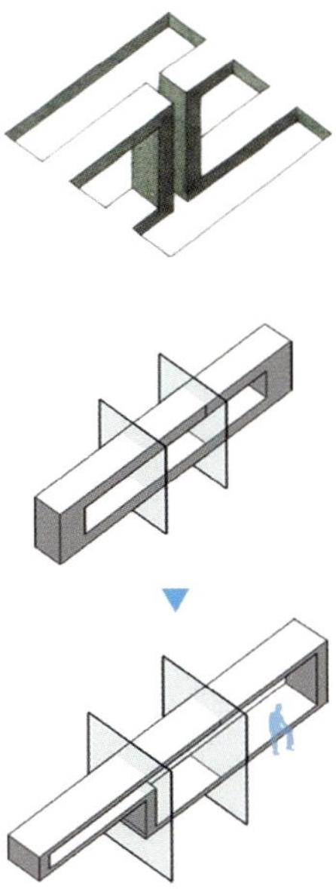
다의도형 착시
(ambiguous figure illusion)

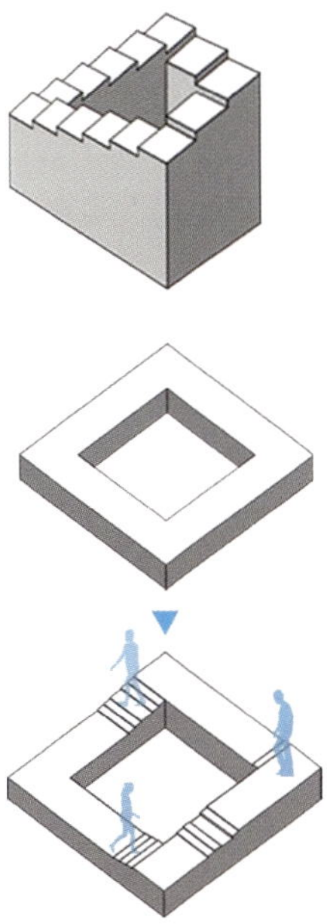
역리도형 착시
(paradoxical figure illusion)

공간 착시
(space illusion)

MOTIVE SPACE

기하학적 착시 : 물체의 크기나 방향, 각도, 모양등 평면 도형의 성질이 실제와 다르게 보이는것
Geometric illusion: having a different view of flat figure from its reality such as size, direction, angle, and shape.

기하학적 착시의 각도와 모양, 평면등 도형의 성질을 이용하여 책꽂이의 형태로 변형하였다. 수직으로된 책꽂이에서 책을 꺼내어 수평으로된 바닥이나 의자에 앉는것이 일반적이지만 이러한 개념을 수평적 요소에서 책을 꺼내어 수직적요소에 앉는 개념으로 변경하였다.

1F

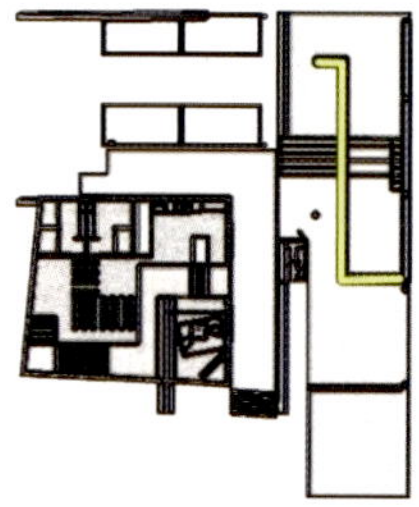

Using features of the figure, its shape is changed to bookshelf based on geometric illusion of angle, shape, and flatness. It is normal to take a book out from vertically placed bookshelf and sit on a horizontal floor or chair for reading. However, it changed to vice versa, taking book out from horizontal bookshelf and reading on vertical object.

다의도형 착시 : 동일한 도형이 2종류 이상의 형태로 보이는 것
Ambiguous figure illusion: having view of more than 2 shapes of the same product.

다의적 착시의 특징을 기능적으로 의미를 두어 조형물로 전환시켰다. 다의적 착시의 도형을 모티브로 조형물의 외관을 디자인하였고, 기능적으로는 천장강화유리의 햇빛을 일부 가려줄 수 있는 그늘역활, 그리고 계단역활과 의자역활로 구성하여 하나의 조형물에 3가지 기능을 넣었다.

2F

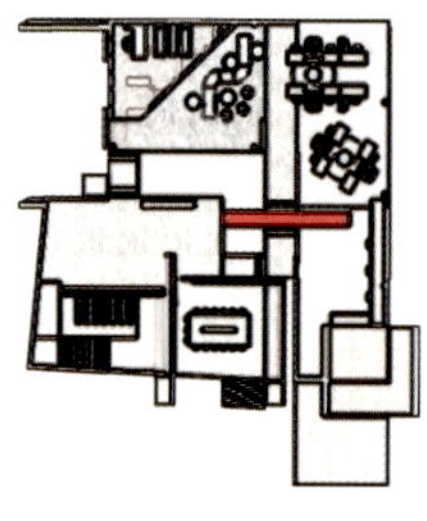

Based on the functional meaning of this ambiguous illusion, it is changed to another object. The external side is designed based on the motive of ambiguous illusion of a figure. Also, 3 more functions are added into one object; function of shadow for covering sunlight, function of stair, and function of chair.

역리도형 착시 : 전체적으로는 모순적이지만, 부분적으로는 가능한 도형
Paradoxical figure illusion: paradoxical in general but possible in part.

역리도형의 모순적인 특징을 담아 복도 계단의 단차이를 표현하였다. 역리도형착시의 대표적인 예시인 상대성 표현효과를 모티브로 전반적으로 모순적이지만, 부분적으로는 가능한 형태의 복도 계단을 완성하였다.

2F

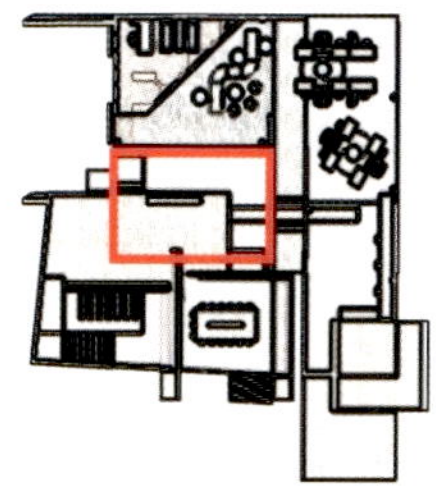

Based on paradoxical features of the figure, height difference of stair is represented. With the motive obtained from one good example of paradoxical figure illusion which is relative expression effect, hallway stair with paradox in general but possible in part is completed.

공간 착시 : 착시의 시각 자극을 인지하는 과정에서 원래의 사물에 대한 시각적인 착각을 일으키는 개념을 공간에 대입시킨 착시이다.
Space illusion: illusion combining concept of causing visual illusion on the object in the process of recognizing visual stimulation into the space.

기본적인 착시개념을 공간에 접목시켜 표현하여 벽과 문의 시각적 느낌을 하나로 통일시켰다. 이 외에도 공간착시를 내부와 외부에 적용시켜 실내와 실외의 개념을 재질의 표현으로 공간적 착시의 효과를 주었다.

1F

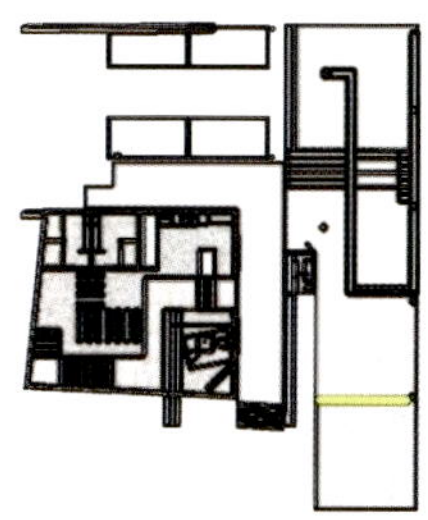

The basic illusion concept is combined into the space to make wall and door into one visual feeling. Along with this, space illusion is applied to internal and external parts to deliver space illusion effect based of inside and outside based on difference in materials.

AXONO METRIC

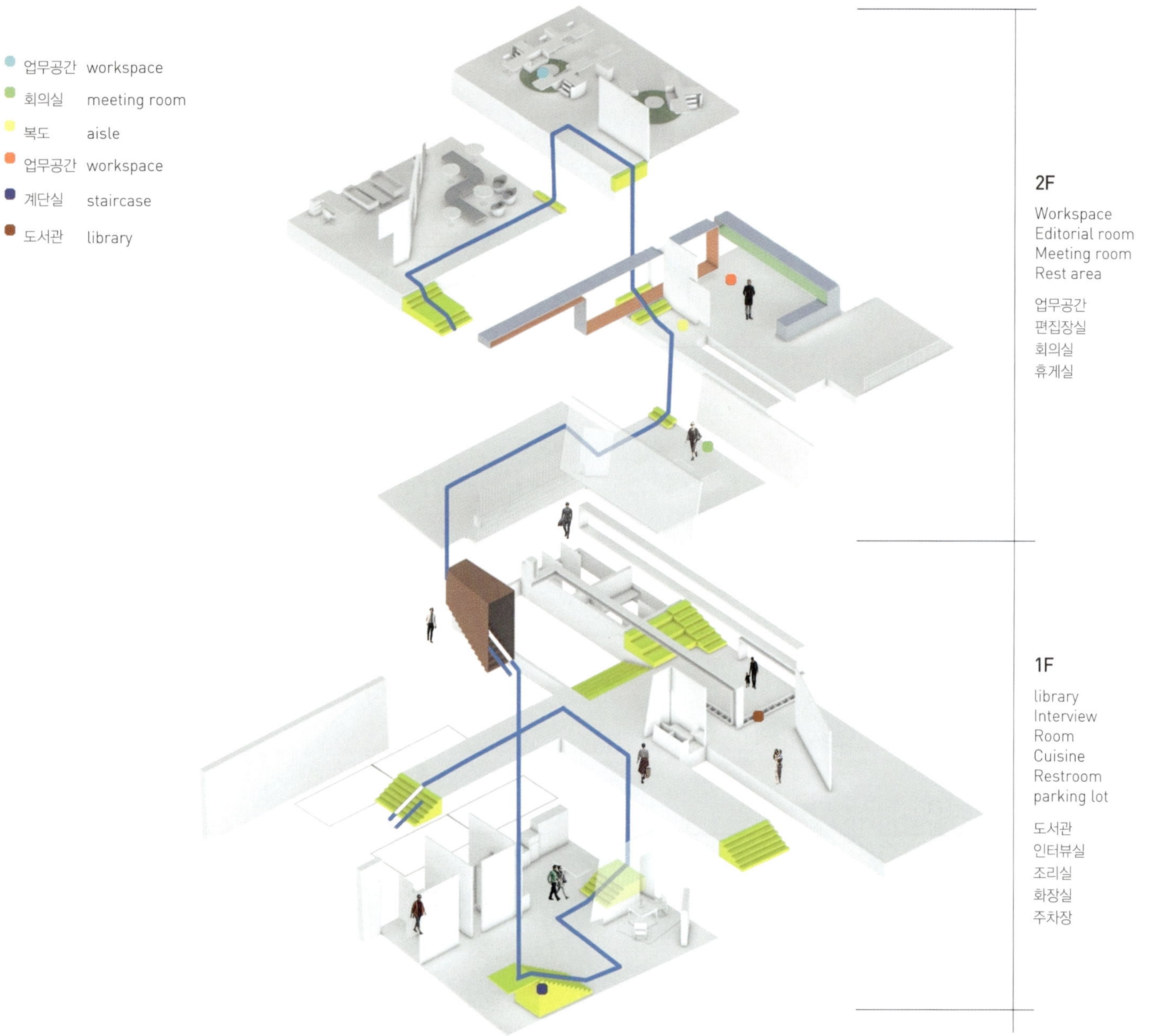

SECTION VIEW

2F - OFFICE

업무용 책상과 쉴 수 있는 소파를 붙여두어 업무와 휴게가 공존 할 수 있는 공간이다.

This is a space where both working and relaxing co-exist by placing desk for working and sofa for relaxing together.

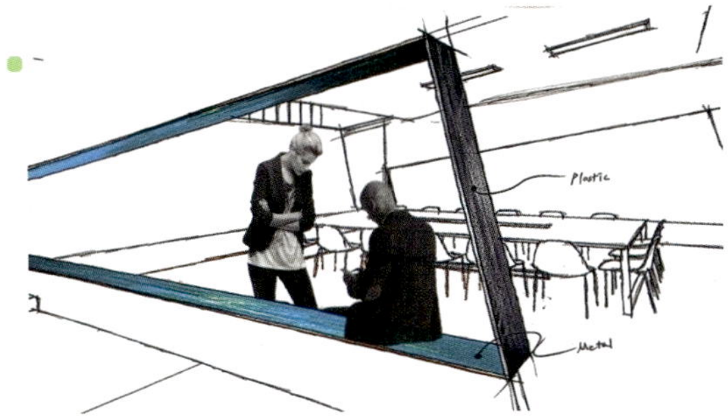

회의실의 창을 통 해 앉을 수 있으며 회의실에 답답한 느낌을 없애고 개방감을 주었다.

One can seat through the meeting room window and it delivers openness, removing closed feeling of the typical meeting room.

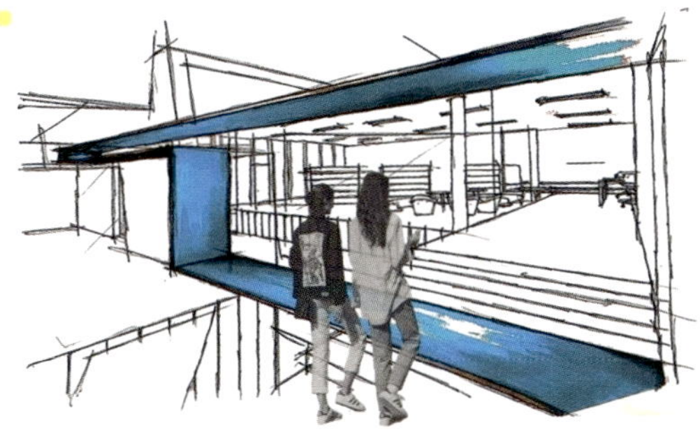

2층에 메인 프레임을 중심에 두어 여러가지 기능을 할 수 있게 하였다.

By placing the main frame in the center of the second floor, we enabled it to serve various purposes.

2층 메인 프레임을 업무공간에선 의자로서의 기능으로 사용한다.

The main frame of the second floor functions as a chair in the work space.

1F - LIBRARY

1층 공용공간에서 2층으로 올라가는 계단으로 공간적 착시효과를 볼 수 있다.

The stairs leading from the communal area on the first floor to the second floor creates a spacial optical illusion.

1층의 소통공간으로 작은 도서관을 연출하였다.

We have created a small library as a communicative space on the first floor.

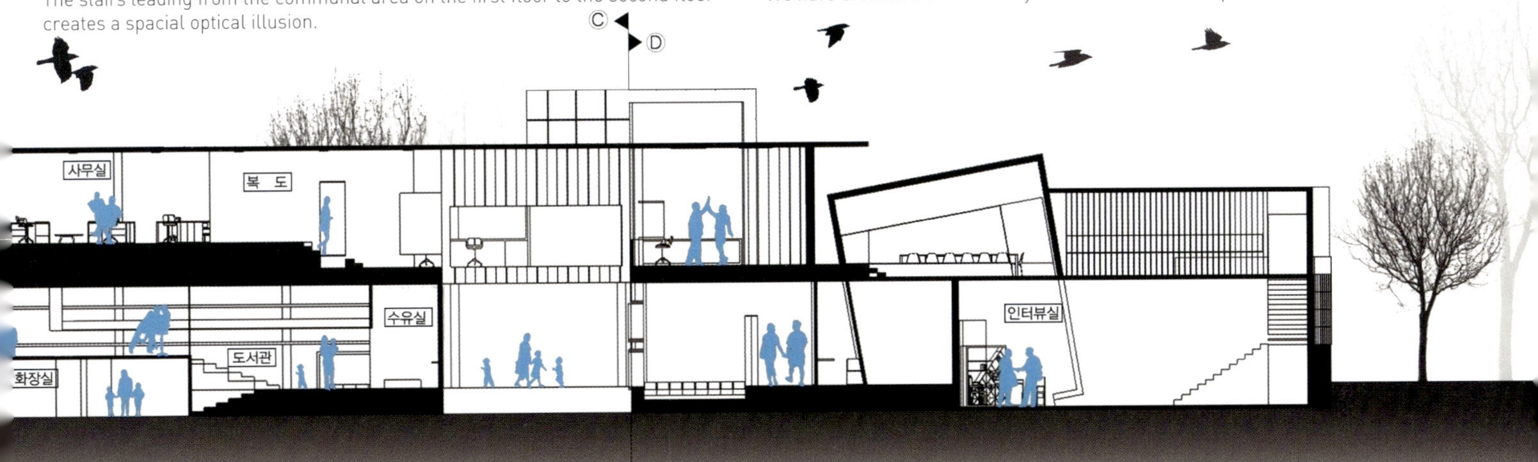

FLOOR PLAN

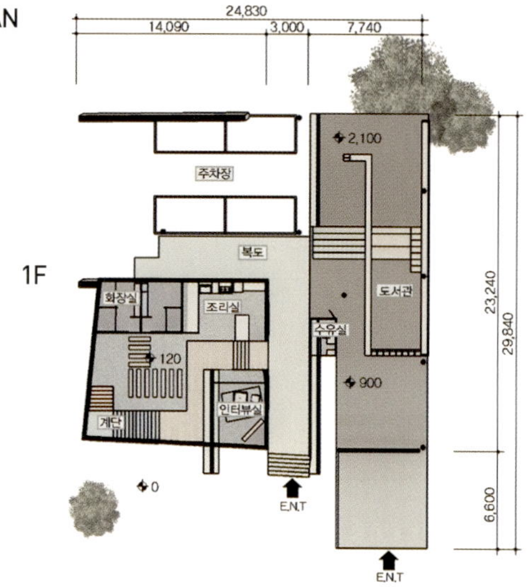

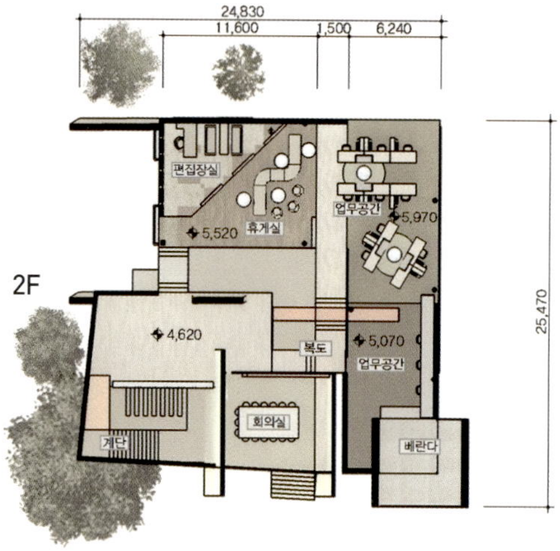

DETAIL VIEW

2F 회의실

비틀린 공간을 회의실로 정하였다. 직원들과 편집장실의 동선을 고려하여 위치를 선정하였고, 전체 사원이 들어갈 수 있게 테이블과 의자를 배치하였다. 또한 실내이지만 실외의 느낌을 주고 싶어 잔디카페트를 사용하였고, 밖과의 경계를 유리를 사용하여 공간을 구분지었다. 복도쪽에서도 밖을 바라 볼 수 있게 안쪽에도 같은 창을 만들어 원근의 느낌을 주었다.

The twisted space is set as a meeting room. It is selected in consideration to pathway of staff and chief editor. Also, it is equipped with table and chair for all staff. What is more, to give outdoor feeling to the inside space, grass carpet is used and boundary between in and outside is separated by using glass. To allow visible view of outside from aisle area, similar windows are made inside to offer feeling of distance.

1F 복도

복도는 주차공간에서 들어올 수 있는 통로이다. 주차를 한뒤 들어와 업무공간으로 올라 갈 수 있는 유일한 동선이다.

The aisle is a pathway entering the place from parking lot. This is the only entrance from parking lot to working space.

1F 주차장

건물의 전면 T부분에는 아이들이 통행하여 뒷쪽 공간에 주차장을 만들었다.
주차공간에는 주민들과의 공용공간과는 분리되어 사원들만의 공간으로 만들었다.

As kids walk around the front side of the building, parking lot is moved to the back side. In the parking lot area, place for staff is separated from the community place of residents.

2F 편집장실

편집장실은 각 팀별 팀장들과 회의를 진행 할 수 있게 회의테이블을 두었다. 벽을 돌과 잔디로 마감하여 색다른 외부적 느낌을 주었고 바닥은 화이트오크 재질을 주어 외부와 내부의 경계가 있게 표현하였다.

Inside the office of chief editor, there is meeting table so that team leaders can have a meeting together. The wall part is made up of stone and grass to deliver exotic external feeling and floor is made up of white oak material to show boundary between inside and outside.

업무공간

업무공간에 내부와 외부의 반전을 표현하기 위해서 외부재질을 마감재로 사용하였다. 벽과 바닥의 마감재를 활용하여 외부적인 느낌을 표현하였는데 벽면에는 징크마감을 하여 외부적인 느낌을 주었고, 바닥에는 외부재질과 비슷한 형태의 카페트를 사용하여 표현하였다. 또한 책상의기능을 책상뿐만 아니라 소파의 기능도 하게하여 업무공간내에 작은 회의실을 만들었다.

To show total difference between inside and outside of working space, external material is used as finishing material. The finishing material of wall and floor is used to represent exotic feeling, zinc material is used on the wall for this. For floor, carpet material similar to external material is used. Furthermore, desk is changed to multi-function, not only as a desk but as sofa, making a small meeting room inside the working space.

휴게실

휴게공간은 외부공간인 복도와 바로 이어져 있어, 반대로 내부적인 느낌을 강조하였다. 화이트오크재질을 사용하여 바닥에 내부의 따듯한 느낌을 주었고 가구와 테이블을 배치하였다. 편집장실과 업무공간 사이에 위치해있으며 개방형 공간이라 쾌적함을 준다.

Resting place is directly connected with aisle. As it is located outside, it is designed to give a very strong feeling of inside. White oak material is used on the floor to make warm feeling of inside and furniture and table are allocated. It is located between office of chief editor and working space, delivering fresh feeling with its openness.

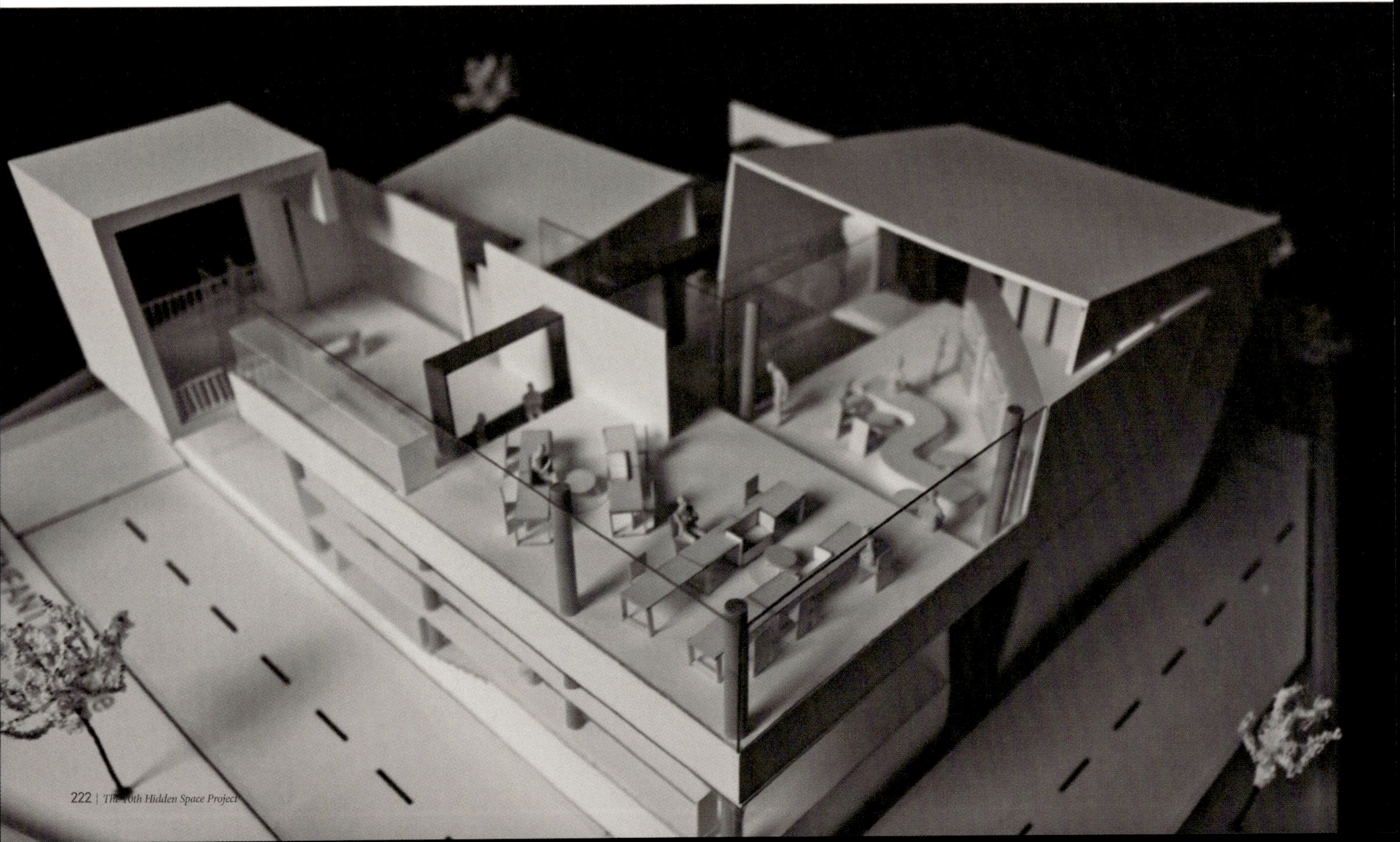

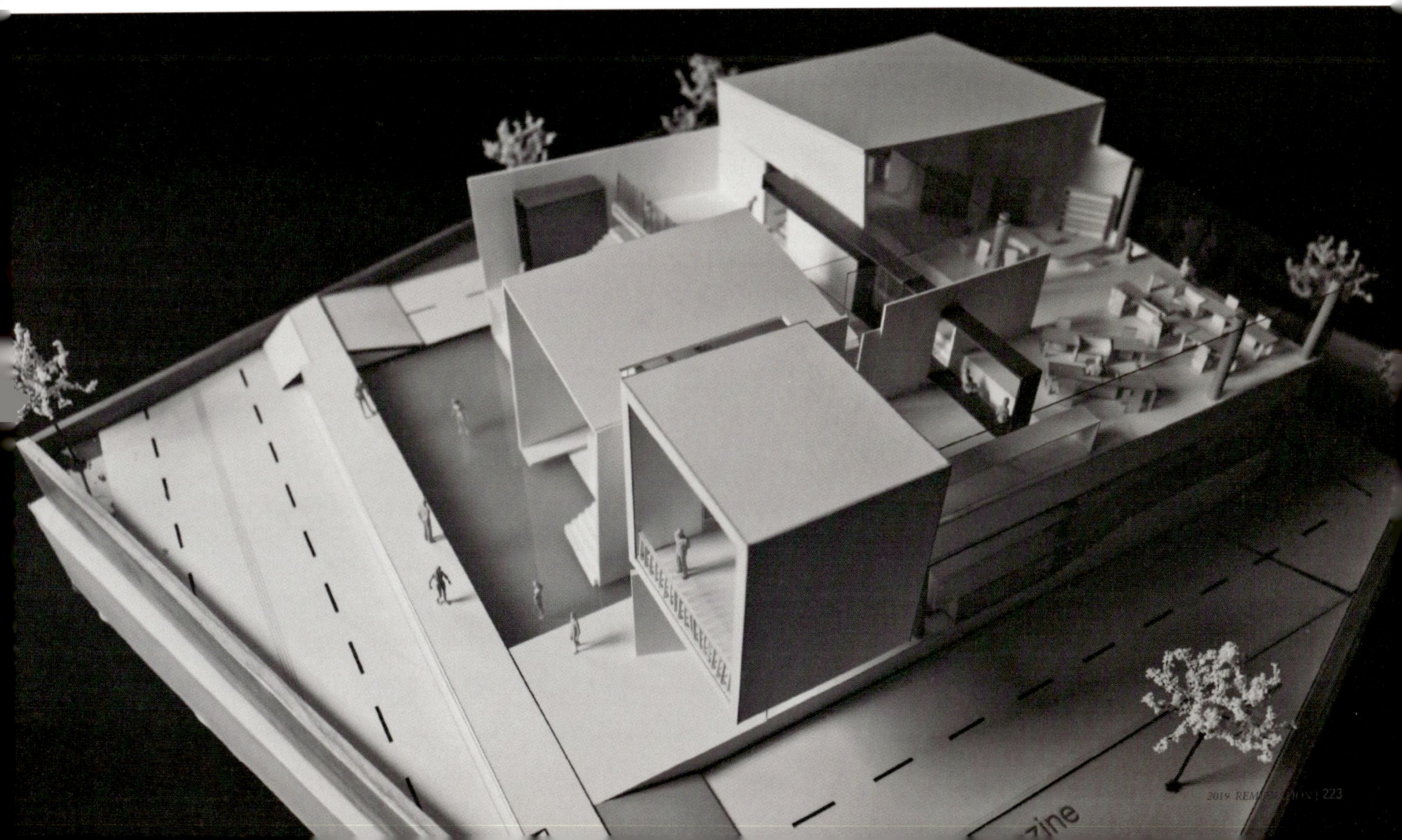

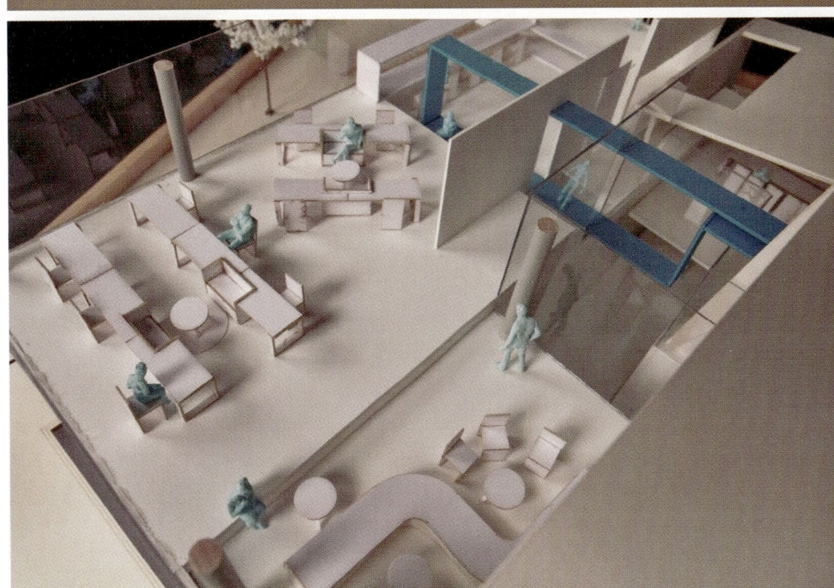

최세령 l Choi, Se Ryeong 최효진 l Choi, Hyo Jin

GARM MAGAZINE

건축 재료 처방전
Building materials prescription.

사람에게 가장 중요한 의식주 중에서 '주'를 중심으로 자신의 공간을 스스로 만들 수 있는 최소한의 방법을 마련하기 위해 건축의 가장 작은 단위인 재료에 대해 고찰한 독립잡지이다. 재료의 역사, 종류, 구성, 응용에 대해 기초부터 설명하는 재료 입문서로 재료의 특징, 판매처, 제작 과정을 알기 쉽게 설명하는 건축재료 가이드북이다.

It is an independent magazine which considered the material, the smallest unit of architecture, to arrange the minimum way to make one's space by oneself focusing on 'shelter' among food, clothing, and shelter which were the most important to person. It is the beginner's book explaining the history, kind, constitution, and application of materials, and is the architecture material guidebook which easily explains about the feature of materials, distributor, and manufacture process.

BACKGROUND

인류의 시작과 함께 발전해 온 건축은 우리의 삶 속 가장 가까운 곳에서 많은 영향을 끼치고 있다. 오랜 시간 인간과 함께 성장한만큼 점진적으로 재료의 한계가 극복되었고 그로 인해 다양한 형태의 건축물을 접할 수 있게 되었다. 하지만 우리는 익숙함에 속아 잘 알지 못하는 건축 재료를 남용하곤 한다. 'GARM MAGAZINE'은 건축 재료의 적절한 사용을 제안하고 정보를 제공하는 매거진으로써 우리가 인식한 문제 의식과 유사한 입장을 지니고 있다.

The architecture which has developed with the beginning of humanity is having a great effect on our life. The limit of materials was gradually overcome as the architecture was growing with the human for a long time and people could contact various kinds of architectures accordingly. However, we usually overuse architecture materials because of the familiarity to them. 'Garm Magazine' is a magazine suggesting the proper use of architecture materials and providing the information, which agrees with our opinion on the problem.

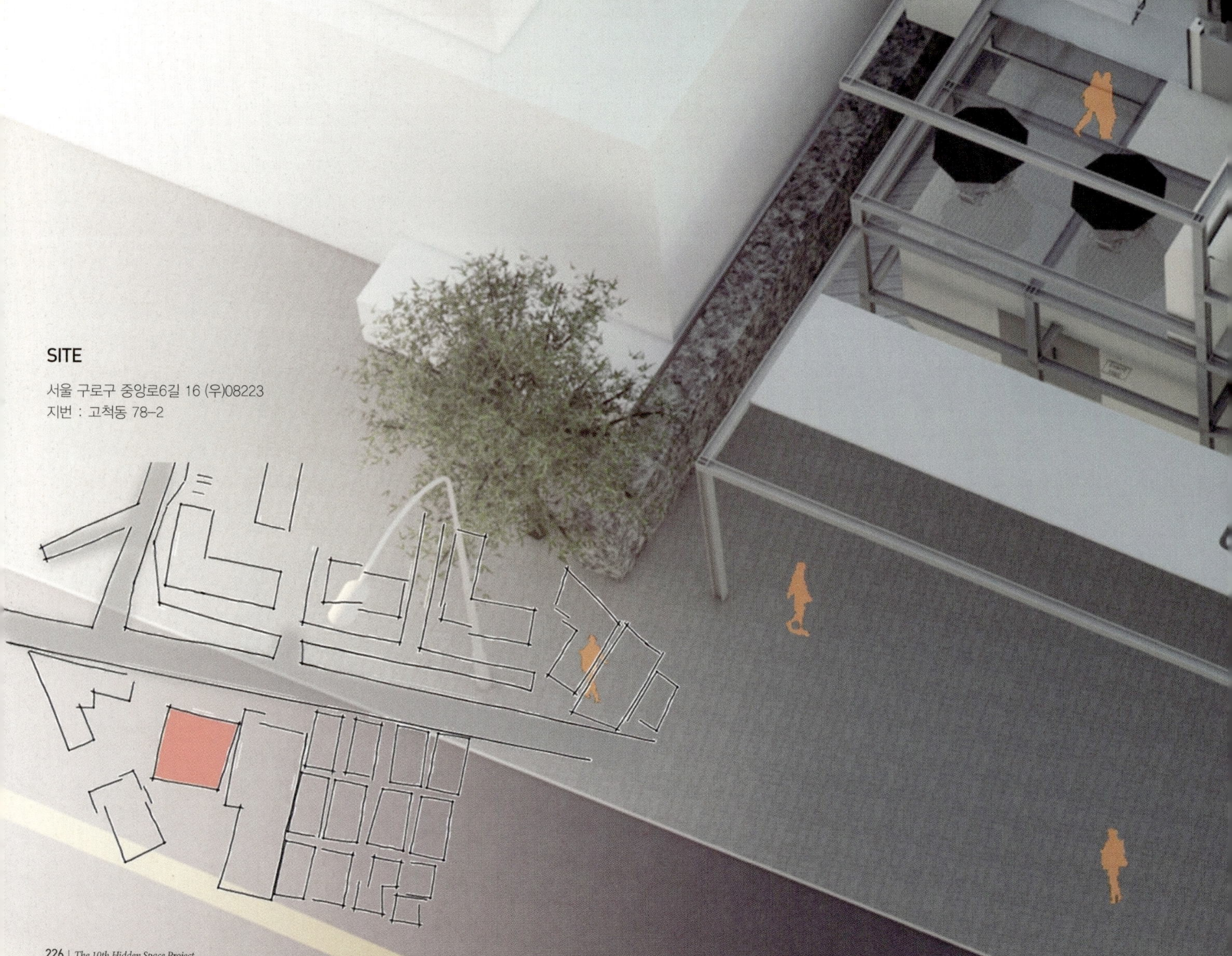

SITE

서울 구로구 중앙로6길 16 (우)08223
지번 : 고척동 78-2

Ensemble of Space

DESIGN CONCEPT

아상블라주 – Assemblage

예술 작품이 아닌 물건의 단편을 모아 만든 작품.
각각의 특징, 특성을 지닌 재료들이 같은 것들끼리, 또 서로 다른 것들끼리 모여 결합되고 조화롭게 이어진다는 점이 '감' 매거진이 다루는 건축재료의 특성과 연결된다.
'전체'로의 조화와 '부분'으로의 기능과 의미가 모두 담길 수 있는 공간을 지향하며 '전체 속 부분의 단편'을 직, 간접적으로 보이려 한다.

The work that is made by gathering scraps of objects not the artworks. That materials having several features and natures are combined harmoniously is connected with the nature of architecture materials dealt with 'Gam' Magazine. It aims for the space that can embody the harmony as a whole and the function as a part, and would show the scrap of a part in the whole directly and indirectly.

DESIGN DIRECTION

물건의 단편이 모여 하나가 된다는 것으로부터 각각의 단편적 모습과 혼합된 작품의 전체적인 모습의 대비, 공존에 집중했다.따라서 단편과 전체의 공존, 그 사이의 혼돈, 그 경계를 보여주는 공간을 표현하는 것을 목적으로 한다.

It concentrated on the contrast and coexistence of the overall image of the work that fragmentary image were mixed with the fact that scraps of objects got together to constitute a whole. Therefore, it is aimed at expressing the coexistence of the scrap and the whole, chaos between them, and space showing the boundary.
M.S1 – The outermost sides of MASS are expressed by the rough texture, the inside by the soft texture.

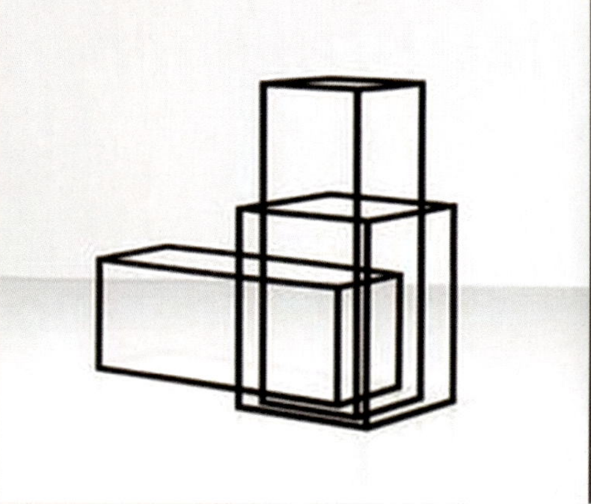

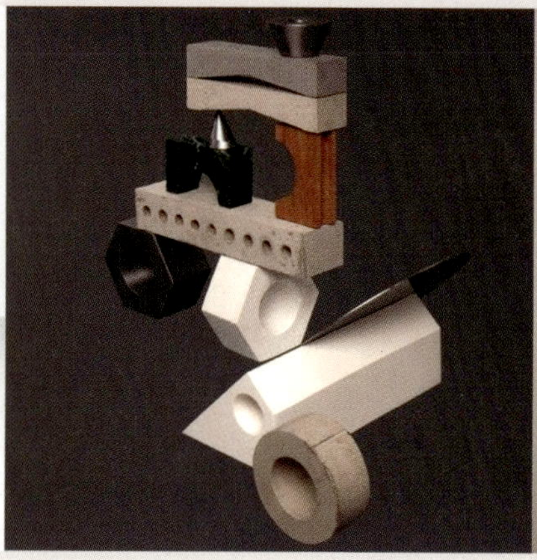

DESIGN PROCESS

1. 전체로서의 틀 규정
Defining the frame as a whole

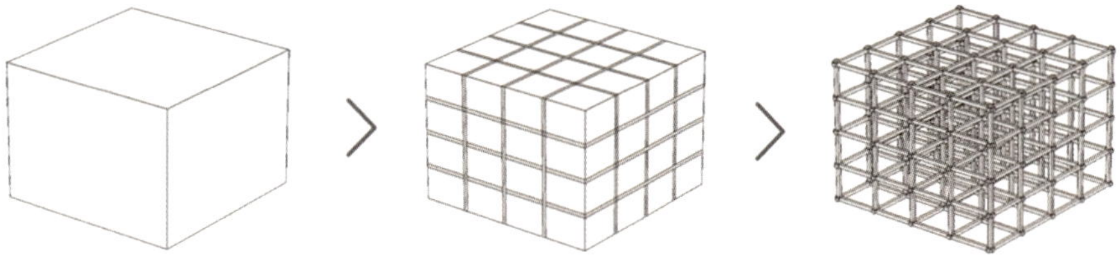

가장 외부의 틀이 되며 건물 구조와 같은 느낌으로 내부에 또 다른 기능을 하는 MASS들이 존재하는 것을 보여준다.

It becomes the outermost frame and shows that there are Masses playing another role on the inside with a feeling like the building structure.

2. 부분으로의 공간을 만들다
Making the space as a part

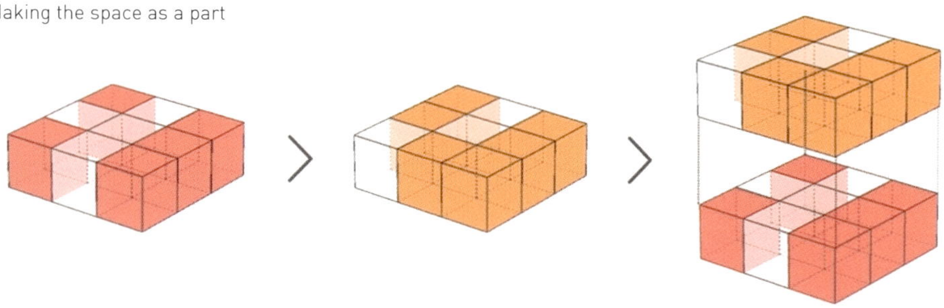

직접적인 기능을 수행하는 부분적 공간을 형성한다.

Forming a partial space which fulfills the direct function.

MATERIAL STRATEGY

1. Rough texture vs Soft texture

MASS의 가장 바깥쪽 면들은 거친 질감으로, 내부로 이어지는 안쪽 면일수록 부드러운 질감으로 표현한다.
The outermost sides of MASS are expressed by the rough texture, the inside by the soft texture.

2. Concrete vs Brick

내부를 훔쳐보는 듯한 느낌을 주는 공간들에서 외부 공간에는 콘크리트를, 창을 통해 보이는 내부 공간에는 벽돌을 사용한다.
In the spaces which look as if peeking the inside, the concrete is used in the outer space and bricks are used in the internal space which is seen through the window.

다른 MASS들과 독립적 MASS인 라이브러리에는 외, 내부의 재질을 반대로 사용함으로써 '뒤집어져 있는 듯한 느낌' 을 부여한다. 이는 '전체의 안에 존재하는 단편' 을 보이는 것이다.
In the library, the independent MASS from other MASSes, the feeling as if it is overturned is given by using the outer/inner materials adversely. It shows the scrap existing inside the whole.

3. Natural element vs Artificial element

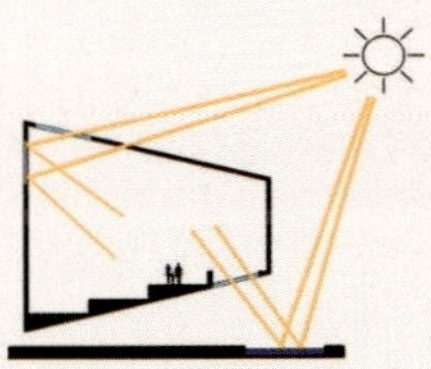

자연적 요소는 외부적 요건으로, 인위적 요소는 내부적 요건으로의 의미를 가지고 인테리어에 도입한다.
Natural element is the external requirement and the artificial element is the internal requirement. They are introduced to the interior design in this way.

사이 공간을 활용해 각 요소가 상반되는 공간에 침범하며 그 경계가 모호하게 된다.
As each element invades the incompatible spaces by using the gap space, the boundary becomes obscure.

3. 전체와 부분을 잇다
Linking the whole to the part

틀에서부터 MASS의 내부로까지 뻗어들어간 빔들은 안에서는 인테리어적 요소로, 사이 공간에서는 전체와 부분을 이어주는 요소로의 기능을 한다.

Beams which stretched out to the inside of Mass from the frame function as an interior-design element on the inside and as an element linking the whole with the part in the gap space.

4. 사이 공간을 드러내다
Revealing the gap space

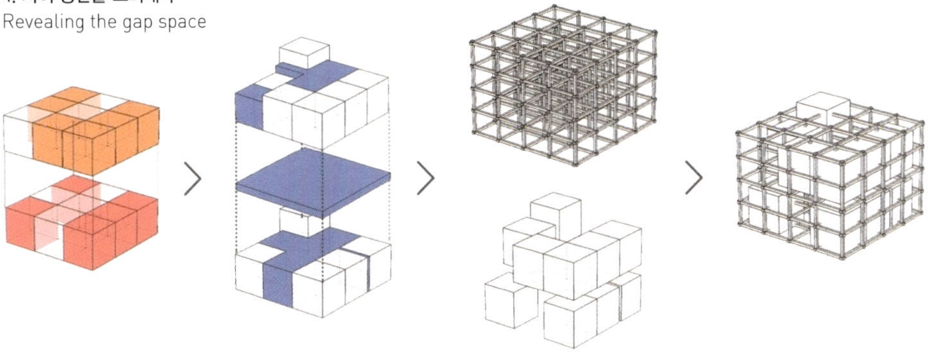

전체와 부분을 이어주는 공간이 드러나며 그 공간은 받아들이는 주체에 따라 서로 다르게 이해하도록 해서 의도적인 인식의 차이를 유도한다.

The space that links the whole with the part is revealed, and the intentional difference of awareness is induced by making the persons accepting the space understand differently.

ISOMETRIC

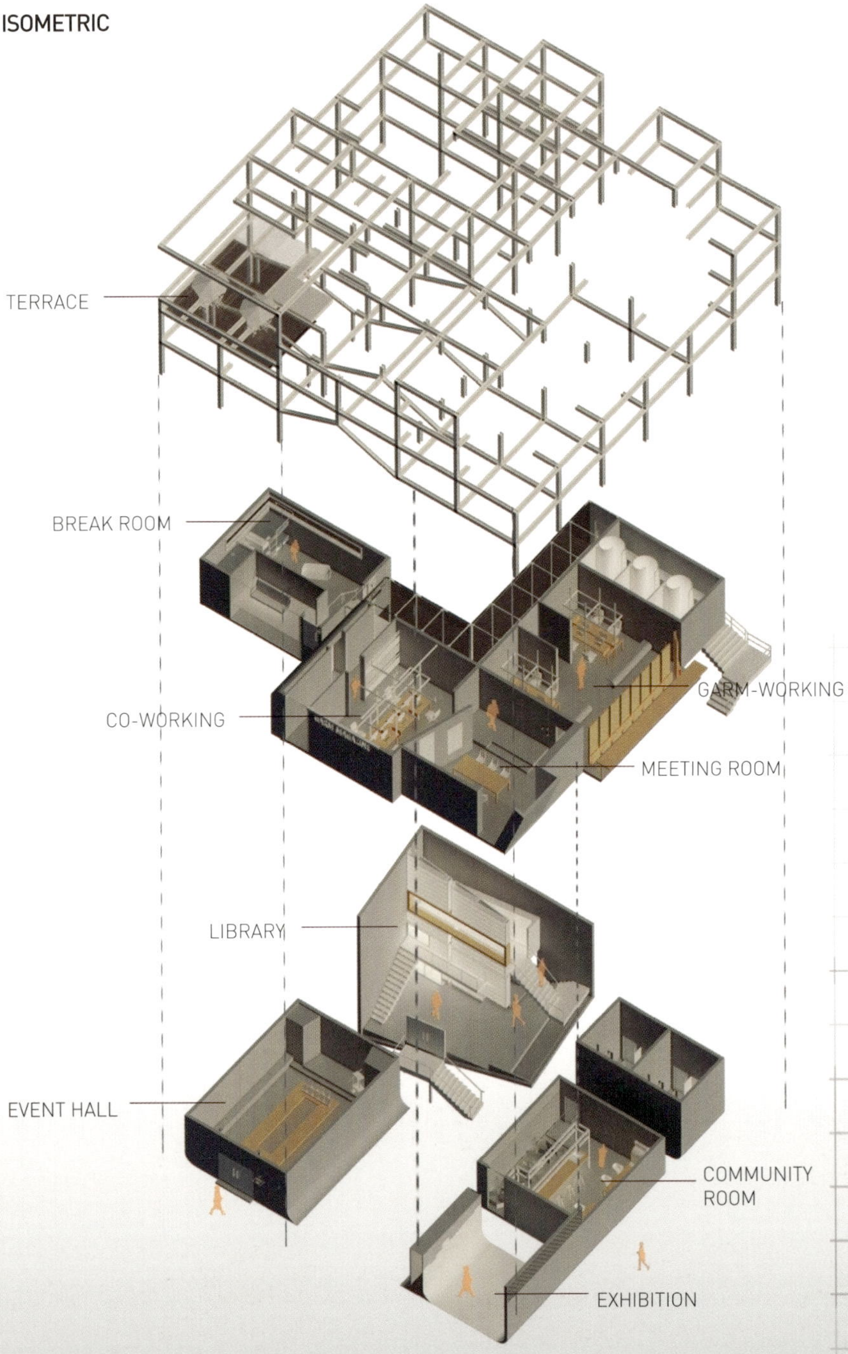

FACADE

MASS 들을 감싸고 있는 H빔을 주차 공간까지 뻗어나오게 하여 외부와 오피스 공간을 구분하고, 지붕을 형성하여 날씨 변화에도 이동이 편리하게 하였다. 지붕에는 'GARM' 을 새겨 낮에는 자연의 빛을, 밤에는 조명을 사용해 주차공간에 그림자가 비춰지도록 한다. 이는 또 하나의 간판이 된다.

Making the H beam surrounding Masses stretch out to the parking lot to distinguish the outside from the office space, and the roof was formed to make the movement convenient even if the weather changed. The word 'GARM' was carved on the roof to make the shadow appear in the parking lot by using the natural light at daytime and the lighting at night. It becomes another signboard.

1st FLOOR

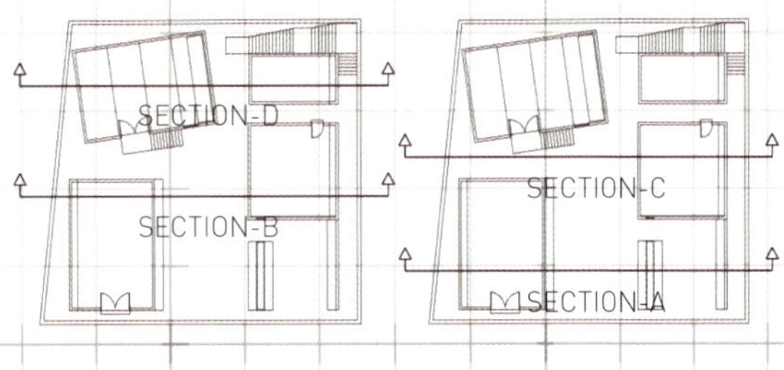

2nd FLOOR

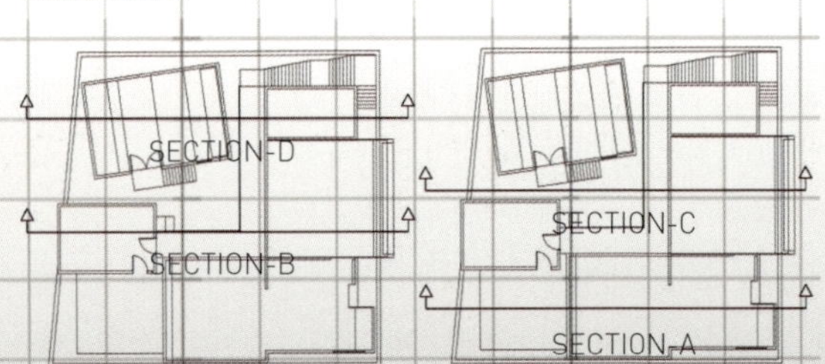

SECTION VIEW

SECTION-A

SECTION-B

SECTION VIEW

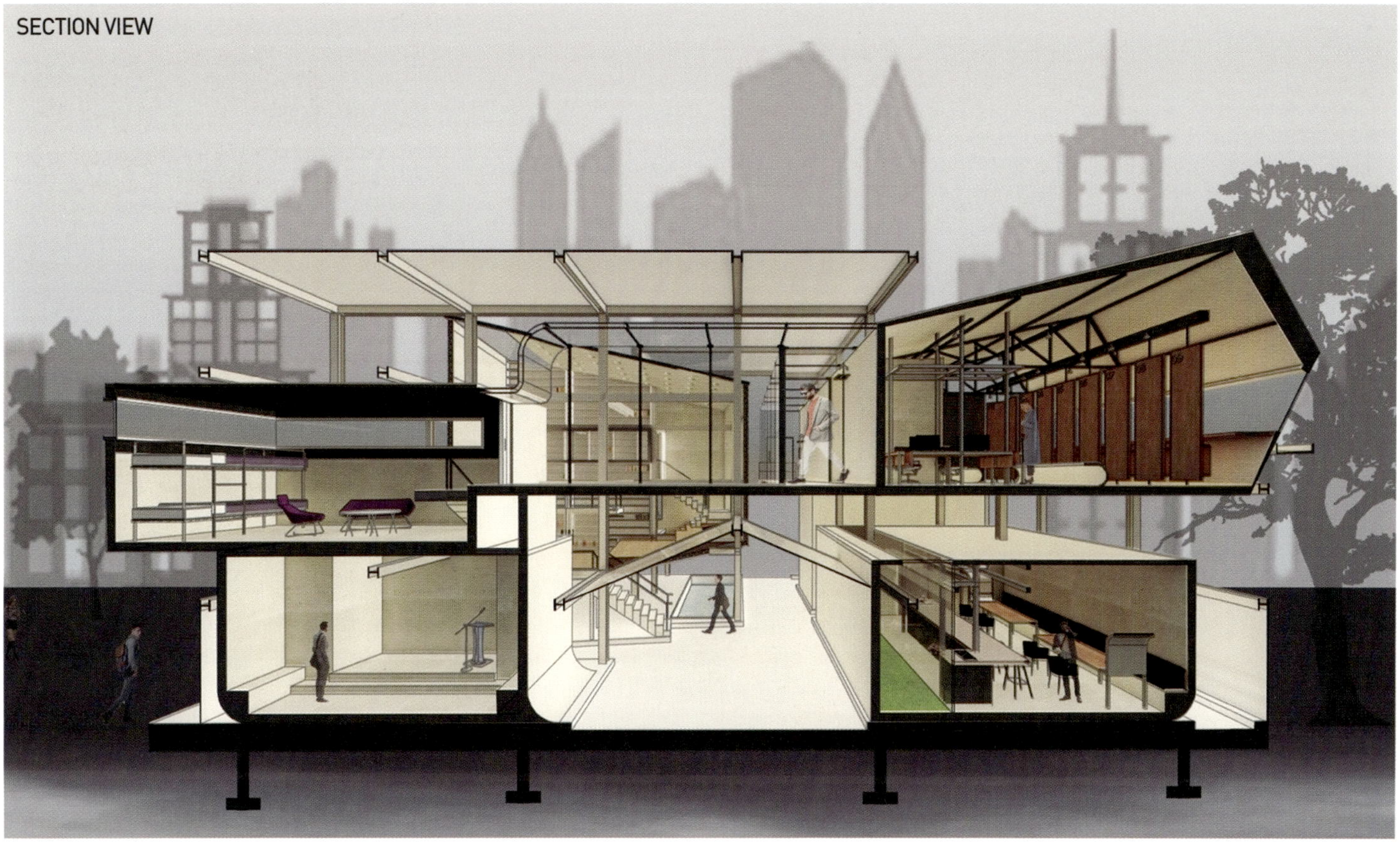

SECTION ISOMETRIC

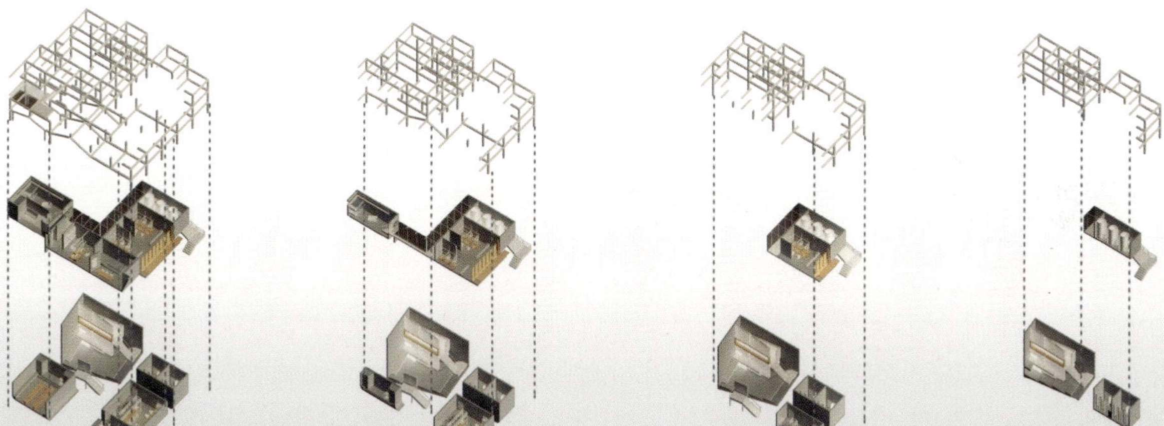

SECTION-A SECTION-B SECTION-C SECTION-D

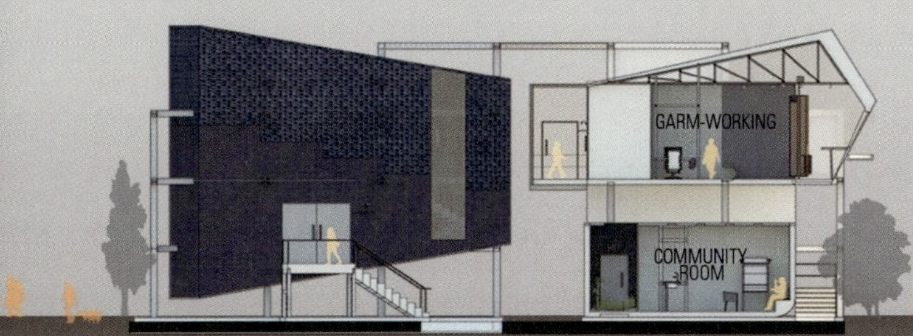

SECTION-C

SECTION-D

PROGRAM

PLAN VIEW

FOR TOGETHER

외부 지역 주민들과 함께 소통하고 공유할 수 있는 공간으로 외부인 공간 이동의 효율을 위해 주로 1층에 배치했다.

It is the space for communicating and sharing with external local residents, which was arranged on the first floor for outsiders to move efficiently in the space.

– Library : 직원들과 외부주민들의 정보 공유를 위한다.
– Event Hall : 잡지 출판행사나 여러가지 이벤트 진행한다.
– Exhibition : 출판된 잡지들을 전시하는 공간으로 잡지의 주재별 다양한 건축재료를 보인다.

- Library : It is intended to enable employees to share the information with outside residents.
- Event Hall : Carrying out the magazine publishing events or various events.
- Exhibition : The space for exhibiting published magazines, which shows various architecture materials depending on each theme of magazine.

FOR EMPLOYEES

직원들만 이용 가능한 공간으로 서로간의 소통과 업무중 휴식을 취할 수 있는 공간이다.

It is the space only available for employees, which is the space for the mutual communication and the rest.

– Community Room : 직원 전용 커뮤니티 룸. 2층 주요 업무공간과는 다르게 자유로운 분위기를 형성하기위해 1층 배치했다.
– Break Room : 장시간 업무가 많아 피로에 쌓인 직원들을 위해 작은 쉼의 공간을 마련했다. Community Room과는 다르게 업무공간과 가까우면서도 소음을 줄일 수 있는 2층 구석진 곳에 있다.

- Community Room : It was arranged on the 1st floor to create a free atmosphere unlike the main business space on the 2nd floor.
- Break Room : The space of rest for employees tried of overwork for a long time was arranged. It is close to the business space unlike the community room and is located in the corner on the 2nd floor to reduce the noise.

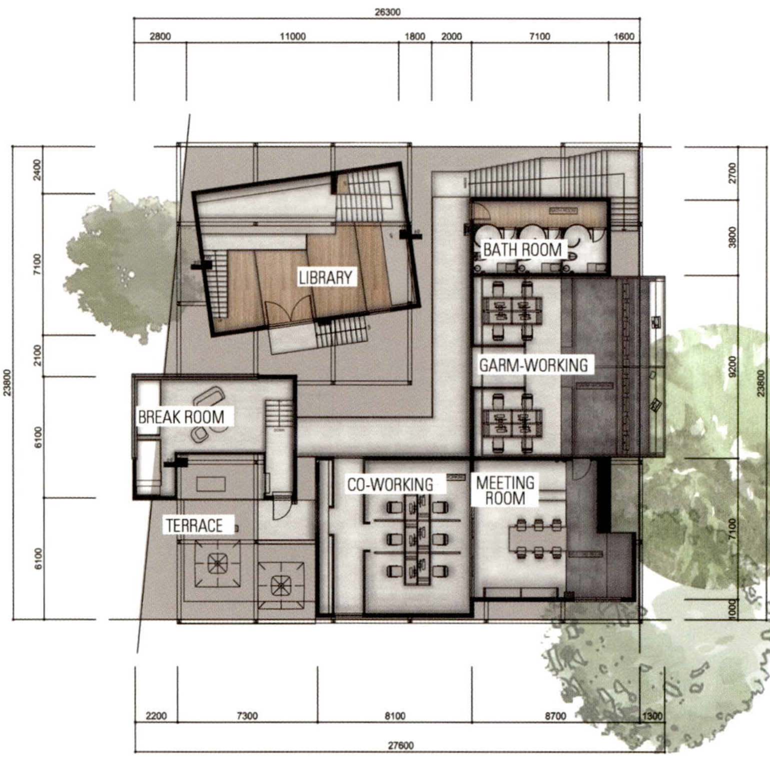

PERSPECTIVE
LIBRARY
1

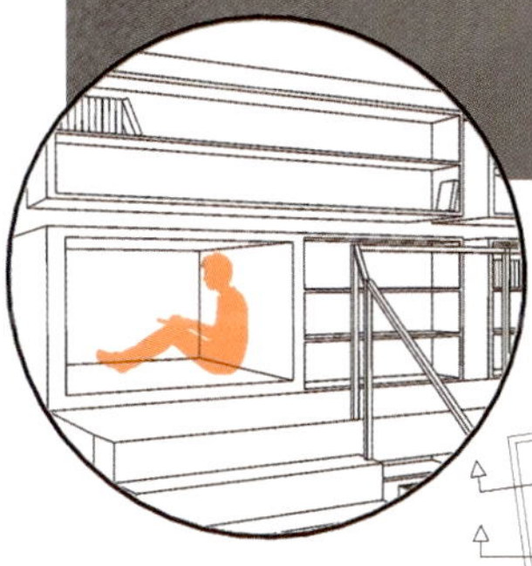

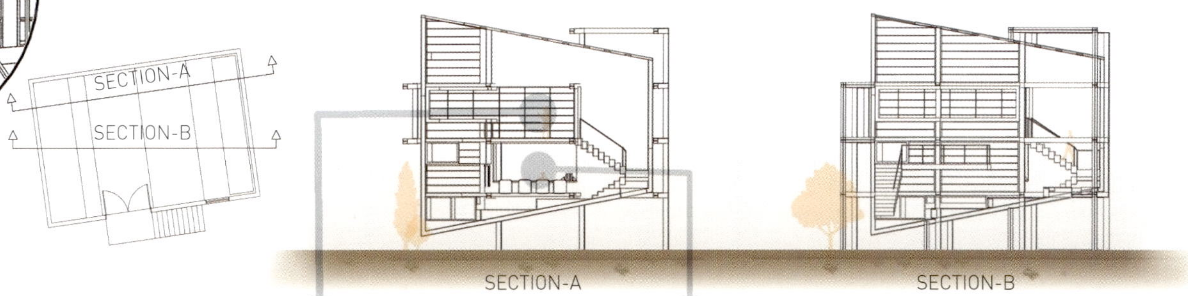
SECTION-A
SECTION-B
SECTION-A
SECTION-B

FOR BUSINESS

서로 도와주며 공유하는 업무 공간이다.

Business space where employees help each other sharing the information.

– GARM–Working : Main 업무공간
– Co–Working : 넓은 공간을 효율적으로 사용하기 위한 오피스 공유 시스템으로 독립적이지만 감과 일정 분야에서 협업하여 업무를 진행한다.
– Meeting Room : Garm–Working과 Co–Working 사용자들이 모두 사용하기 쉽도록 중간에 배치해 동선의 효율성을 높였다.

- GARM - Working : Main business space
- Co-Working : It is the office sharing system for using the wide space efficiently, which is independent but fulfills the work in cooperation with GARM in a certain field.
- Meeting Room : It was arranged in the middle for users from Garm-Working and Co-Working to use easily and to raise the efficiency of traffic line.

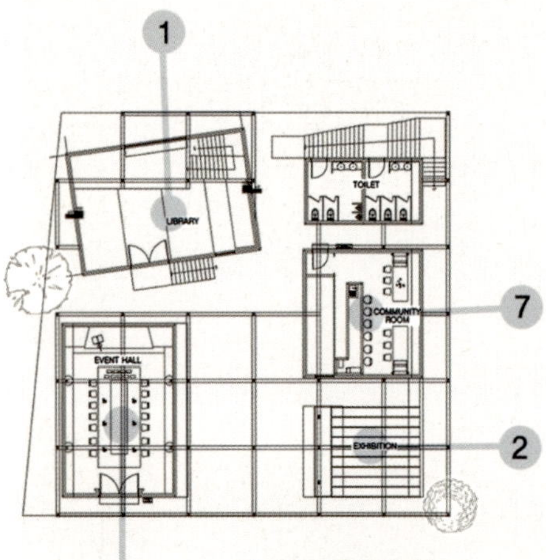

1st FLOOR PLAN

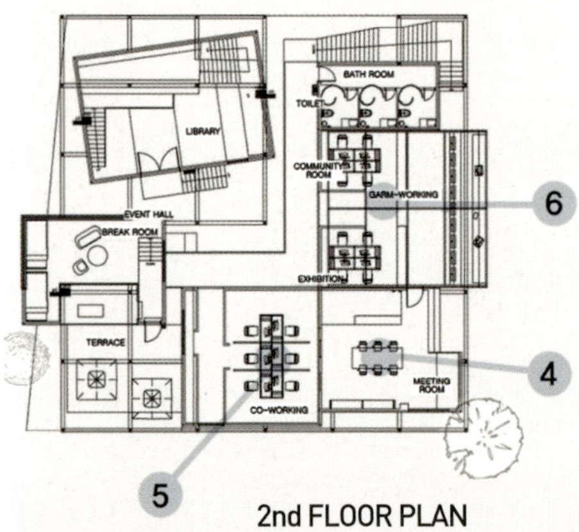

2nd FLOOR PLAN

Furniture
6
08
07
06
05
04
03
02
01
GARM-WORKING

7
COMMUNITY ROOM

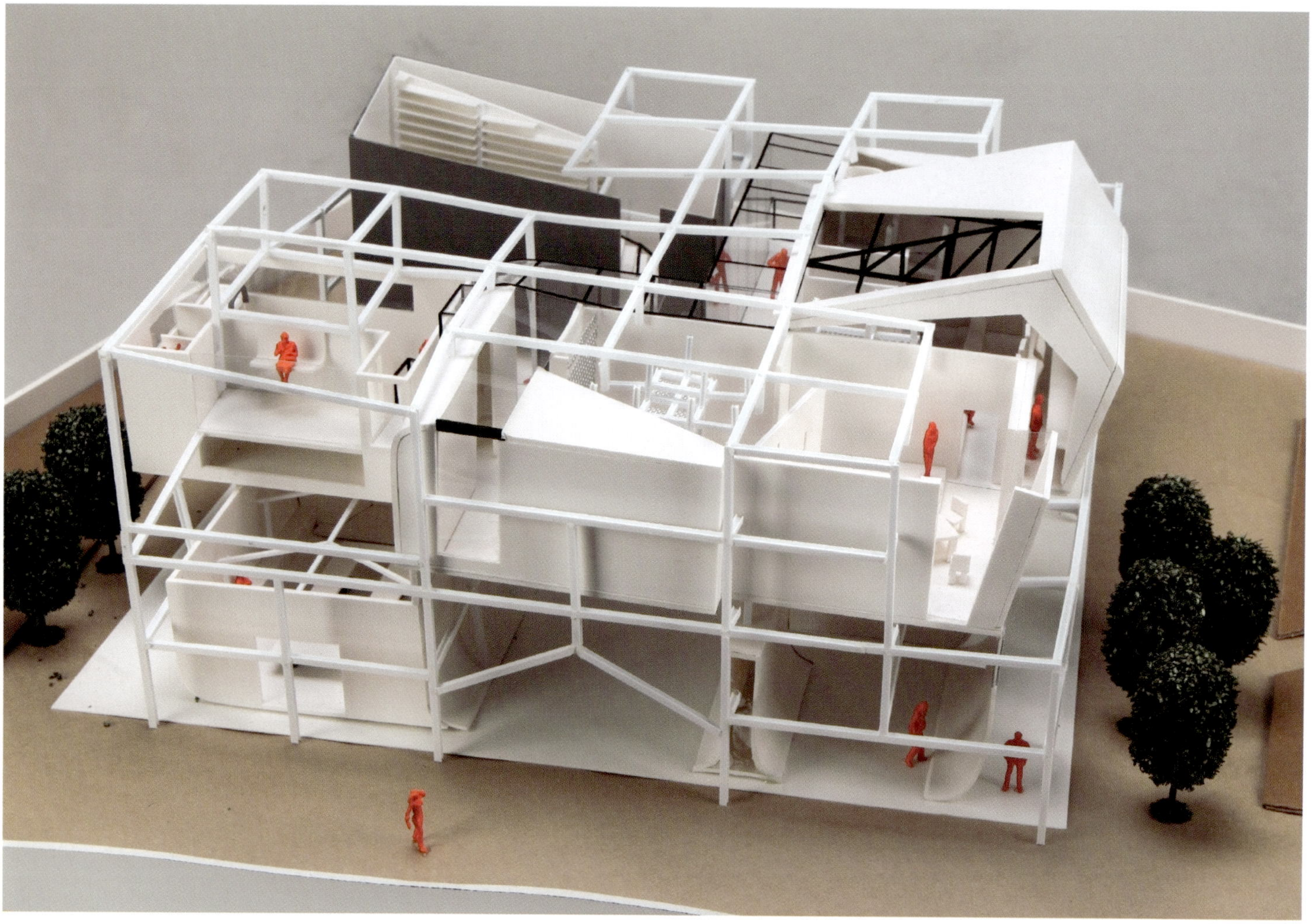

이윤서 I Lee, Yoon Seo 김경곤 I Kim, Kyeong Gon

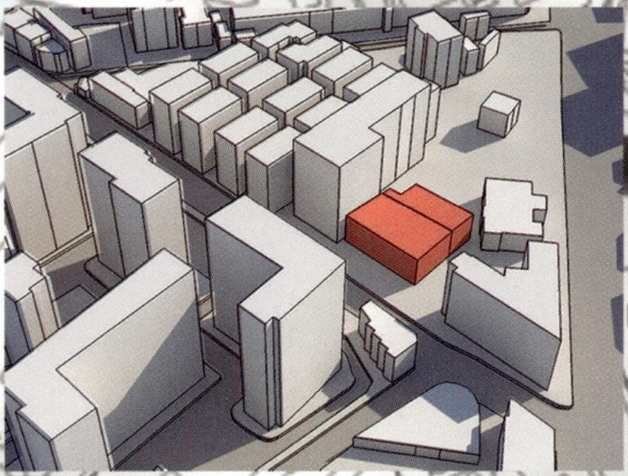

SITE ANALYSIS

서울특별시 구로구 중앙로6길 16(지번주소 고척동 78-2)에 위치한 곳으로 주변에 아파트와 상가건물, 편의시설 등이 위치하고 있으며 동양미래대학교 부속 건물이다.

It is located at 6 Gil 16 (Gocheok-dong 78-2), Guro-gu, Guro-gu, Seoul. Apartments and commercial buildings are located nearby. It is an affiliated building of Dong-yang Mirae University.

sense of contrast

:가벼움과 무거움이라는 감각을 각각 1층, 2층 공간에 대비되게 형상화하여 '가벼움과 무거움' 이라는 컨셉을 공간화.

The concept of lightness and heaviness is expressed in contrast to the first and second floor spaces, respectively.

BEERPOST (독립잡지사–맥주)

국내 외 맥주 이야기를 다루는 잡지사로 맥주 전문가들로 구성된 에디터 들이 맥주의 스타일, 재료에 대한 설명부터 발효과정, 맥주 관리 방법 등의 다양한 정보를 공유한다. 2016년부터는 한계레 교육 문화 센터와 함께 한계레 맥주 학교를 운영하며 맥주의 기초 지식과 펍투어, 푸드 페어링 등의 여러 가지 프로그램을 진행하고 있다.

As a magazine dealing with domestic and foreign beer stories, editors made up of beer experts share various information such as the description of beer style, ingredients, fermentation process, and beer management method. Starting in 2016, we have been operating the Hankyoreh Beer School with the Hankyoreh Education and Cultural Center and have been conducting various programs such as basic knowledge of beer, pub tours, and food pairing

CONCEPT

가벼움 속 무거움

맥주를 잔에 따르면 하얀 거품이 맥주 위를 가득 덮고 있다. 부드럽고 가벼운듯한 거품은 깊은 향을 담고 있는 맥주와 상반되는 느낌이지만 맥주를 덮고 있는 하얀 거품은 맥주의 목넘김을 부드럽게 도와주면서도 맥주의 청량감과 깊은 향을 극대화 시킨다. 이 거품으로 인해 맥주의 향과 맛은 한층 더 좋게 받아들여진다. 대비되는 두 감각의 조화를 통해 공간을 구성하고자 하였다.

When you pour beer into a glass, the white foam is covering the top of the beer. Though the soft and light bubbles give the feeling contrast to the deep-flavor of the beer, white bubbles covering the beer not only help soften your swallowing, but also maximize the refreshing sensation and deep flavor of the beer. Because of this bubble, the aroma and flavor of the beer becomes even better. We intend to compose the space through the harmony of the two contrast senses.

맥주거품의 가벼움이라는 감각을 1층 공간에 형상화.

The sense of lightness of beer bubbles is expressed in the first floor space.

맥주거품 밑에 맥주의 목넘김,풍미 등의 감각을 2층 공간에 형상화.

Under the bubble of beer, the sense of the beating of the beer, the flavor and the like in the second floor space.

MOTIVE

무거움의 표현으로 건물이 밀집된 마을을 떠올렸다. 좁은 땅에 많은 건물들이 들어서기 위해서 빼곡하게 들어서 있는 마을의 모습을 통해 무거움을 표현했다.

As an expression of heaviness, a built-up village came to my mind. I expressed the heaviness through the image of the village which is filled with lots of buildings in the cramped land.

통로는 분리된 두 공간을 연결해 주는 요소이다. 통로를 통한 연결은 사람의 이동에 편의를 주기도 하지만 빛, 공기, 소리 등 여러 가지 자연 요소들도 이 통로를 통하여 두 공간을 자유롭게 이동한다. 1층은 개방된 느낌으로 이러한 통로를 사용하여 표현했다.

The Passage is the element connecting the two separated spaces. The connection through the passage not only makes the movement of the person convenient but also various natural elements such as light, air, and sound freely move the two spaces through the passage. The openness feeling of the first floor was expressed using this passage.

USER NEEDS

2~3명이 팀을 이뤄
작업하는 경우가 많음

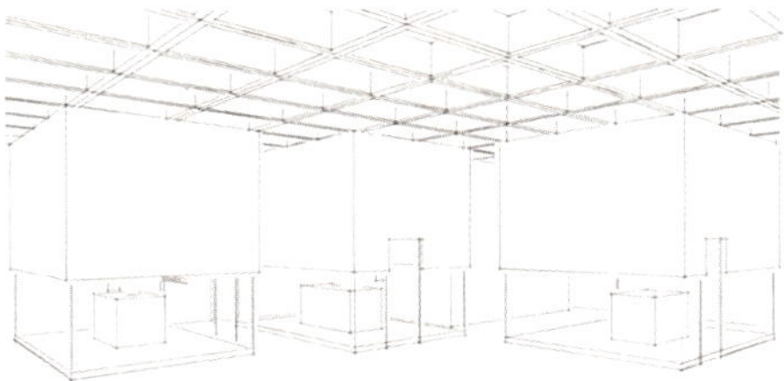

공동 작업 공간 필요

월간호 특성상 매달
아이디어 회의가 이루어짐

직원 전체 회의공간 필요

매달 잡지의 주제 선정을 위한 직원 전체 회의가 이루어지고 모든 작업에서 각각의 부서가 2~3명씩 팀을 이루어 작업하기 때문에 많은 팀 작업 공간과 전 직원을 수용할 수 있는 회의실이 필요하다. 또한 외부 객원 기자와의 미팅도 자주 있기 때문에 접근성이 좋은 위치에 미팅공간을 필요로 한다.

Because a full staff meeting is held each month to select a theme for the magazine, and all the works are done by many teams of two or three people from each department, there is a need for a large work space and a meeting room that can accommodate all employees. Also, meetings with external guest reporters are frequent, so meeting space is needed in a good accessible location.

PROBLEM & SOLUTION

수직적 조직문화로 자신의 의견이나 아이디어를 제대로 표현할 수 없음.

Vertical organizational culture can not express your opinions or ideas properly.

평등한 업무공간 배치와 자유롭게 의견을 낼 수 있는 오픈 스페이스를 형성.

Create an open space and place equal work spaces and freely comment.

항상 같은 공간, 같은 자리에서 업무를 처리하는 단조로움 속에선 창의적인 사고가 어렵움.

It is always difficult to think creatively in the same space, in the monotony of handling tasks in the same place.

한 공간 속에 독립적인 MAS들을 배치시켜 다양한 공간의 형성, 다양한 동선을 형성하여 단조로움을 피함.

By arranging independent mass in a space, various spaces are formed, various copper lines are formed, and monotony is avoided.

COMMUNITY

지역 주민들과 소통의 공간으로 Bar를 계획했다. 주변에 아파트와 오피스텔, 대학교로 둘러싸인 주변 환경을 고려하여 주민들과 학생들이 일을 마치고 집으로 돌아가기 전 간단히 친구 또는 가족들과 맥주를 마시며 하루의 피로를 조금이나마 털어놓을 수 있는 공간을 마련했다.

A bar was planned as a communication space of local residents. Considering the environment surrounded by apartments, officetels, and universities, a space was arranged, where residents and students could be relieved of a day's fatigue, drinking beer lightly with friends or family before returning home after working.

MASS PROCESS

밀집된 Mass의 배치로 마을과 같은 골목길 형성

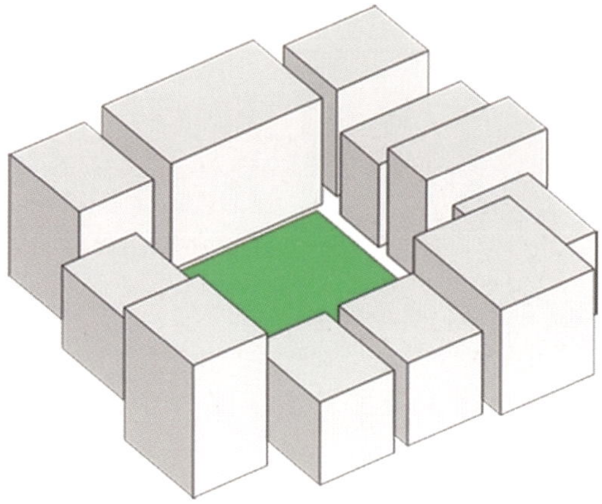

공터를 중심으로 Mass밀집

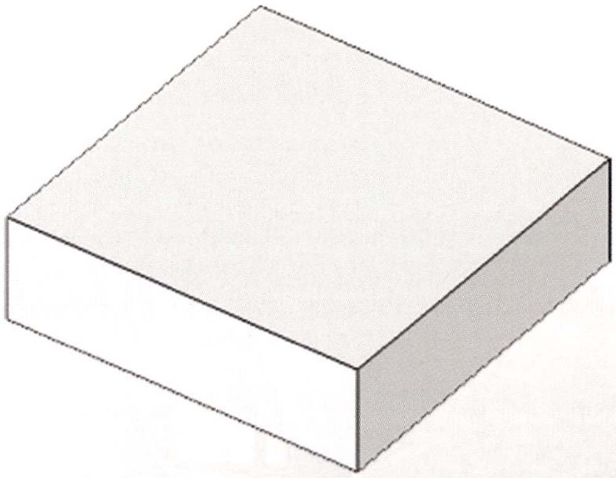

하나의 Mass 생성

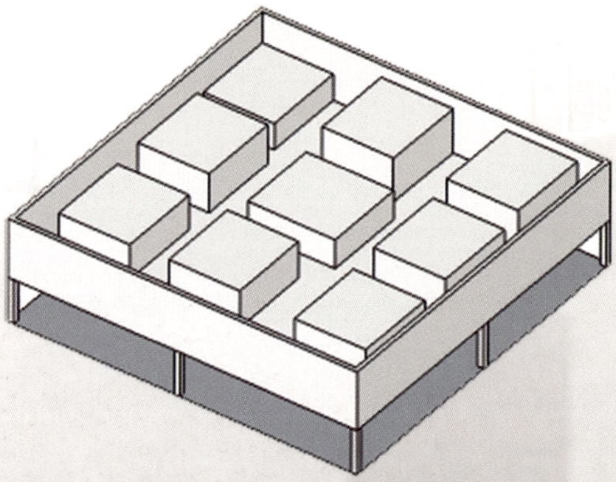

Mass를 여러개로 분할

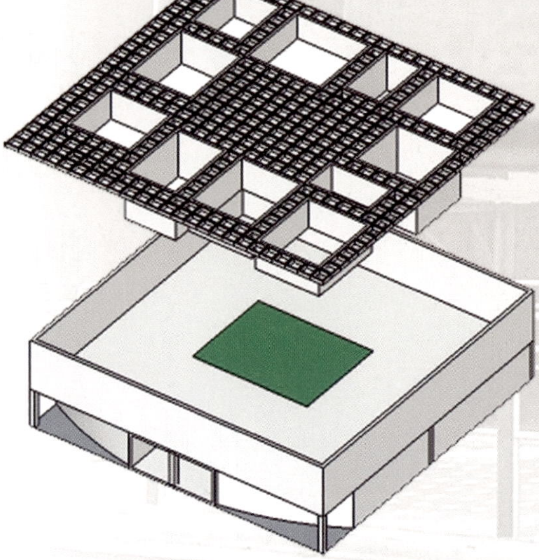

격자에 Mass들을 배치하여 내림
마을 분위기 형성

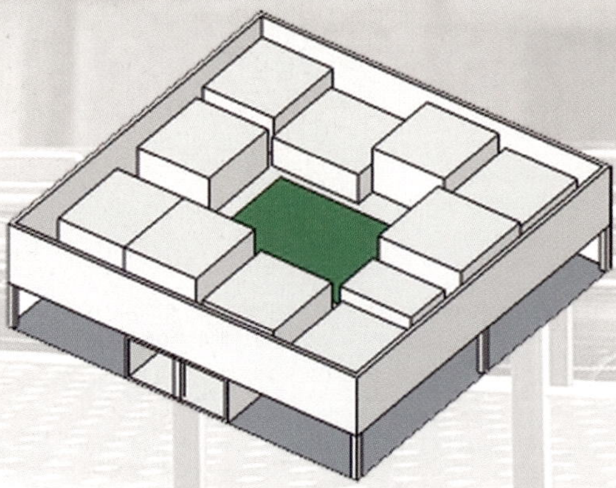

공터와 같은 공간 형성

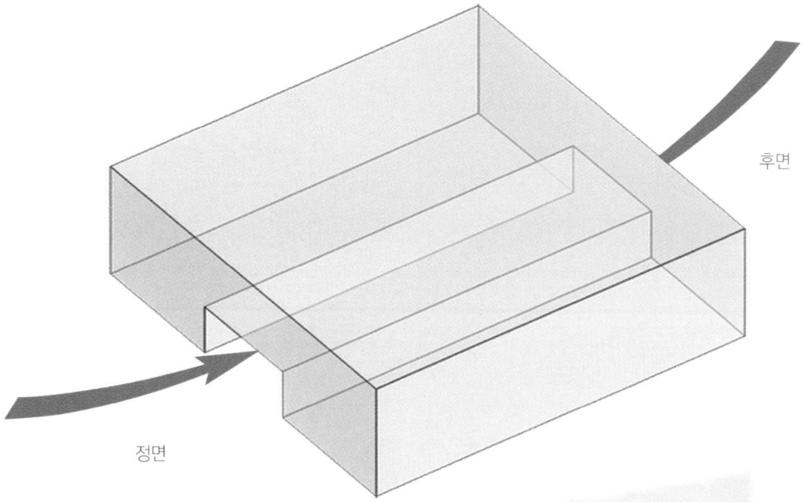

1층은 2층의 거대한 사각형 콘크리트를 받치고 있는 기둥이 주를 이루어 개방감을 준다. 중앙에는 건물의 뒤편까지 비어있는 긴 통로가 있어 개방감을 더한다. 통로의 뒤편에는 정원에서 뿜어져 나오는 자연의 빛이 통로를 감싸고 있고, 통로를 통해 건물의 앞까지 전달되어 온 빛은 그늘져 있는 건물의 FACADE와 대비되며 통로를 강조하고 있다. 통로는 건물의 앞과 뒤를 최대한 연결하여 방문객으로부터 부담 없이 들어오기를 권유하고 있다.

On the first floor, the pillars that support the huge rectangular concrete of the second floor are made up, which give a sense of openness. In the center, there is a long open passage to the back of the building, which adds a sense of openness. Behind the passage is a natural light pouring out of the gardens surrounding the passage, and the light transmitted to the front of the building through the passage contrasts with the FACADE of the shaded building, emphasizing the passage. The passage connects the front and back of the building as much as possible, which encourages visitors to feel free to come in.

2층은 무거움을 큰 매스들을 여러 곳에 배치하여 밀집된 마을느낌으로 표현했다. 각각의 매스들은 하나의 집, 건물처럼 각자의 천장을 갖고 독립된 공간을 형성한다. 각각의 매스들을 통해 개인 업무 공간, 회의실, 팀 스페이스 등 다양한 공간을 독립적으로 형성하여 단조로움을 피했다.

On the second floor, the heaviness was expressed with a dense village feel by placing large masses in various places. Each mass has its own ceiling and forms a separate space like a house or a building. Through individual masses, various spaces such as individual work space, conference room, and team space were independently formed to avoid monotony.

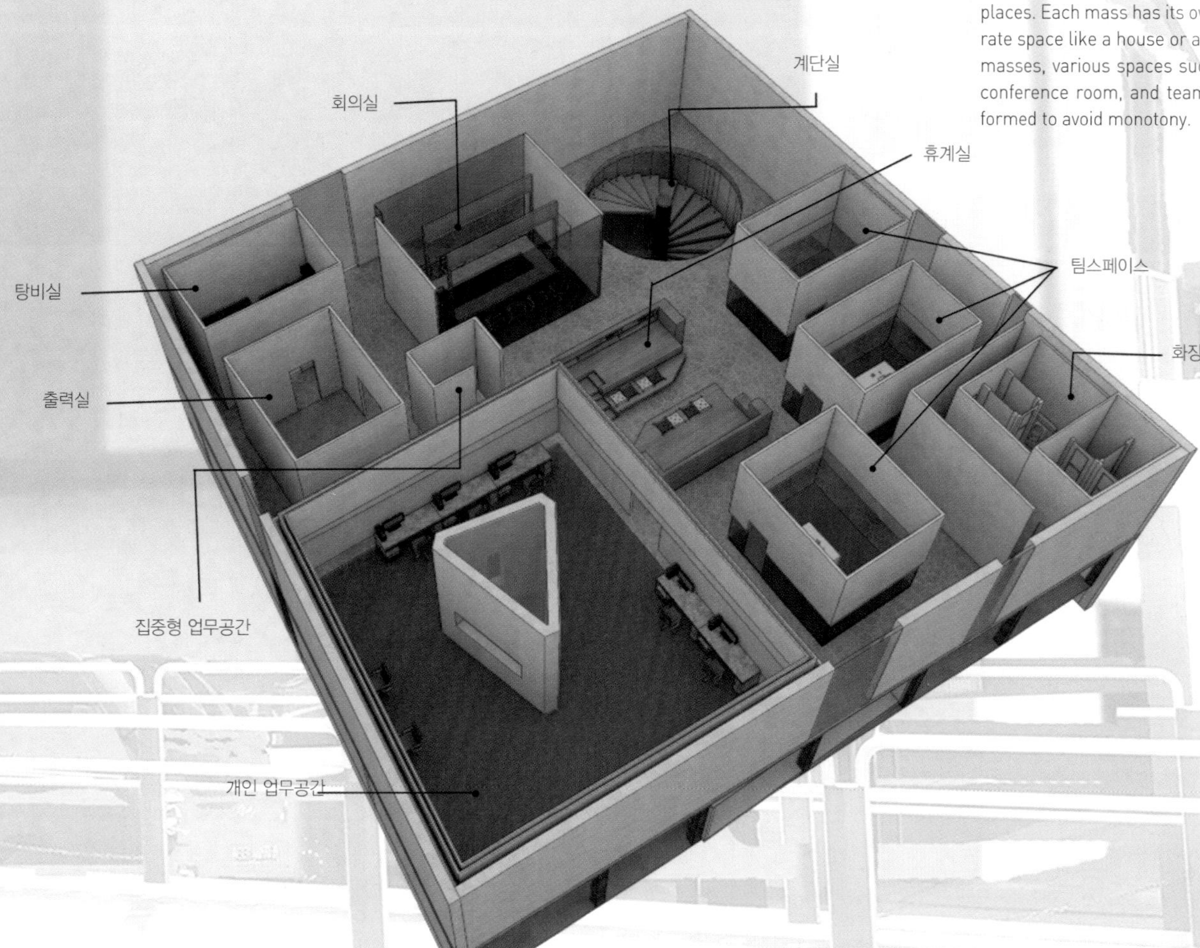

2층의 휴게공간은 마을의 광장과 같은 역할을 한다. 휴게 공간을 중심으로 공간을 배치하였고 모든 동선은 휴게공간과 연결되어 있다. 공간과 공간을 이동하는 중심에 휴게 공간을 배치함으로서 휴게공간의 접근성을 쉽게 하고 전체적인 업무 공간의 분위기를 한층 더 부드럽게 완화시켜 주는 역할을 한다.

The rest area on the second floor functions like the plaza of the village. The space of the second floor was arranged centering around the rest area, and all the moving lines are connected to the rest area. By arranging the rest area in the center of moving from a space to space, it facilitates access to the rest area and softens the atmosphere of the whole work space more smoothly.

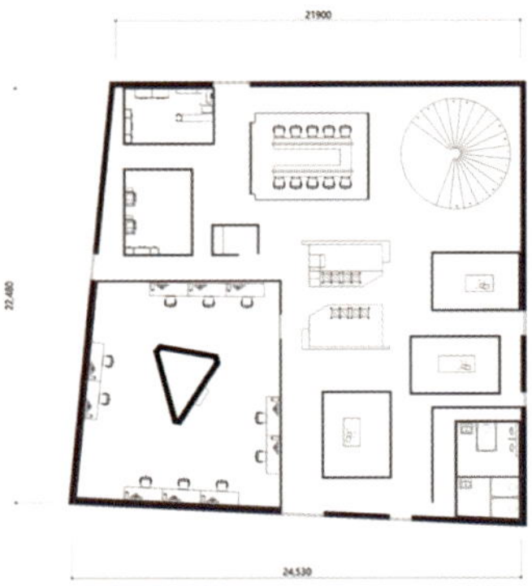

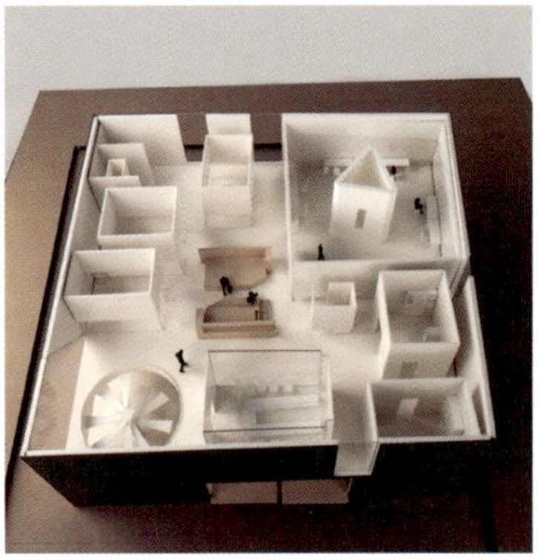

모든 직원들이 같이 일하는 개인 업무공간은 서로의 소통을 방해하지 않으면서도 분리된 공간을 형성시키고자 했다. 큰 사각형 형태의 공간 중앙에 삼각형 기둥 모양의 벽은 개인 공간의 시선을 차단해 주는 역할을 하는 동시에 천장에서부터 자연광을 끌어들여 업무공간에 생기를 불어넣어 준다.

The personal work space where all the employees work together was intented to create a separate space without obstructing communication with each other. The triangular pillar-shaped wall at the center of the large rectangular space blocks the gaze of the individual space, and at the same time brings in the natural light from the ceiling and invigorates the workspace.

매달 직원 전체 회의가 이루어지는 회의실은 중요한 공간이지만 자주 이용되지는 않는 공간이다. 회의실의 양 옆을 막힌 벽이 아닌 투명한 유리로 하여 회의를 할 때 공간 내부에서는 외부와 차단된 독립족인 공간이 되어 회의에 집중을 돕지만, 평상시에 회의실을 사용하지 않을 때에는 시선을 막지 않아 전체적인 공간의 답답함을 줄여준다.

Meeting rooms where monthly staff meetings are held are important but often unused. When a meeting room has a transparent glass rather than a closed wall on both sides, it becomes an independent space that is blocked from the outside in the inside of the space, thereby helping to focus on the meeting, but when the meeting room is not used normally, It reduces the crampedness of the overall space by not obstructing your eyes.

회의실의 테이블은 끝이 사선으로 깎인 대리석의 상판을 천장에 매달아 두는 방식으로 되어있다. 대리석의 사선으로 깎인 모서리는 무거운 대리석의 상판을 가벼워 보일 수 있게 하여 새로운 감각과 지각을 경험하도록 의도했다.

The table in the meeting room is made by hanging the marble top with the diagonal-cut-edge from the ceiling. The diagonal-cut-edge of the marble were intended to look the heavy marble top light to experience new senses and perceptions.

1층은 지역 주민들과 회사 직원들이 모두 사용할 수 있는 공간이다. 중앙 통로를 중심으로 맥주를 판매하는 Beer post bar와 미팅룸, 계단실이 배치되어 있다. 시선을 막는 벽을 모두 없애고 기본 골격인 기둥과 보를 모두 노출시켜 가벼움을 표현한 공간이며, 중앙의 통로와 곡선의 긴 bar 테이블은 1층 공간이 건물의 뒤편으로 방향성을 가지게 하고 있다.

The first floor is a space that can be used by local residents and company employees all together. There are a beer post bar selling beer, a meeting room and a staircase centering around the central passage. It is a space where lightness is expressed by exposing both the columns and beams that are basic skeleton by removing the walls obstructing sight lines, and the central passage and the curved long bar table makes the first floor space have the direction to the backside of the building.

건물의 1층 중앙을 가로지르는 통로는 건물의 앞과 뒤를 이어주며, 건물의 뒤편에서 들어오는 햇빛이 통로의 바닥과 천장을 타고 건물의 앞까지 전달되도록 도와준다. 통로를 통해 들어온 자연광은 그늘져 있는 건물의 FACADE를 밝혀주며, 동시에 지역 주민들이 좀 더 부담 없이 접근할 수 있게 도와준다.

The passage through the center of the first floor of the building connects the front and rear of the building, and helps the sunlight from the back of the building pass through the floor and ceiling of the passage to the front of the building. The natural light entering through the passage lights up the shaded FACADE of the building and at the same time helps the local residents to access more freely.

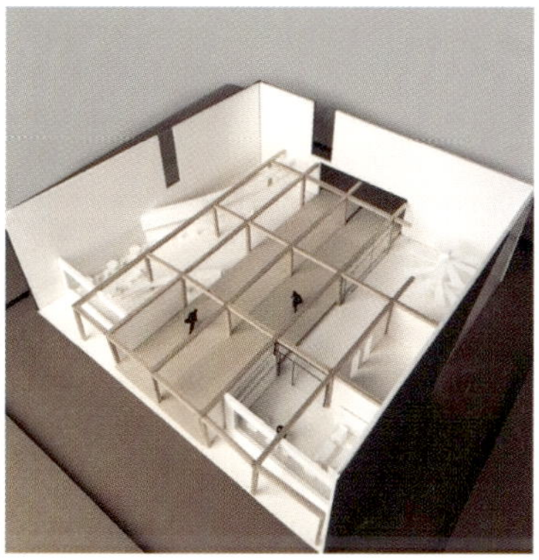

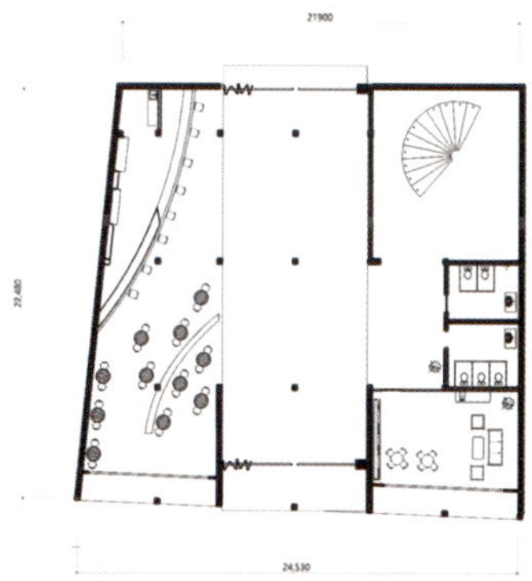

ISOMETRIC

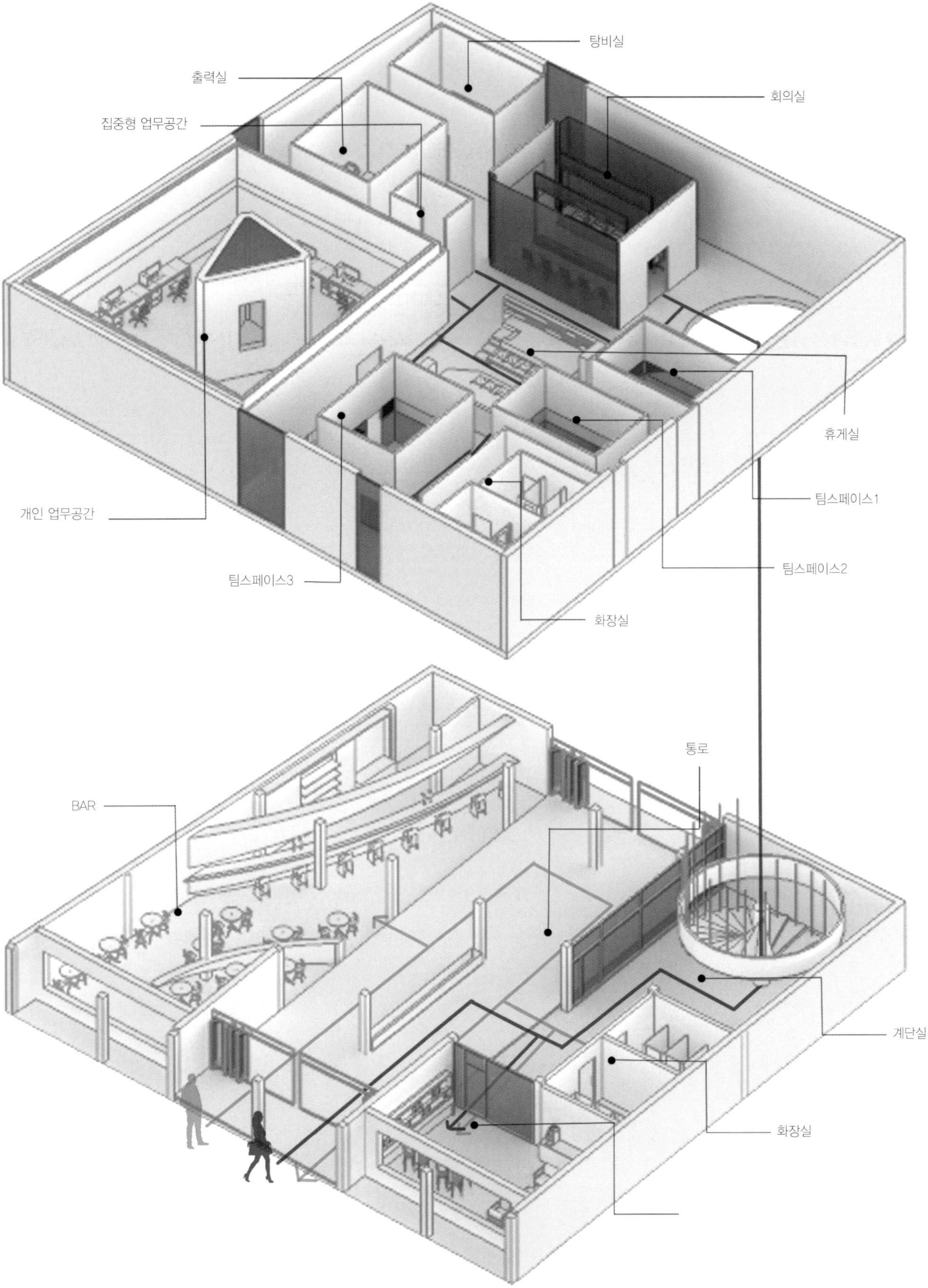

탕비실
출력실
회의실
집중형 업무공간
휴게실
팀스페이스1
개인 업무공간
팀스페이스2
팀스페이스3
화장실
통로
BAR
계단실
화장실

SECTION PLAN

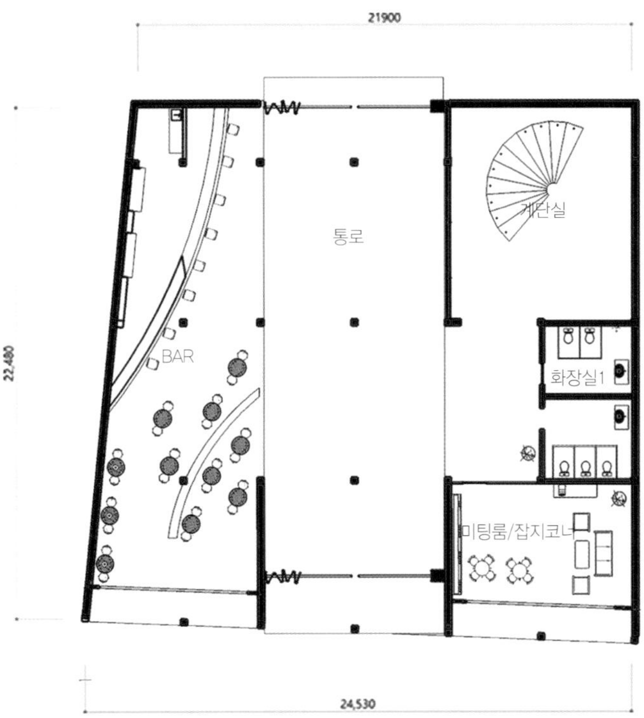

1FLOOR

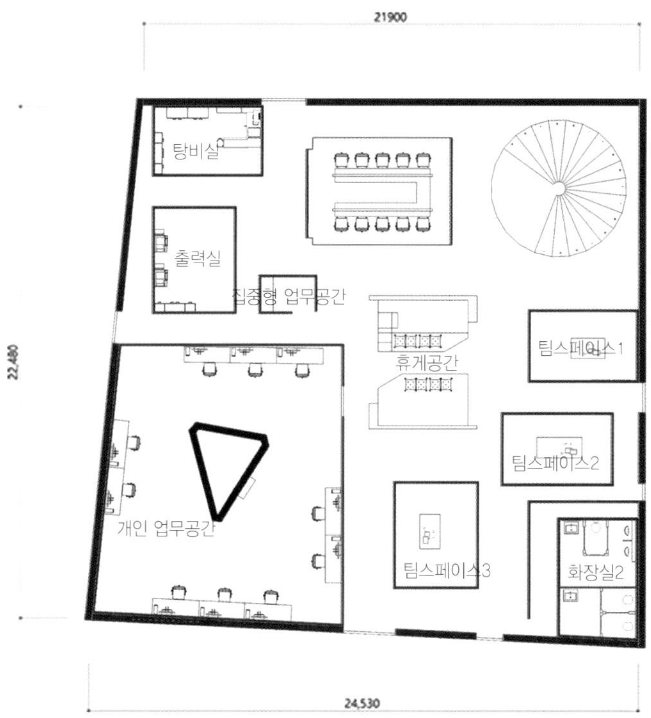

2FLOOR

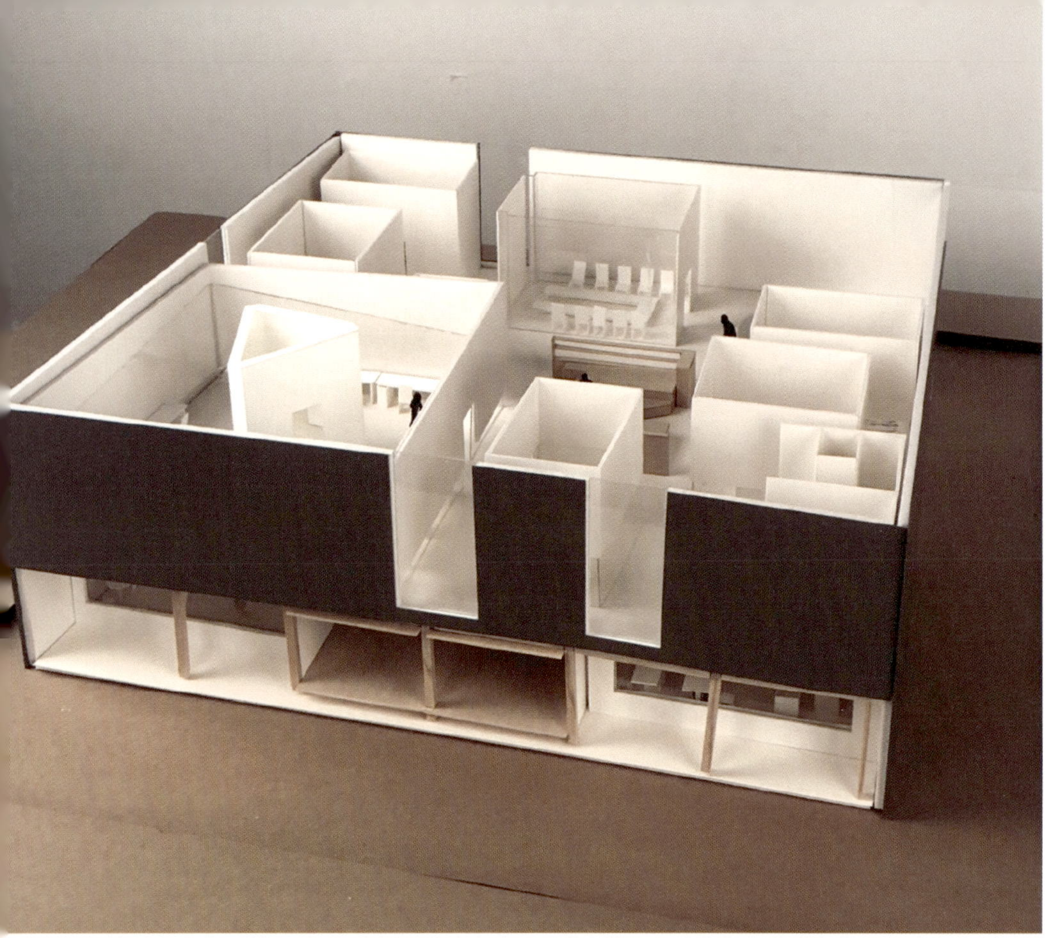

최다연 I Choi, Da Yeon 김서호 I Kim, Seo Ho

SCENT OF SCENT

BACKGROUND

우리의 곁엔 항상 냄새가 따라 다닌다. 눈에 보이지도 않고 잡히지도 않는 냄새는 홀연히 어디론가 움직여 사라진다. 사라진 냄새에 대한 기억이나 태도는 사람마다 다르며, 'SCENT' 는 이 기억들을 모아 사람들에게 한 권의 잡지로 추억과 즐거움을 전달한다. SCENT의 직원뿐만 아니라 지역주민을 포함한 외부인들이 공간 안에서 다양한 '시각' 의 기억을 가질 수 있는 공간을 제안하고자 한다.

There are always scents all around us. Indeed, these scents are invisible, but they move away and disappear suddenly. The memories and attitudes about the disappearing smell vary from person to person. 'SCENT' collects these memories people have and conveys these recollections and pleasures to people through a magazine. SCENT attempts to propose a space where not only the staff, but also

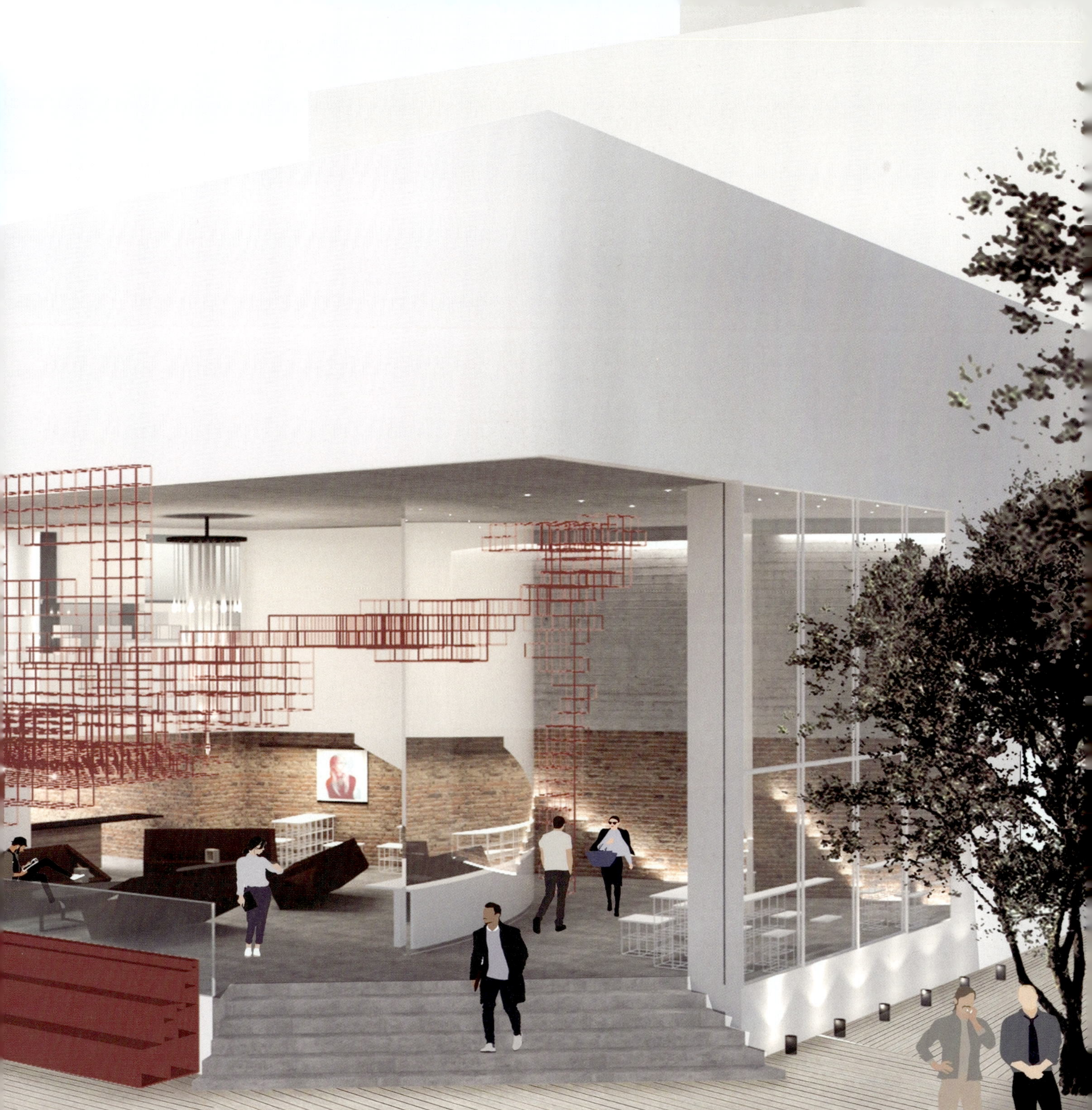

SITE

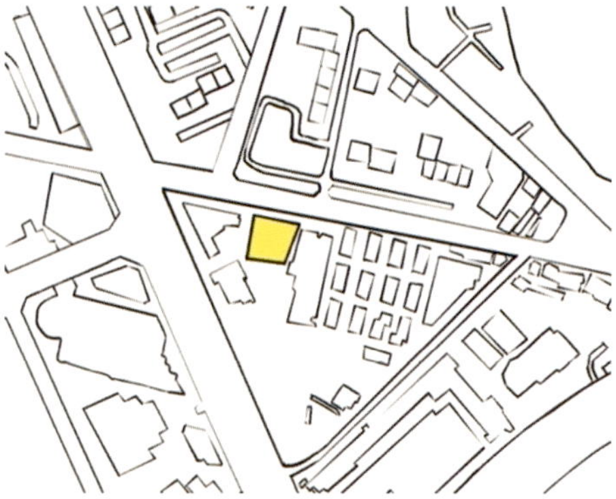

서울특별시 구로구 중앙로6길 16에 위치하고 있는 두드림 센터는 주변에 대학교가 위치하고 있으며, 양옆, 앞뒤로 아파트와 오피스텔 등 주거공간이 밀집된 지역이다. 큰 야구경기장이 두블럭 사이를 두고 위치하고 있어 경기시즌 및 행사 시 다소 소란스러움이 느껴지는 지역이다.

Located at 6th street 16, Jungro-gu, Guro-gu, Seoul, the doodle center is located in the surrounding univer sity area, and is a densely populated residential area including apartments and officetels. A big baseball st adium is located between the two blocks, making it a bit crowded during the games and events.

Organization chart

a manager editor

design team	shooting team	marketing team
the whole range of a magazine planning &Design (5)	Take and edit pictures (2)	promote sns and on-line and manage customers (2)

SCENT가 공간에서 원하는 점?

주위의 무심코 지나쳤던 냄새들을 모아 다른 표현 방법으로 사람들에게 알리며, 향이라는 것을 실제로 본인이 가지고 있고 기억을 떠올리며 꺼내 맡을 수 있게하여 그 기억이 오래 지속되는 것이 최종목표이다.

The ultimate goal of SCENT is to collect the smells that we have missed and to express these fragrances through a different approach. That is, the ultimate goal is to let people be reminiscent of the memories they have and let these memories last long.

Space required?

Workspace: Design Room / Laboratory / Conference Room / Public Relations Room / Lounge / Studio
Public space: exhibition room / library / cafe / outdoor space

SCENT는 냄새에 관한 다양한 시각과 경험을 소개하는 잡지로 기획자와 디자이너가 주요 역할을 한다. 그에 필요한 부서로 촬영과 편집을 하는 제작부와 홍보를 위한 마케팅부가 있다.

SCENT introduces a variety of views and experiences about smell. Planners and designers play a major role as magazines. Accordingly, the production department that shoots and edits, the PR department, There is a marketing department for this.

CONCEPT IDEA

'PIPE MESH'

파이프 매쉬' 는 보는 높이, 각도 등에 따라 그 밀도와 느낌을 달리하며, 냄새의 불명확한 형태가 파이프 매쉬로서 상징적으로 표현될 수 있다.

'Pipe Mesh' differs in density and feeling depending on from which height orangle it is viewed upon. The indefinite form of the smell can be symbolically expressed as Pipe Mesh.

CONCEPT PROCESS

이를 파이프를 이용해 매쉬를 만들어 움직임을 나타내었다. 매쉬 흔적들은 공간의 곳곳에서 나타나며 가구 / 구조/소품에서 모두 찾아볼 수 있다.

Thus, I used the pipe to make it into Mesh to represent the movement. These Mesh traces appear throughout the space and can be found in furniture / structures / props.

매쉬 결합 -> 냄새를 형상화 변화가 일어나려는 조짐을 보이며 공간에 서서히 달라 붙고있다.

Combination of Mesh --> Symbolization of Smell The mesh is showing the signs of change and is gradually adhering to the space. found in furniture / structures / props.

매쉬의 이동 -> 냄새의 속성대로 이동하면서 생긴흔적이며, 모양대 로 매쉬가 구조로 스며들어 공간을 구성하게된다.

Movement of the Mesh --> The traces are left as the scent moves according to its properties. The Mesh permeates through the strcture according to the shape and forms the space

매쉬의 융합 -> 공간을 구성하던 큰 매쉬가 지나가고 남은 매쉬들이 가구와 사람의 동선 주위에 달라붙어 냄새의 흐름을 정리해준다.

Fusion of Mesh --> The large Mesh that constituted the space disappears, and the remaining Mesh is adhered to the furniture and movement path of people. This remaining Mesh organizes the flow of the smell.

냄새의 불특정적인 움직임으로 매쉬가 흐트러지고 깨지며 생긴 흔적 사이를 바람이 지나가며 공명, 진공이 생거 공간에 소리를 생성한다.

The unspecified movement of the smell makes traces in the Mesh, as it is continuously disturbed and broken. The wind will pass through these traces, creating resonance and vacuum, and finally the sound in the space.

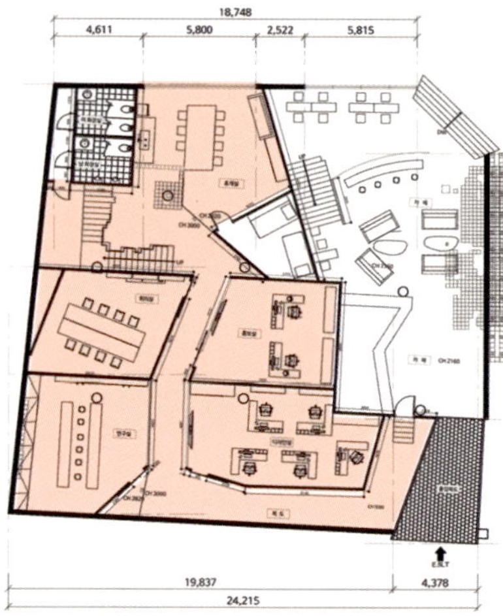

OFFFICE SPACE

디자인실, 연구실,회의실,영업부실,휴게실 위치함 직원들의 통로이며 고객에게 노출되지 않음 가운데 직선통로를 한번 꺾으면서 들어가게 만들어 오픈공간인 휴게실에 바로 시선이 가지 않도록 함.

There are design room, lab room, meeting room, business meeting room, sales department room and the lounge. / The employees' passageway is not exposed to the clients. It is located at a place swerving to the straight passageway, so that the gaze does not lie immediately on the lounge, an open space.

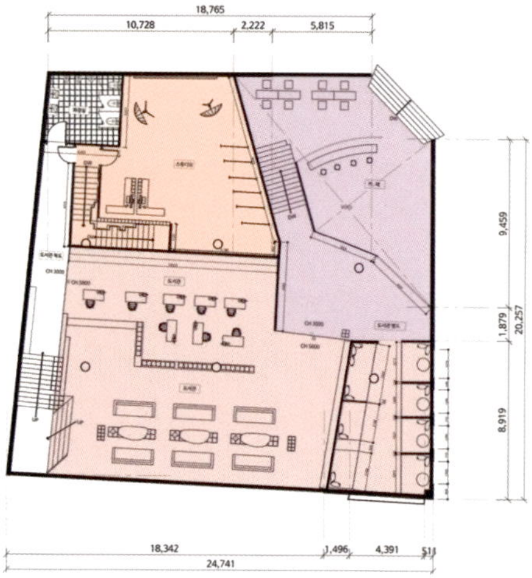

좌측은 직원의 이동통로 구간

위와 같이 SCENT는 궁극적으로 회사라는 공간이 사람에게 주어진역할과 주어진 용도라고 규정을 하는 것을 원하지않는다. 자신이 이야기를 만들어가며 다양한 소통을 시도할 수 있도록 노력하며 이를 공간에 드러내고자 한다.

FRONT VIEW

도서관_디자인실_연구실_상영관_입구

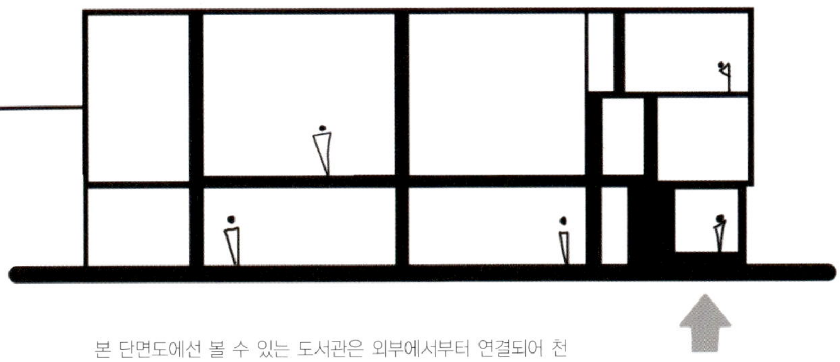

본 단면도에선 볼 수 있는 도서관은 외부에서부터 연결되어 천정 구조물이 된 매쉬이다.1층과 다르게 2~3층까지 높은 천정고를 가지고 있는데 업무공간은 집중을 요하는 공간임에 반해 공공 장소인 도서관은 열린 공간이 되어 닫힌공간과 소통하고 이어주는 역할을 하고있다.

As can be seen in this cross-sectional diagram, the library is a ceiling structured Mesh that is connected from the outside. Unlike the first floor, it has a high ceiling that ranges about two to three floors high. Whereas the work space is a space that requires concentration, the library is a public place. Thus, this ceiling serves as the bridge to the connection and communication between the open space and the closed space.

OFFFICE SPACE

스튜디오, 도서관(개별용, 단체 라운지, 총 3종류), 카페가 이어짐. 직원 전용 숨은 계단을 두어 문이 닫히면 무의식적으로 길게 이어진 벽에 시선을 두지 않으면서 문을 벽으로 인식하게 했다. 도서관은 높은 천장을 두어 자유로운 이동과 착석이 가능하도록 가구를 퍼뜨려 배치했다.

Studio, library (individual use, group lounge, total three kinds), cafe are connected. There is a separate hidden stair for the employees so that the people can perceive the door as a wall without unconsciously keeping their eyes to the long, continuous wall. The library has a high ceiling, and the furniture is spread out to enable free movement and seating.

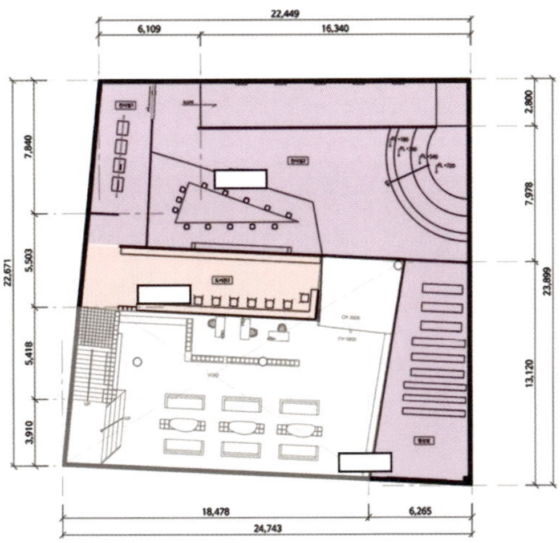

EXHIBITION SPACE

도서관에서 연결되는 계단을 타고 올라가면 전시장 1에서 scent의 잡지와 관련 정보를 얻을 수 있다. 전시장 2에서는 구조적으로 최소화하고 공간 안에서 scent의 향을 즐기며 소통할 수 있도록 하였다.

If you climb the stairs connected to the library, you can get information about the SCENT magazine at the Exhibition Hall 1. In the Exhibition Hall 2, the library is placed to structually minimize the space, so that visitors can communicate and enjoy the scent in the space.

우측은 고객의 이동통로 구간

ISOMETRIC

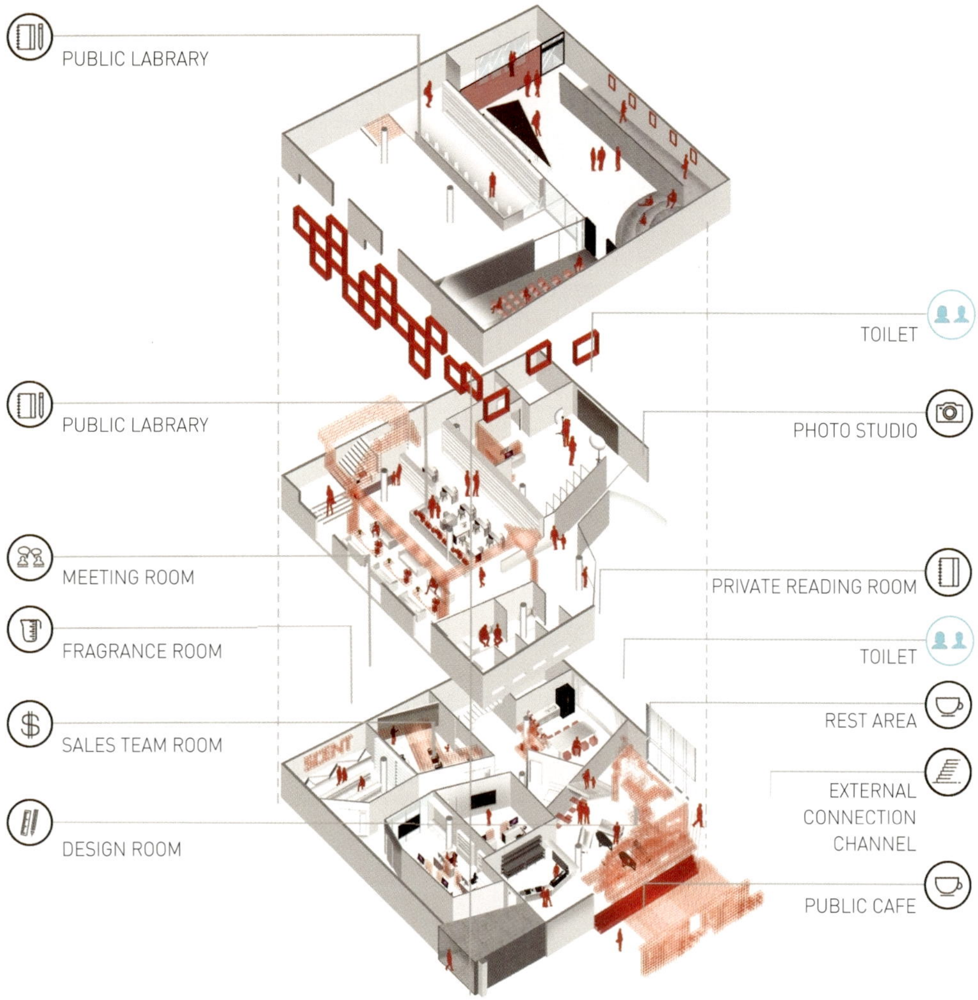

DESIGN ROOM

카페에 설치된 매쉬와 연결되어 센트의 이미지를 보여주고있다.

It is connected to the mash installed in the cafe to show the image of the cent.

MEETING ROOM

평면상으로 사선의 벽에 의해 형성된 공간에 어울리게 파이프 매쉬가 구조를 형성하였다.

Pipe mash formed a structure to match the space formed by the wall of the oblique line on the plane.

PUBLIC CAFE

넓게 오픈된 카페 공간에 구조적 및 가구적 요소로 파이프 매쉬가 적용되었다.

Piping mash has been applied as structural and household elements in the wide open cafe space.

FURNITURE PROCESS

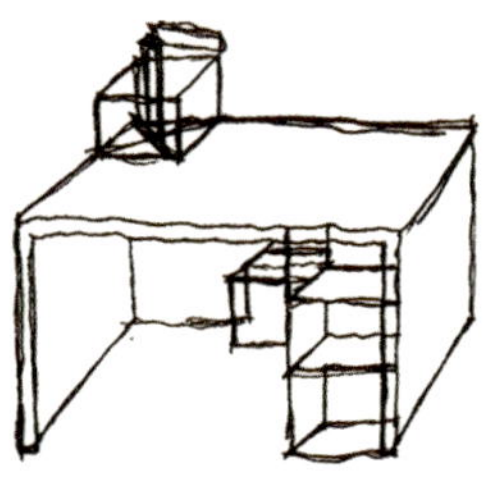

사람의 손이 주로 닿는 부위에 냄새의 입자가 끼어들어가 테이블, 조명, 수납함이 하나가 되었다. 이 테이블은 한명만 사용할 수 있는데 그래서 매쉬의 입자가 크게 들어갈 수 있었다.

The particles of the smell are caught in the area where the human hand mainly touches, making the table, the lighting, and the storage box to become as one. This table can only be used by one person, so that the Mesh particles can enter in large quantity.

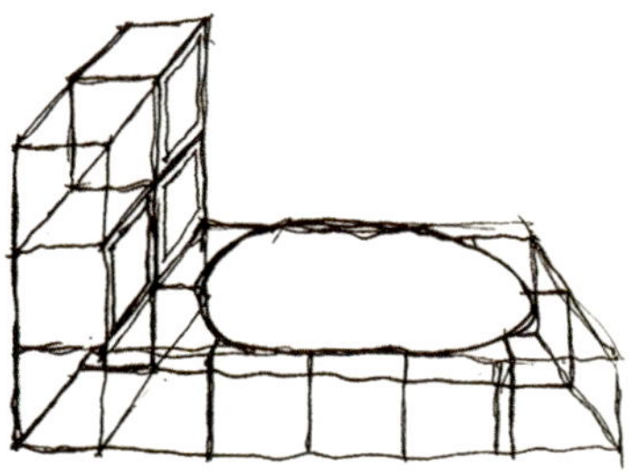

외부매쉬가 천장을 타고 들어와 흘러 굳어져 테이블과 책장을 자연스레 연결시키면서 떨어진 가구로만 존재하던 것의 경계를 모호하게 만들었다.

The outer Mesh comes in through the ceiling and is hardened. The table and the bookshelf are naturally connected, thus obscuring the boundaries of what existed only as distinct furnitures.

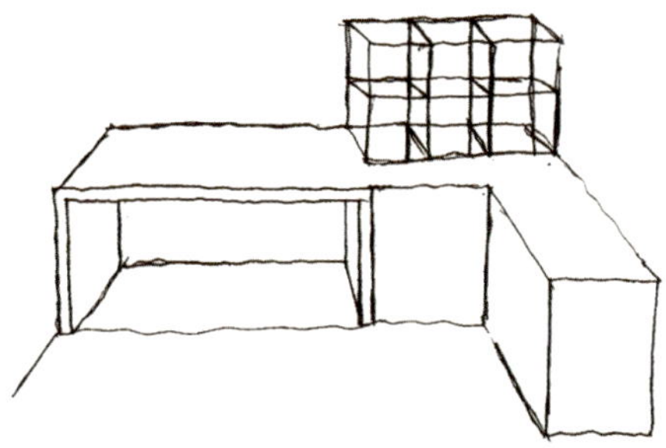

개인 책상에 수납, 메모 등 사용자의 개성을 그대로 들어낼 수 있도록 만든 가구이다. 개성이 만들어내는 변화는 비어있는 매쉬에 더하기를 해주는 것과 같다.

This furniture is designed to wholly display the personality of users such as storage and memo on personal desk. The change that personality of the user makes is like adding from the empty Mesh.

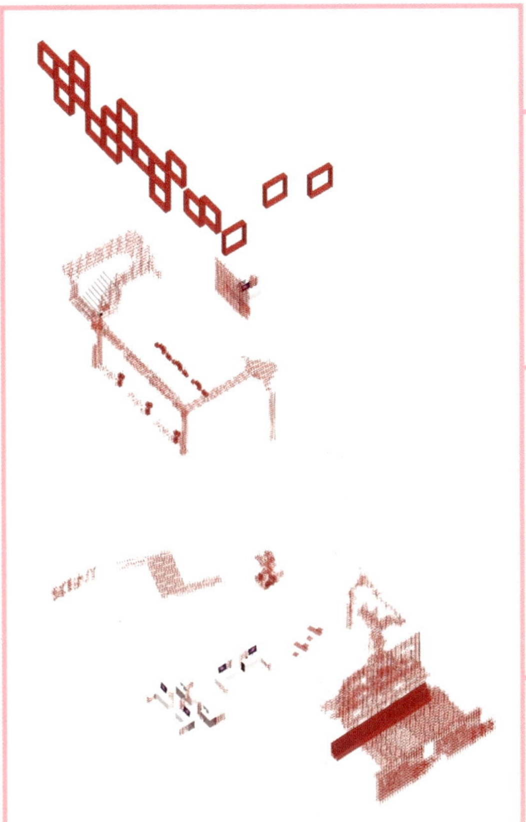

FACADE

카페에 설치된 매쉬와 연결되어 센트의 이미지를 보여주고있다.

It is connected to the mash installed in the cafe to show the image of the cent.

PUBLIC LIBRARY / EXHIBITION SPACE / PUBLIC CAFE

길고 얇게 연결되어 연속성을 보여주고 큰 스케일로 공간의 부피감을 더욱 극대화 시키고 있다.

Long and thin connection shows continuity and volume of space at large scale It is further maximizing the feeling.

FRAGRANCE ROOM / SALES TEAM ROOM / DESIGN ROOM

크고 강렬한 매쉬와 작고 얇은 밀도감의 매쉬가 대비를 이루어 공간의 재미를 보여주고 있다.

A large, intense mesh and a small, thin, dense mesh provide contrast It shows the fun of space.

매쉬로 글자 디자인을 블럭처리 하여 SCENT의 이름을 표현 하였다.

Blocks character design with mesh The name was expressed.

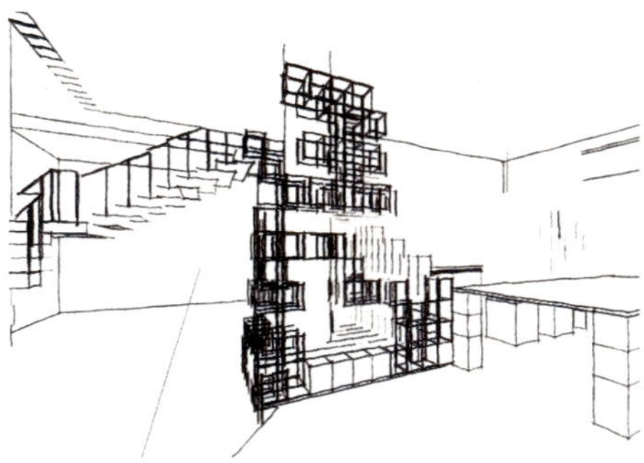

휴게실에서 음식도 해먹으면서 수납공간이 필요한데 프레임을 선반화 시켜 물건을 올려둘 수 있게 만들었다.

I need some storage space in the break room. The frame was shelved to make things stand up.

회의실에 설치된 매쉬의 경우 조명을 매쉬로 만들어 그 끝에 연결하고 기둥에 영향을 주는 구조체인 것을 모호하게 만든다.

As for the Mesh installed in the conference room, the lighting is changed into the Mesh. The end of the lightbulb is connected, thus making it ambiguous that this structure affects the column of the space.

REST ROOM

업무공간사이의 좁은 복도가 냄새가 흐르면서 생기는 리듬의 불규칙한 파동을 표현했고 점차 응축되면서 직원들의 휴게 공간에서 응축됨이 크게 확장하여 개방된 공간으로 나타나 오픈휴게실의 형태를 가지게 되었다. 크게 확장하여 흩어진 냄새의 리듬은 곧 창문을 통해 밖으로 자유롭게 연결되어 밖과 안을 연결하고 있다.

The irregular ripples of the rhythm are used to represent the flowing of the smell through the narrow corridor between the work spaces. As the smell is condensed, the condensation of the employee lounge gradually expands, and appears in the open space to become the open lounge. The rhythm of the smells that is expanded broadly soon connectes freely with the outside through the window, thus connecting both the inside and the outside.

OPEN CAFE

크게 오픈된 카페에 파이프 매쉬가 다양한 모습으로 분포되어 있다. 앉을 수 있는 벤치 역할을 하기도 하고, 동선을 알려주는 역할을 하기도 한다.
Pipe mesh is distributed in a wide open cafe. They serve as benches to sit on, or as a guide to traffic.

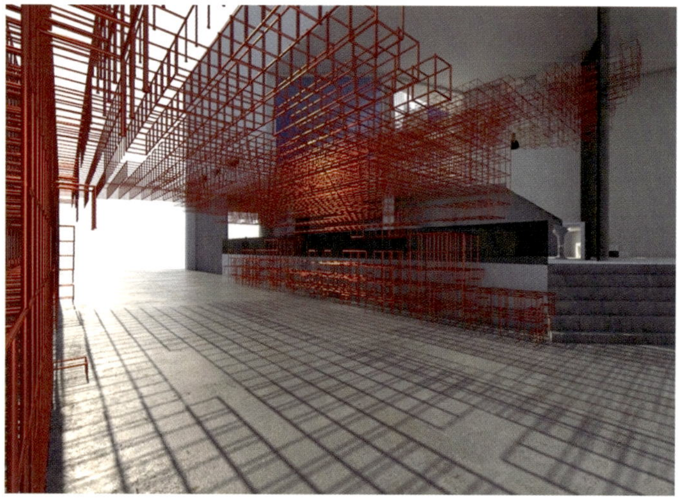

DESIGN OFFICE

평면상으로 각도가 조금씩 변화하며 세워지는 벽으로 인해 지루한 느낌을 없애주며, 다양한 가구배치가 가능하다. 가구 또한 파이프 매쉬와 결합되어 파이프 매쉬의 다양한 모습을 보여준다.

The wall that gradually stands still as the angle changes in subtle amount in the plane removes the feeling of triteness, allowing for various arragements of furniture. The furniture is also combined with the Pipe Mesh to show various aspects of it.

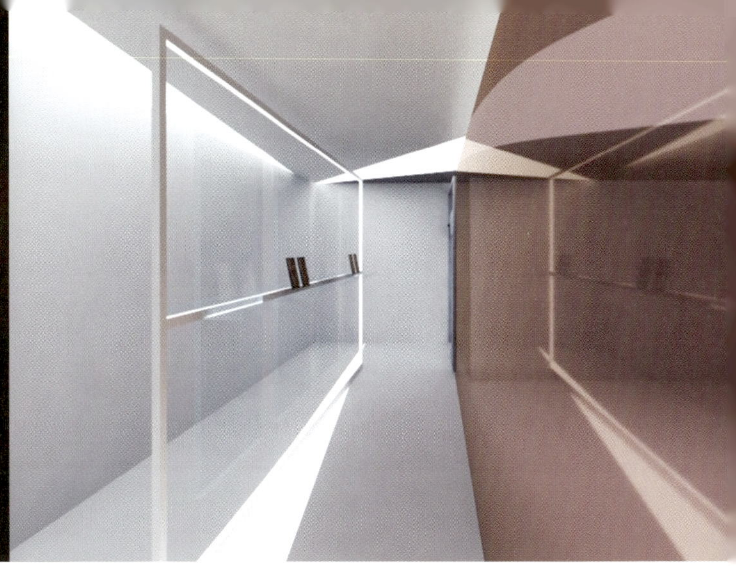

CUSTOMER SPACE

소통과 개인의 공간을 동시에 제공할 수 있을까란 고민을 하며 구성한 공간이다. 도서관의 경우 높은 책장과 낮은 책장으로 시선을 사람의 키나 앉은 위치에 따라 구분하고 낮은 바닥 레벨변화를 주어 약간의 방향성과 변화를 주어 시선을 끌면 카페에서 전시장까지 하나의 시퀀스를 생성하였다. 시퀀스는 SCENT가 주고자 하는 감성과 향의 의미를 후각 – 지적 요소(센트가 가진 지식의 양, 혹은 그동안 이뤄온 노력의 산물) – 결과물로 보여주고있다.

This is a space that is constructed by contemplating whether communicative and individual space can be provided simultaneously. As for the library, the gaze is differentiated according to the heights or the location of the visitors through the high and low bookshelves. A level change was made to the lower floor to give a subtle direction and change, so that a single sequence from the cafe to the exhibition hall is created. This sequence displays the significance of sensibility and fragrance that SCENT attempts to convey tthrough olfactory-cognitive elements (the amount of knowledge that the scents have, or the fruit of the efforts that have been made).

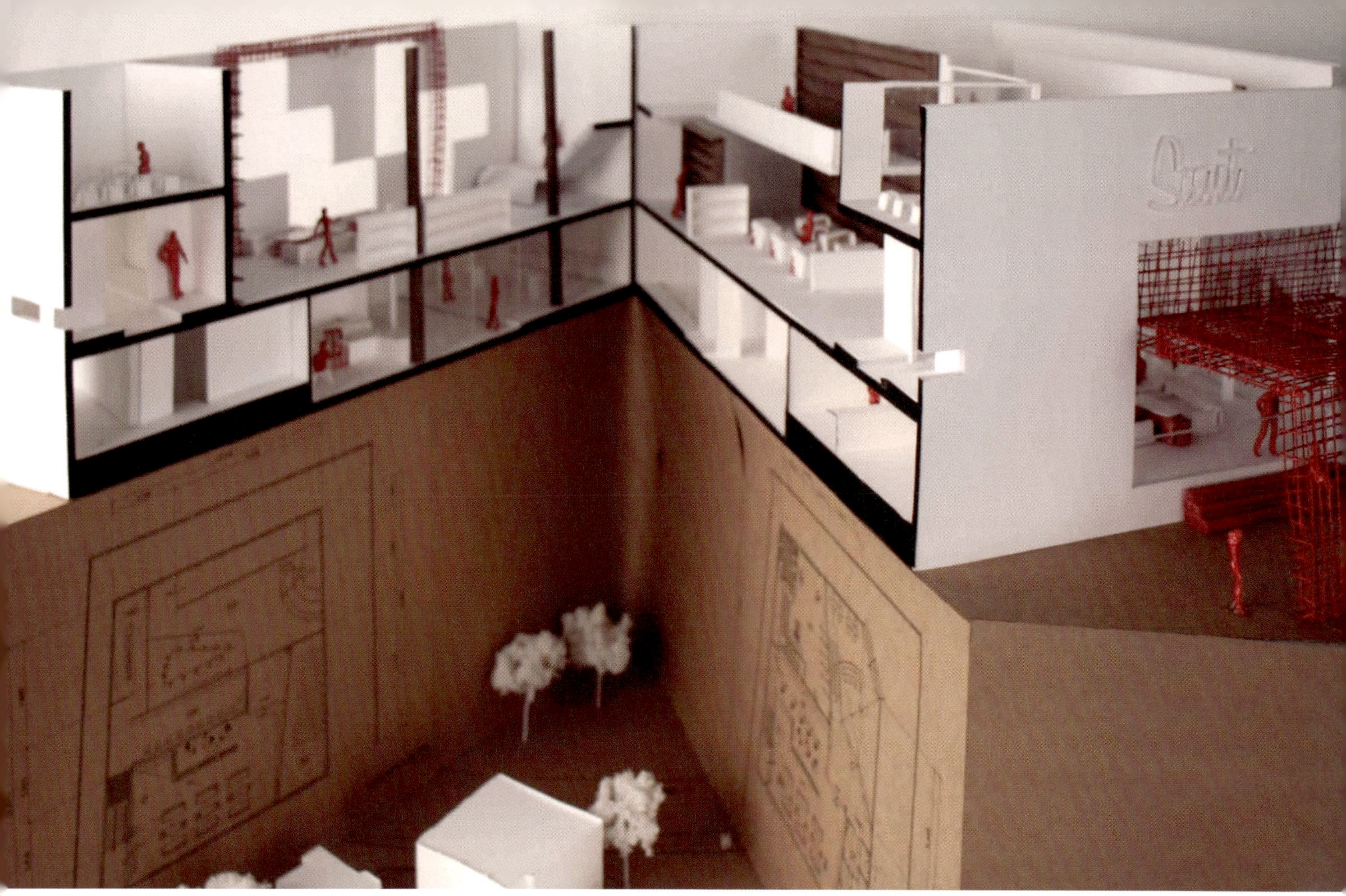

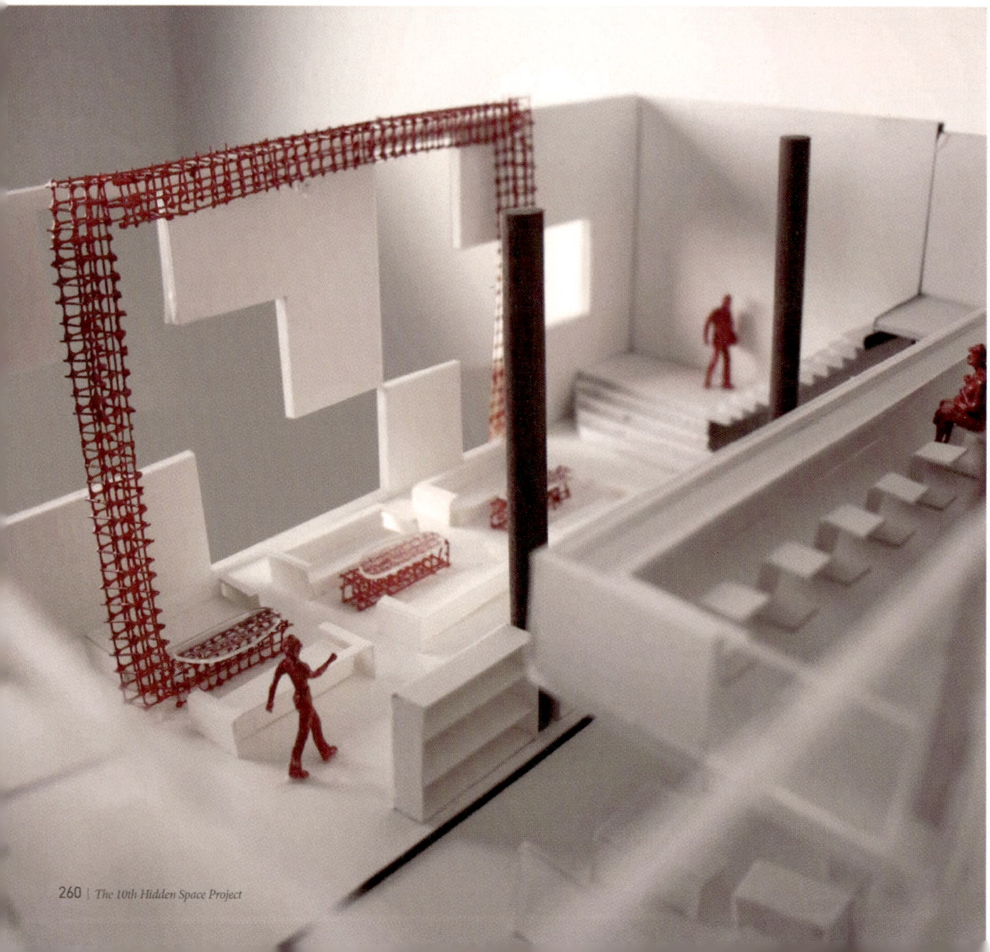

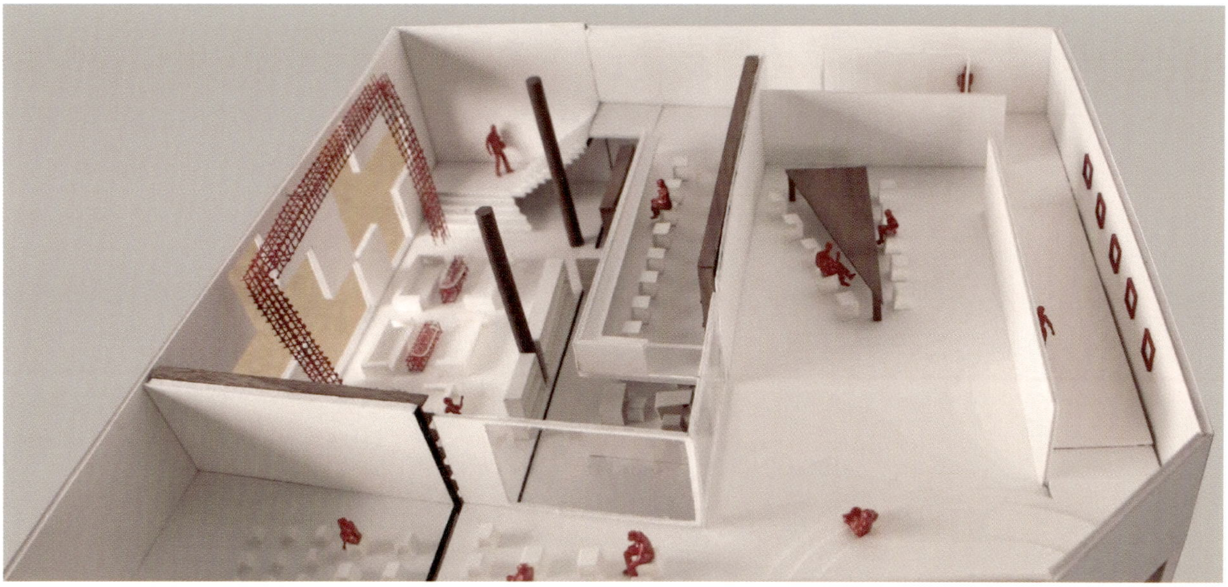

2019
INTERNATIONAL DESIGN EXCHANGE PROJECT

The 10th Hidden Space Project

REMEDIATION

The Cass London Metropolitan University
The Sir John Cass Faculty of Art, Architecture and Design
Interior Design BA (Hons)
UK

Dongyang Mirae University
Dept. of Interior Environmental Design, in School of Living and Environmental Engineering
Rep. of KOREA

19 June
4 July 2010
HUN
IAS
IDN
Intl Architectural Showcase
Presented by foreign embassies and cultural institutes in collaboration with The British Council and The Architecture Foundation for the London Festival of Architecture 2010

JPN
KOR
LVA
Living Bridges
A UK and Korea collaboration
26. Korean Cultural Centre and London Metropolitan University
31
16 Goulston Street, London E1 7TP
22 June– 27 June
10am– 5pm
Private View: 24 June, 6–8pm
An exhibition of work by 25 students from three collaborative universities; London Metropolitan, Kyung Won and Dong Yang. The project brief focused on urban and community issues shared by the UK and Korea and asked the students to create a 'living bridge' structure that is both an architectural metaphor for community support and a space for the communities to enjoy, integrate and communicate. Each student has built a model and will show the structure, its location and context.
The work from this project has been compiled in a book that will be published alongside the exhibition. Living Bridges has given us a valuable insight into how young creative minds work, and how they deal with physical evidence and community challenges.
Image: Night Visual Designer, 2010. © Anelika Ghouralal

2010
INTERNATIONAL DESIGN EXCHANGE PROJECT

The 1st Hidden Space Project

LIVING BRIDGES

The Cass London Metropolitan University
Dongyang Mirae University
Kyungwon University

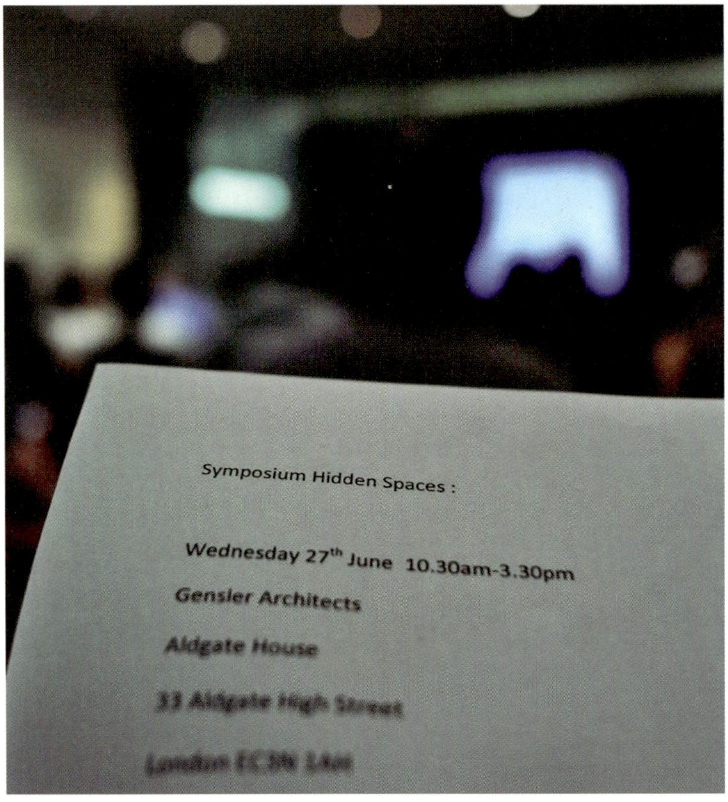
Symposium Hidden Spaces :
Wednesday 27th June 10.30am-3.30pm
Gensler Architects
Aldgate House
33 Aldgate High Street

2012

INTERNATIONAL DESIGN EXCHANGE PROJECT

The 3rd Hidden Space Project

HIGH STREET

The Cass London Metropolitan University
Dongyang Mirae University
Kyungwon University

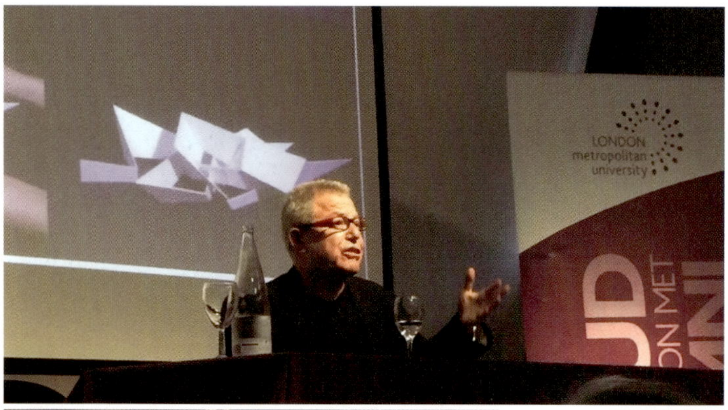
LONDON
metropolitan
university

Gerard Byrne: 1984 and Beyond

2014
INTERNATIONAL DESIGN EXCHANGE PROJECT

The 5th Hidden Space Project

THE RE-ENERGISING SPACE

The Cass London Metropolitan University
Dongyang Mirae University
Hunan University

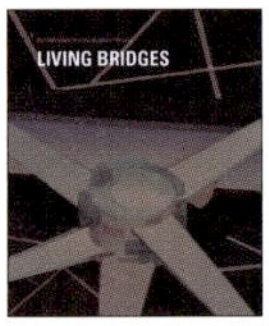

FHHS2010 - *The 1st Hidden Space Project* **"LIVING BRIDGES"**

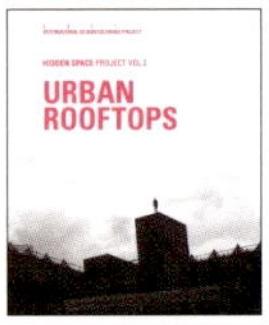

FHHS2011 - *The 2nd Hidden Space Project* **"URBAN ROOFTOPS"**

FHHS2012 - *The 3rd Hidden Space Project* **"HIGH STREET"**

FHHS2013 - *The 4th Hidden Space Project* **"THE RISE OF THE CREATIVE DISTRICT"**

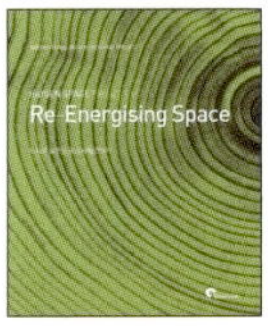

FHHS2014 - *The 5th Hidden Space Project* **"RE-ENERGISING SPACE"**

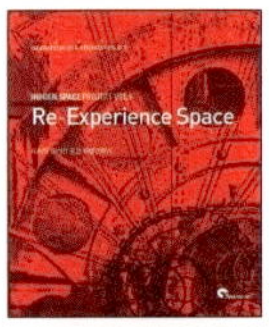

FHHS2015 - *The 6th Hidden Space Project* **"RE-EXPERIENCE SPACE"**

FHHS2016 - *The 7th Hidden Space Project* **"SPATIAL HAPPINESS"**

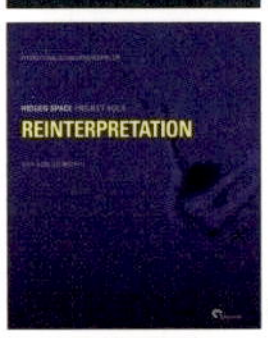

FHHS2017 - *The 8th Hidden Space Project* **"REINTERPRETATION"**

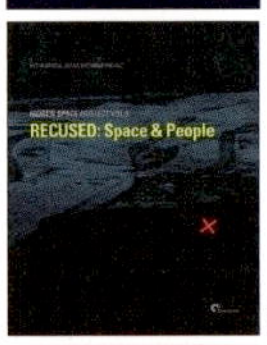

FHHS2018 - *The 9th Hidden Space Project* **"RESCUED: Space & People"**

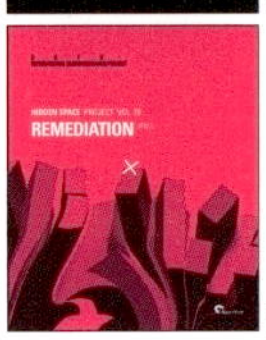

FHHS2019 - *The 10th Hidden Space Project* **"REMEDIATION"**

* FHHS (Future Horizon for Hidden Space)

2010년 [LIVING BRIDGES]주제로 영국 런던 메트로폴리탄대학과 한국의 가천대학교, 한국의 동양미래대학교가 공동으로 런던 아키텍 페스티벌 참가하면서 시작된 FHHS는 우리시대 숨겨진 공간들을 10년간 10개의 주제로 교류전을 기획, 현재 9회까지 진행했으며 이후 주제를 현재 기획중이다.
FHHS는 이후 계속해서 우리주변의 숨겨진 공간들을 찾아 열정과 지혜, 그리고 성찰의 노정(路程) 을 계속해 나갈 예정이다.

2019

INTERNATIONAL DESIGN EXCHANGE PROJECT

The 10th Hidden Space Proect

REMEDIATION 再媒介

Published By 2019 REMEDIATION
(Kaye Newman / Park Youngtae)

Total Design
Park Youngtae | DONGYANG MIRAE University
만든이 | 동양미래대학교 실내환경디자인과 박영태
편집, 기획, 표지 디자인 | 박영태 / 편집보조 | 염은진, 서정인

펴낸날 | 2019. 8. 20. 1쇄 발행
발행인 | 김기현
발행처 | spacetime · 시공문화사
등 록 | 1993년 3월12일 1-1493
주 소 | 120-796 서울시 서대문구 독립문공원길 13 극동프라자 5층
전 화 | 02-3147-1212
팩 스 | 02-3147-2626
ISBN | 978-89-5592-430-5

정가 | 20,000원